Nonlinear Interpolation and Boundary Value Problems

TRENDS IN ABSTRACT AND APPLIED ANALYSIS

Series Editor: John R. Graef
The University of Tennessee at Chattanooga, USA

Trends in Abstract
and Applied Analysis
Volume **2**

Nonlinear Interpolation and Boundary Value Problems

Paul W Eloe
University of Dayton, USA

Johnny Henderson
Baylor University, USA

World Scientific

NEW JERSEY · LONDON · SINGAPORE · BEIJING · SHANGHAI · HONG KONG · TAIPEI · CHENNAI · TOKYO

Published by

World Scientific Publishing Co. Pte. Ltd.

5 Toh Tuck Link, Singapore 596224

USA office: 27 Warren Street, Suite 401-402, Hackensack, NJ 07601

UK office: 57 Shelton Street, Covent Garden, London WC2H 9HE

Library of Congress Cataloging-in-Publication Data
Names: Eloe, Paul W., author. | Henderson, Johnny, author.
Title: Nonlinear interpolation and boundary value problems / Paul W. Eloe
 (University of Dayton, USA), Johnny Henderson (Baylor University, USA).
Description: Singapore ; Hackensack, NJ : World Scientific, [2016] | 2016 |
 Series: Trends in abstract and applied analysis ; volume 2 |
 Includes bibliographical references and index.
Identifiers: LCCN 2015040530| ISBN 9789814733472 (alk. paper) |
 ISBN 9814733474 (alk. paper)
Subjects: LCSH: Boundary value problems. | Differential equations.
Classification: LCC QA379 .E435 2016 | DDC 515/.35--dc23
LC record available at http://lccn.loc.gov/2015040530

British Library Cataloguing-in-Publication Data
A catalogue record for this book is available from the British Library.

In-house Editors: V. Vishnu Mohan/Kwong Lai Fun

Typeset by Stallion Press
Email: enquiries@stallionpress.com

Printed in Singapore

Dedicated to the memory of Mary JoAnn Kramer Eloe.

Dedicated to David Ray Strunk III and Jana Elisabeth Strunk, and to the memory of their sister, Kathryn Madora Strunk.

Preface

The questions most often addressed concerning dynamical systems (such as differential equations, finite difference equations, dynamic equations on time scales, and so on) involve existence of solutions and uniqueness of solutions. The topics dealt with in this book address those two questions for boundary value problems for dynamical systems, but for the greater part, in the stated opposite order. That is, given a dynamical system

$$Ly = f, \tag{1}$$

satisfying boundary conditions,

$$l_i y = c_1, \quad 1 \leq i \leq n, \tag{2}$$

where $L : C^{(n)}(I) \to C(I)$ is an nth order linear differential operator (or difference operator or dynamic differential operator, etc.), I is an interval of the reals, f is nonlinear in $y^{(i-1)}$, $1 \leq i \leq n$, $l_i : C^{(n-1)}(I) \to \mathbb{R}$ are continuous and bounded, $c_i \in \mathbb{R}$, $1 \leq i \leq n$, and $n \geq 2$, the questions dealt with are **(i)** "When are solutions of (1)–(2) unique, when solutions exist?"; and **(ii)** "When does uniqueness of solutions of (1)–(2) imply existence of solutions of (1)–(2)?" Questions of type (ii) are called "uniqueness implies existence" questions. If (1) is a nonlinear ordinary differential equation, the answers to questions of type (i) and type (ii) are quite technical. If (1) is the simple equation,

$$y^{(n)}(x) = 0, \quad x \in I,$$

then the questions of type (i) and type (ii) are precisely those questions that motivate the study of polynomial interpolation. In the case of polynomial interpolation, a linear problem, the questions of type (i), when are solutions unique when they exist, or type (2), uniqueness implies existence, are equivalent.

An interpolation problem is the problem of selecting a unique function from a family of functions that satisfies prescribed conditions. Hence, to consider boundary value problems for ordinary differential equations as interpolation problems is appropriate and Hartman [43] employed the phrase, interpolation problems for nonlinear differential equations. For the purposes of the title of this book, we adopted a phrase, **nonlinear interpolation**. Specifically, for this book, we mean the following: the family of functions is the set of solutions of the nonlinear equation, (1), and the prescribed conditions are those given by the linear boundary conditions, (2). Since type (i) questions and type (ii) questions are apparently independent questions for nonlinear ordinary differential equations, we intend that the one phrase, nonlinear interpolation, contains both questions.

In routes taken to establish "uniqueness implies existence" for solutions of (1)–(2), uniqueness of solutions of one type of boundary value problem for (1) may imply uniqueness of solutions for related types of boundary value problems for (1), which can be ultimately useful in obtaining existence of solutions of (1)–(2). These types of results are sometimes referred to as "uniqueness implies uniqueness" results. These questions for nonlinear differential equations have a long history, with some motivational papers originating with Hartman [41], Lasota and Luczynski [86, 87] and Lasota and Opial [88] in the 1950's and 1960's. Soon after those papers, the most prominent such results for the case of (1) nonlinear were established by P. Hartman and L. Jackson, as well as by L. Jackson's students, throughout the 1960's and 1970's.

For a period in excess of 30 years, this book's authors have occupied the center of research attention devoted to "uniqueness implies uniqueness" and "uniqueness implies existence" questions for solutions of the boundary value problem (1)–(2). This book pulls together much of their work, along with closely related work by other authors, as well as includes some of the classical results as historical motivation. (The book by Agarwal [3], the paper by Agarwal [4] and the survey paper by Mawhin [90] contain documentation of some of the important historical works.)

In Chapter 1, conjugate boundary value problems and right focal boundary value problems are introduced for when (1) is an nth order ordinary differential equation. One goal of the chapter will be directed toward "uniqueness implies uniqueness" results for both conjugate boundary value problems and right focal boundary value problems. Yet, before those questions are addressed, the chapter is focused on a Kamke Theorem, a historically significant "Compactness Condition" on solutions of (1), and a result concerning solutions of boundary value problems and their continuous dependence on the boundary conditions. Each of these plays a fundamental role in the "uniqueness implies uniqueness" results of the chapter, as well as throughout many of the other chapters of the book. Some of the "uniqueness implies uniqueness" results for the cases of $n = 3, 4$, are necessary and sufficient results.

Chapter 2 is devoted to the "uniqueness implies existence" question for the boundary value problems introduced in Chapter 1; that is, the conjugate boundary value problems and the right focal boundary value problems. Historically motivational results for the conjugate boundary value problems for second and third order (1) comprise the first two sections of Chapter 2, with generalizations in the third section. The last two sections of the chapter focus on right focal problems. Throughout the chapter, extensive use is made of the "Compactness Condition" and continuous dependence of solutions on boundary conditions in conjunction with shooting methods, along with induction based on the arrangement indexing pattern for each family of boundary value problems.

Nonlocal boundary value problems for (1) are considered in Chapter 3. The first and third sections of this chapter deal with uniqueness of solutions of certain types of nonlocal boundary value problems leading to uniqueness of solutions for other nonlocal boundary value problems. The arguments for these "uniqueness implies uniqueness" results take different paths. Then, for each of the paths taken, in the second and fourth sections, use is made of the respective preceding uniqueness results in establishing "uniqueness implies existence" for the nonlocal boundary value problems. Again, the methods for existence include continuous dependence of solutions on boundary conditions, the "Compactness Condition" and shooting adapted to the unique nature of nonlocal boundary value problems.

The results of Chapter 4 are concentrated on the uniqueness of solutions implying their existence for discrete boundary value problems for finite difference equations. The addressed problems are, in some sense, discrete analogues of the results from Chapter 2, in that, discrete conjugate boundary value problems and discrete right focal boundary value problems are the focus. Discrete Lidstone boundary value problems are also dealt with for fourth order difference equations. Uniqueness hypotheses for solutions are viewed in terms of Hartman's *generalized zeros* and *discrete Rolle's Theorem*. The methods in each context rely strongly on induction and shooting, yet the inductions depend on the spacing of the discrete boundary points.

Chapter 5 contains results on "uniqueness implies existence" for solutions of boundary value problems for Hilger's dynamic equations on time scales. Given a nonempty closed subset of the reals, boundary value problems are defined on the subset in terms of the *Hilger delta derivative* (a general differentiation concept that unifies ordinary differentiation and discrete differences, with resulting equations called *dynamic equations*). Then, for dynamic equations of orders $n = 2, 3, 4$, uniqueness of solutions assumptions are imposed in terms of Bohner and Eloe's *generalized zeros* and *Rolle's Theorem on a time scale*. Shooting methods are adapted to the boundary value problems of the chapter, with the topological nature of the boundary points (such as *right dense boundary points*, *left dense boundary points*, *isolated boundary points*, and so on) playing major roles in the details of the existence arguments.

Chapter 6 contains a few remarks about "uniqueness implies uniqueness" and "uniqueness implies existence" for other boundary value problems along with some citations to works involving sufficient conditions for uniqueness of solutions, when solutions exist.

When many years of research result in a book such as this, there are always those to thank, who in one way or another, led to its writing. Paul W. Eloe thanks his wife, Laura. He also thanks his co-author, Johnny Henderson, for first, his extensive contributions in the study of nonlinear interpolation and boundary value problems and second, for the opportunity to contribute to this project which in many ways is a tribute to (the late) Lloyd K. Jackson. Johnny Henderson thanks first his wife, Darlene, and then others to whom he owes his thanks include David L. Skoug, Allan C. Peterson, James R. Wall, John V. Baxley, Overtoun M. Jenda, Jerry A. Veeh, August J. Garver, (the late) Wyth S. Duke, (the late) George M. Randall, (the late) Carroll Bo Sanderlin, (the late) Edward N. Mosely, (the late) Louis J. Grimm, (the late) Willam N. Hudson, (the late) William G. Leavitt, and (the late) Lloyd K. Jackson. Of course, he also thanks his co-author, Paul W. Eloe, not only for his collaboration on this book, but also for his enthusiastic and enlightening thoughts and comments while tossing around the baseball.

The authors especially want to thank their editors at World Scientific Publishing: Trends in Abstract and Applied Analysis Series Editor John R. Graef, Executive Editor Rochelle Kronzek, and Desk Editors V. Vishnu Mohan and Lai Fun Kwong.

Paul W. Eloe
Johnny Henderson

Contents

Chapter 1

Uniqueness Implies Uniqueness

This chapter is devoted primarily to uniqueness of solutions for boundary value problems for an ordinary differential equation. In particular, focus will be on when uniqueness of solutions for one type of boundary value problem leads to uniqueness of solutions for other types of boundary value problems. In addition to interest in the results themselves, such results sometimes play a role in establishing the existence of solutions (the topic of some of the subsequent chapters of this book). For example, in the case of linear boundary value problems for linear ordinary differential equations, uniqueness of solutions is equivalent to their existence.

The "uniqueness implies uniqueness" results in this chapter will be concentrated on conjugate boundary value problems and right focal boundary value problems. Boundary value problems, such as nonlocal boundary value problems, and some others will be dealt with in other chapters.

1.1 Some preliminaries

Throughout this chapter, we will consider boundary value problems for an nth order ordinary differential equation,

$$y^{(n)} = f(x, y, y', \ldots, y^{(n-1)}), \quad a < x < b. \tag{1.1.1}$$

And throughout, we will use the following hypotheses concerning the differential equation:

(A) $f(x, r_1, \ldots, r_n) : (a, b) \times \mathbb{R}^n \to \mathbb{R}$ is continuous.
(B) Solutions of initial value problems for (1.1.1) are unique and extend to (a, b).

Given $2 \leq k \leq n$, $m_1, \ldots, m_k \in \mathbb{N}$ such that $\sum_{j=1}^{k} m_j = n$, points $a < x_1 < \cdots < x_k < b$, and $y_{ij} \in \mathbb{R}$, $1 \leq i \leq m_j$, $1 \leq j \leq k$, a boundary value problem for (1.1.1) satisfying

$$y^{(i-1)}(x_j) = y_{ij}, \ 1 \leq i \leq m_j, \ 1 \leq j \leq k, \qquad (1.1.2)$$

will be called either a *k-point conjugate boundary value problem*, or an $(m_1, \ldots, m_k)$ *conjugate boundary value problem*.

   Also, given $2 \leq r \leq n$, $m_1, \ldots, m_r \in \mathbb{N}$ such that $\sum_{j=1}^{r} m_j = n$, $s_0 := 0$, $s_k := \sum_{j=1}^{k} m_j$, $1 \leq k \leq r$, points $a < x_1 < \cdots < x_r < b$, and $y_{ik} \in \mathbb{R}$, $s_{k-1} \leq i \leq s_k-1$, $1 \leq k \leq r$, a boundary value problem for (1.1.1) satisfying

$$y^{(i)}(x_k) = y_{ik}, \ s_{k-1} \leq i \leq s_k - 1, \ 1 \leq k \leq r, \qquad (1.1.3)$$

will be called either an *r-point right focal boundary value problem*, or an $(m_1, \ldots, m_r)$ *right focal boundary value problem*. (We remark that, historically, such conditions were sometimes called *right* $(m_1, \ldots, m_r)$ *focal point boundary conditions*, but we will not use that terminology.)

   Some of the results of this chapter, as well as the subsequent chapters on existence, rely on a type of "compactness condition" on sequences of solutions of (1.1.1). In particular, it was conjectured for many years that hypotheses (A) and (B), along with a uniqueness condition on $(1, 1, \ldots, 1)$ conjugate boundary value problems for (1.1.1) on (a, b), implied the following "compactness condition" on solutions:

   (CP) If $\{y_k(x)\}$ is a sequence of solutions of (1.1.1) such that for some $[c, d] \subset (a, b)$ and some $M > 0$, $|y_k(x)| \leq M$ on $[c, d]$, for all $k \geq 1$, then there exists a subsequence $\{y_{k_j}(x)\}$ such that $\{y_{k_j}^{(i)}(x)\}$ converges uniformly on each compact subinterval of (a, b), for each $0 \leq i \leq n - 1$.

The conjecture was based in part on an existing proof by Jackson and Schrader [82] for the case of $n = 3$. In 1985, Schrader [102] presented a paper, during a Sectional Meeting of the American Mathematical Society held at the University of Missouri-Columbia, in which he revealed that he and L. Jackson had established the validity of (CP) under the assumptions of (A) and (B) and the uniqueness condition:

   (C) There exists at most one solution of each $(1, 1, \ldots, 1)$ conjugate boundary value problem for (1.1.1) on (a, b).

Jackson and Schrader never published their result, yet written notes taken by A. C. Peterson of Jackson's presentation of the result in a seminar were communicated to J. Henderson. Because of its fundamental importance for this book, we will present the proof of the compactness result in its form as presented by Jackson in that seminar. (We remark that Agarwal [4] did publish the Jackson and Schrader proof.)

The proof that Jackson and Schrader gave made major use of results from a paper by Agronsky, Bruckner, Laczkovich and Preiss [5]. We now state some of the definitions and results from [5].

Definition 1.1.1. [[5], Agronsky *et al.*, p. 660] *Let*

$$P_n := \{p(x) : \mathbb{R} \to \mathbb{R} \mid p(x) \text{ is a polynomial of degree at most } n\}.$$

Definition 1.1.2. [[5], Agronsky *et al.*, p. 660] *Given* $E \subseteq [c, d]$, x_0 *is a* bilateral accumulation point *of* E, *in case* x_0 *is an accumulation point of both* $E \cap [c, x_0]$ *and* $E \cap [x_0, d]$.

Definition 1.1.3. [[5], Agronsky *et al.*, p. 666] *A function* $g : I \to \mathbb{R}$, *with* I *an interval, is said to be* n-convex (n-concave) *on* I *in case, for any distinct* $x_0, \ldots, x_n \in I$, $\sum_{i=0}^{n} \frac{g(x_i)}{w'(x_i)} \geq 0$ (≤ 0), *where* $w(x) = \prod_{i=0}^{n}(x - x_i)$, *(so* $w'(x_j) = \prod_{i=0, i \neq j}^{n}(x_j - x_i)$*).*

Theorem 1.1.1. [[5], Agronsky *et al.*, p. 666] *Suppose* $g \in C^{(n)}(I)$. *Then* g *is* n-convex *on* I *if, and only if,* $g^{(n)}(x) \geq 0$, *on* I.

Theorem 1.1.2. [[5], Agronsky *et al.*, p. 666] g *is* n-convex (n-concave) *on* I *if, and only if,* $g \in C^{(n-2)}(I)$ *and* $g^{(n-2)}$ *is convex (concave).*

Theorem 1.1.3. [[5], Agronsky *et al.*, Thm. 13] *Let* $g \in C[c, d]$ *and assume that, for each* $p \in P_n$, *the set,* $\{x : p(x) = g(x)\}$, *does not have a bilateral accumulation point in* (c, d). *Then, there exists a subinterval* $I \subseteq [c, d]$ *on which* g *is either* $(n+1)$-convex *or* $(n+1)$-concave.

One final fundamental result is given prior to stating and proving the Jackson and Schrader "compactness condition." This result is known as the Kamke convergence theorem for solutions of initial value problems; the reader is referred to [42, Theorem 3.2, p. 14] or [77].

Theorem 1.1.4. [Kamke] *Assume that in a sequence of differential equations*

$$y^{(n)} = f_k(x, y, y', \ldots, y^{(n-1)}), \quad k = 1, 2, \ldots, \qquad (1.1.4)$$

the functions $f_k(x, r_1, r_2, \ldots, r_n)$ *are continuous on* $I \times \mathbb{R}^n$, *where* I *is an interval of the reals, and assume*

$$\lim_{k \to \infty} f_k(x, r_1, r_2, \ldots, r_n) = f_0(x, r_1, r_2, \ldots, r_n)$$

uniformly on each compact subset of $I \times \mathbb{R}^n$. Assume that $\{x_k\}_{k=0}^{\infty} \subset I$ converges to x_0. For each $k = 1, 2, \ldots$, let y_k denote a solution of (1.1.4), for the respective k, defined on a maximal interval $I_k \subset I$ where $x_k \in I_k$. Assume $\lim_{k \to \infty} y_k^{(i-1)}(x_k) = y_i$ for each $i = 1, 2, \ldots, n$. Then there is a subsequence $\{y_{k_j}\}$ of $\{y_k\}$ and there is a solution y_0 of

$$y^{(n)} = f_0(x, y, y', \ldots, y^{(n-1)})$$

defined on a maximal interval $I_0 \subset I$ such that $x_0 \in I_0$, $y_0^{(i-1)}(x_0) = y_i$, $i = 1, 2, \ldots, n$, and such that for any compact interval $[c, d] \subset I_0$, then $[c, d] \subset I_{k_j}$ eventually and $\{y_{k_j}^{(i-1)}\}$ converges uniformly to $\{y_0^{(i-1)}\}$ on $[c, d]$, $i = 1, 2, \ldots, n$.

Now, we provide the Jackson and Schrader result known as the "compactness condition."

Theorem 1.1.5. [Jackson and Schrader] *If, with respect to (1.1.1), conditions (A)–(C) are satisfied, then condition (CP) is satisfied.*

Proof. Assume that $\{y_k(x)\}$ is a sequence of solutions of (1.1.1) which is uniformly bounded on some $[c, d] \subset (a, b)$. Then, by Helly's Selection (or Choice) Theorem [95, Helly's First Theorem, p. 222], there exists a subsequence $\{y_{k_j}(x)\}$ and a function $h \in BV[c, d]$ such that

$$\lim_{j \to \infty} y_{k_j}(x) = h(x) \text{ pointwise on } [c, d].$$

Set

$$H(x) := \int_c^x h(t) dt.$$

Then $H \in C[c, d]$, and by Theorem 1.1.3, either

(i) There exists a $p \in P_{n+1}$ such that $\{x : p(x) = H(x)\}$ has a bilateral accumulation point in (c, d),

or

(ii) H is $(n + 2)$-convex or $(n + 2)$-concave on some $[c_1, d_1] \subseteq [c, d]$.

We now relabel the subsequence $\{y_{k_j}(x)\}$ as $\{y_k(x)\}$.

And we first consider case (ii). It follows from Theorem 1.1.2 that $H \in C^{(n)}[c_1, d_1]$ and $H^{(n)}$ is convex, (or concave). Then, $H' \in C^{(n-1)}[c_1, d_1]$ and $H'(x) = h(x)$ a.e. on $[c_1, d_1]$. As a consequence of an application of the Schauder-Tychonoff fixed point theorem, $(1, \ldots, 1)$ conjugate boundary value problems for (1.1.1) are locally solvable to the extent that, there exists a $\delta = \delta(H', d_1 - c_1) > 0$ such that, for fixed $c_1 \leq x_1 < \cdots < x_n \leq d_1$, with $x_n - x_1 \leq \delta$, there exists an $\epsilon_0 > 0$ such that the boundary value problem for (1.1.1) satisfying

$$y(x_j) = H'(x_j) + \epsilon_j, \quad 1 \leq j \leq n,$$

where $|\epsilon_j| \leq \epsilon_0$, $1 \leq j \leq n$, has a solution $y(x)$. Moreover, this solution has bounds on $|y^{(i)}(x)|, 0 \leq i \leq n-1$, depending only on H' and $d_1 - c_1$; call these respective bounds $N_0 + 1, \ldots, N_{n-1} + 1$.

Next, choose points $c_1 \leq x_1 < \cdots < x_n \leq d_1$, with $x_n - x_1 \leq \delta$ and such that $H'(x_j) = h(x_j), 1 \leq j \leq n$. Then, there exists a $K \in \mathbb{N}$ such that, for $k \geq K$,

$$|y_k(x_j) - h(x_j)| \leq \epsilon_0, \quad 1 \leq j \leq n.$$

For $k \geq K$ and $1 \leq j \leq n$, let

$$\epsilon_{kj} := y_k(x_j) - h(x_j).$$

Then, for $k \geq K$ and $1 \leq j \leq n$,

$$y_k(x_j) = h(x_j) + \epsilon_{kj} = H'(x_j) + \epsilon_{kj},$$

and it follows from condition (C) that $y_k(x)$ is the solution referred to above from the local solvability. As a consequence, we have, for $k \geq K$,

$$|y_k^{(i)}(x)| \leq N_i + 1 \text{ on } [x_1, x_n],$$

for each $0 \leq i \leq n - 1$. An application of the Kamke Convergence Theorem yields a further subsequence $\{y_{k_\ell}(x)\}$ such that $\{y_{k_\ell}^{(i)}(x)\}$ converges uniformly on each compact subinterval of (a, b), for each $0 \leq i \leq n - 1$.

Now, we proceed with case (i). Hence, there exists a $p \in P_{n+1}$ such that $\{x : p(x) = H(x)\}$ has a bilateral accumulation point $x_0 \in (c, d)$. Assume $\{x_r\}_{r=1}^{\infty} \downarrow x_0$ is such that

$$p(x_r) = H(x_r), \text{ for each } r \geq 1.$$

If we assume on some subinterval $[x_{j+1}, x_j]$, that we have

$$H'(x) = h(x) \geq p'(x) \text{ a.e.,}$$

then

$$
\begin{aligned}
H(x_j) - H(x_{j+1}) &= \int_{x_{j+1}}^{x_j} h(t)dt \\
&\geq \int_{x_{j+1}}^{x_j} p'(t)dt \\
&= p(x_j) - p(x_{j+1}) \\
&= H(x_j) - H(x_{j+1}),
\end{aligned}
$$

so that the above inequality is an equality. From $h(x) - p'(x) \geq 0$ a.e. on $[x_{j+1}, x_j]$, whereas $\int_{x_{j+1}}^{x_j} (h(t) - p'(t))dt = 0$, we conclude $h(x) - p'(x) = 0$ a.e. on $[x_{j+1}, x_j]$; that is

$$H'(x) = h(x) = p'(x) \text{ a.e. on } [x_{j+1}, x_j].$$

Similarly, if we had assumed on some subinterval $[x_{j+1}, x_j]$, that

$$H'(x) = h(x) \leq p'(x) \text{ a.e.,}$$

then, we would arrive at

$$H'(x) = h(x) = p'(x) \text{ a.e. on } [x_{j+1}, x_j].$$

Now, $p'(x) \in C^{(n-1)}[c, d]$, and so again by the local solvability of solutions of $(1, \dots, 1)$ conjugate boundary value problems for (1.1.1), as an application of the Schauder-Tychonoff fixed point theorem, there exists a $\delta = \delta(p', d - c) > 0$ such that , for $c \leq t_1 < \cdots < t_n \leq d$, $t_n - t_1 \leq \delta$, there exists an $\epsilon_0 > 0$ such that the boundary value problem for (1.1.1) satisfying

$$y(t_j) = p'(t_j) + \epsilon_j, \quad 1 \leq j \leq n,$$

where $|\epsilon_j| \leq \epsilon_0, 1 \leq j \leq n$, has a solution $y(x)$, and as in the case (ii), $|y^{(i)}(x)|$, $0 \leq i \leq n - 1$, has bounds depending only on p' and $d - c$; again, call these bounds $N_0 + 1, \dots, N_{n-1} + 1$.

We now choose points $x_{\ell+n} < x_{\ell+n-1} < \cdots < x_\ell$ from $\{x_r\}_{r=1}^\infty$ such that $x_\ell - x_{\ell+n} \leq \delta$. There are two subcases to consider.

(a) For some $1 \leq \nu \leq n$, on the interval $[x_{\ell+\nu}, x_{\ell+\nu-1}]$, we have $h(x) \geq p'(x)$ a.e., or $h(x) \leq p'(x)$ a.e. Using arguments from above, we have $h(x) = p'(x)$ a.e. on $[x_{\ell+\nu}, x_{\ell+\nu-1}]$. We now repeat the argument of case (ii); that is, choose points $x_{\ell+\nu} \leq t_1 < \cdots < t_n \leq x_{\ell+\nu-1}$ such that $h(t_j) = p'(t_j), 1 \leq j \leq n$. Then, there exists a $K \in \mathbb{N}$ such that, for $k \geq K$ and $1 \leq j \leq n$,

$$|y_k(t_j) - h(t_j)| \leq \epsilon_0.$$

For $k \geq K$ and $1 \leq j \leq n$, let

$$\epsilon_{kj} := y_k(t_j) - h(t_j).$$

Then, for $k \geq K$ and $1 \leq j \leq n$,

$$y_k(t_j) = h(t_j) + \epsilon_{kj} = p'(t_j) + \epsilon_{kj},$$

and so from (C), $y_k(x)$ is the solution of the local problem referred to above. Hence, for $k \geq K$,

$$|y_k^{(i)}(x)| \leq N_i + 1 \text{ on } [t_1, t_n],$$

for each $0 \leq i \leq n - 1$. Again, an application of the Kamke Convergence Theorem yields a further subsequence $\{y_{k_\ell}(x)\}$ such that $\{y_{k_\ell}^{(i)}(x)\}$ converges uniformly on each compact subinterval of (a, b), for each $0 \leq i \leq n - 1$.

(b) For each $1 \leq \nu \leq n$, there exist positive Lebesgue measurable subsets A_ν and B_ν of $[x_{\ell+\nu}, x_{\ell+nu-1}]$, such that

$$h(x) > p'(x) \text{ on } A_\nu,$$
$$h(x) < p'(x) \text{ on } B_\nu.$$

Since $\lim_{k \to \infty} y_k(x) = h(x)$, there exists a $K \in \mathbb{N}$ such that, for $k \geq K$,

$$y_k(x) > p'(x) \text{ for some } x \in A_\nu,$$
$$y_k(x) < p'(x) \text{ for some } x \in B_\nu.$$

And so by continuity, for each $1 \leq \nu \leq n$, there exists $t_\nu \in (x_{\ell+\nu}, x_{\ell+\nu-1})$ such that $y_k(t_\nu) = p'(t_\nu)$.

In particular, there exist points $x_{\ell+n} \leq \tau_1 < \cdots < \tau_n \leq x_\ell$ (so $\tau_n - \tau_1 \leq \delta$), so that, for some $L \geq K$,

$$y_k(\tau_j) = p'(\tau_j) + \epsilon_{kj}, \quad 1 \leq j \leq n,$$

where $|\epsilon_{kj}| \leq \epsilon_0, 1 \leq j \leq n$ and $k \geq L$.

Again, it follows from (C) that $y_k(x)$ is the solution to the local problem referred to before subcase (a). So, for $k \geq L$,

$$|y_k^{(i)}(x)| \leq N_i + 1 \text{ on } [\tau_1, \tau_n],$$

for each $0 \leq i \leq n - 1$. An application of the Kamke Convergence Theorem provides a subsequence $\{y_{k_\ell}(x)\}$ such that $\{y_{k_\ell}^{(i)}(x)\}$ converges uniformly on each compact subinterval of (a, b), for each $0 \leq i \leq n - 1$.

This completes the proof. $\qquad\qquad\qquad\qquad\qquad\qquad\qquad\square$

Another major tool used throughout this book involves continuous dependence with respect to boundary conditions for solutions of boundary value problems for (1.1.1), within the context of uniqueness hypotheses on such solutions. Arguments used in proofs for this type of continuous dependence can be found in, to cite a couple, [62] and [76]. We offer as a model a typical such continuous dependence result for solutions of $(1, \ldots, 1)$ conjugate boundary value problems for (1.1.1).

Theorem 1.1.6. [Continuous Dependence] *Assume that with respect to* (1.1.1), *conditions* (A)–(C) *are satisfied. Let* $u(x)$ *be an arbitrary, but fixed, solution of* (1.1.1) *on* (a, b). *Then, for any* $a < x_1 < \cdots < x_n < b$, *any* c *and* d, *with* $a < c < x_1$ *and* $x_n < d < b$, *and any* $\epsilon > 0$, *there exists a* $\delta > 0$ *such that, for* $c < t_1 < \cdots < t_n < d$ *with* $|t_i - x_i| < \delta, 1 \leq i \leq n$, *and for* $v_i \in \mathbb{R}$ *with* $|u(x_i) - v_i| < \delta, 1 \leq i \leq n$, *there exists a solution* $v(x)$ *of* (1.1.1) *on* (a, b) *satisfying* $v(t_i) = v_i, 1 \leq i \leq n$, *and*

$$|u^{(j-1)}(x) - v^{(j-1)}(x)| < \epsilon, \quad c \leq x \leq d, \quad j = 1, \ldots, n.$$

Proof. First, let $\alpha \in (a, b)$ be fixed. Next, let

$$G := \{(t_1, \ldots, t_n, c_1, \ldots, c_n) \mid a < t_1 < \cdots < t_n < b \text{ and } (c_1, \ldots, c_n) \in \mathbb{R}^n\}.$$

G is an open subset of $\mathbb{R}^{2n}$. Now, define a mapping $\phi : G \to \mathbb{R}^{2n}$ by

$$\phi(t_1, \ldots, t_n, c_1, \ldots, c_n) := (t_1, \ldots, t_n, v(t_1), \ldots, v(t_n)),$$

where $v(x)$ is the solution of (1.1.1) satisfying the initial conditions, $v^{(i-1)}(\alpha) = c_i, 1 \leq i \leq n$.

Condition (B) implies the continuity of solutions of initial value problems for (1.1.1) with respect to initial conditions, which, in turn, implies the continuity of ϕ. Moreover, by condition (C), ϕ is one-to-one on G. Since G is an open subset of $\mathbb{R}^{2n}$, it follows from the Brouwer Theorem on Invariance of Domain that $\phi(G)$ is an open subset of $\mathbb{R}^{2n}$, that $\phi : G \to \phi(G)$ is a homeomorphism, and that ϕ^{-1} is continuous on $\phi(G)$.

With $u(x)$ the fixed solution in the hypotheses of the theorem, now let $a < x_1 \cdots < x_n < b$ be chosen, and let $a < c < x_1, x_n < d < b$, and $\epsilon > 0$ be chosen. By continuity with respect to initial conditions, there exists an $\eta > 0$ such that $|u^{(i-1)}(\alpha) - c_i| < \eta$, for $i = 1, \ldots, n$, implies $|v^{(j-1)}(x) - u^{(j-1)}(x)| < \epsilon$ on $[c, d]$, where $v(x)$ is the solution of (1.1.1) with $v^{(i-1)}(\alpha) = c_i, i = 1, \ldots, n$.

Now $(x_1, \ldots, x_n, u(x_1), \ldots, u(x_n)) \in \phi(G)$ and $\phi(G)$ is open, and $\phi^{-1} : \phi(G) \to G$ is continuous. So, there exists a $\delta > 0$ such that $|t_i - x_i| < \delta$, $i = 1, \ldots, n$, and $|v_i - u(x_i)| < \delta$, $i = 1, \ldots, n$, imply $(t_1, \ldots, t_n, v_1, \ldots, v_n) \in \phi(G)$, and by the continuity of ϕ^{-1}, $\phi^{-1}(t_1, \ldots, t_n, v_1, \ldots, v_n)$ belongs to the open cube of half-edge η and centered at $\phi^{-1}(x_1, \ldots, x_n, u(x_1), \ldots, u(x_n)) = (x_1, \ldots, x_n, u(\alpha), \ldots, u^{(n-1)}(\alpha))$. It follows from the preceding paragraph that (1.1.1) has a solution $v(x)$ with $v(t_i) = v_i, 1 \leq i \leq n$, and $|v^{(j-1)}(x) - u^{(j-1)}(x)| < \epsilon$ on $[c, d]$, for $j = 1, \ldots, n$. $\square$

It is also important to mention that alternative proofs have been given for continuous dependence of solutions of boundary value problems, and prominent among these would be the results in the paper by Vidossich [105].

Remark 1.1.1. *In the above result, the importance of the uniqueness condition in* (C) *for the continuous dependence of solutions is clear. In future contexts, continuous dependence of solutions on boundary conditions will be in play in analogy to Theorem 1.1.6 from certain uniqueness assumptions that will have been made then.*

Each of the results, Theorem 1.1.5 and Theorem 1.1.6, will be referred to frequently throughout much of this book.

1.2 Conjugate boundary value problems: for $m > k$, uniqueness of m-point implies uniqueness of k-point

This section is devoted to uniqueness results for solutions of conjugate boundary value problems (1.1.1), (1.1.2). Namely, we are concerned with uniqueness of solutions of m-point conjugate boundary value problems for (1.1.1), for fixed $2 < m \leq n$, implying the uniqueness of solutions of k-point conjugate boundary value problems for (1.1.1), for $2 \leq k < m$.

Our results involve from Section 1.1 conditions (A), (B), (C) and "compactness condition" (CP), and continuous dependence from Theorem 1.1.6 and Remark 1.1.1, along with other conditions we will impose.

For the case when (1.1.1) is linear, Opial [96] proved the result for $m = n$; that is, when (1.1.1) is linear, uniqueness of solutions of n-point conjugate boundary value problems implies uniqueness of solutions of all k-point boundary value problems, for all $2 \leq k < n$. (Of course, when (1.1.1) is linear, existence of solutions of all conjugate boundary value problems is also present in the Opial paper.) Using methods different from Opial, Sherman [104] also established, in the context of (1.1.1) linear that uniqueness of solutions of n-point conjugate boundary value problems implies the uniqueness of solutions of k-point conjugate boundary value problems, for all $2 \leq k < n$. Proof of that result, for when (1.1.1) is linear, can also be found in the monograph by Coppel [15].

Also, Peterson [97] proved, for (1.1.1) linear, that uniqueness of solutions of $(n-1)$-point conjugate boundary value problems implies the uniqueness of solutions of k-point conjugate boundary value problems, for all $2 \leq k < n-1$. Still in the context of (1.1.1) linear, for fixed $\lceil \frac{n}{2} \rceil \leq m \leq n$, Henderson

[48] proved that uniqueness of solutions of m-point conjugate boundary value problems implies the uniqueness of solutions of k-point conjugate boundary value problems, for all $2 \leq k < m$. We should add at this point that Sherman [103] proved that uniqueness of solutions of 2-point conjugate boundary value problems implies the uniqueness of solutions of n-point conjugate boundary value problems for (1.1.1) linear, and so when coupled with this Sherman result and the Opial result, many of the other above results actually yield uniqueness of solutions of all the conjugate boundary value problems. In fact, Muldowney [94] generalized these results, when he showed, for (1.1.1) linear and any fixed $2 \leq m \leq n$, uniqueness of solutions of m-point conjugate boundary value implied uniqueness of solutions of all k-point conjugate boundary value problems, for all $2 \leq k \leq n$.

For (1.1.1) nonlinear and $n = 3$, Jackson and Schrader [81] assumed conditions (A), (B) and (C) (that is, that solutions of 3-point conjugate boundary value problems for (1.1.1) are unique on (a, b), when they exist). They then proved that solutions of 2-point conjugate boundary value problems for (1.1.1) are unique on (a, b), when they exist. We present that result now.

Theorem 1.2.1. [[81], Jackson and Schrader, Lem. 1] *Let* $n = 3$ *and assume that with respect to* (1.1.1), *conditions* (A)–(C) *are satisfied. Then, each 2-point conjugate boundary value problem for* (1.1.1) *on* (a, b) *has at most one solution.*

Proof. We will establish the proof for the $(2, 1)$ conjugate boundary value problem. So, assume to the contrary that there are distinct solutions $y(x)$ and $z(x)$ of (1.1.1) and points $a < x_1 < x_2 < b$ such that

$$y(x_1) = z(x_1), \ y'(x_1) = z'(x_1) \text{ and } y(x_2) = z(x_2).$$

By (B), $y''(x_1) \neq z''(x_1)$, and so, we may assume, without loss of generality, that $y''(x_1) > z''(x_1)$ and that $y(x) > z(x)$ on (x_1, x_2).

Fix $a < \tau_1 < x_1$ and $x_2 < \tau_2 < b$. By (C) and Theorem 1.1.6, solutions of (1.1.1) depend continuously on 3-point conjugate boundary conditions. In particular, given $\epsilon > 0$, there is a $\delta > 0$ such that there is a solution $z_\delta(x)$ of (1.1.1) satisfying

$$z_\delta(\tau_1) = z(\tau_1), \ z_\delta(x_1) = z(x_1) + \delta, \ z_\delta(x_2) = z(x_2),$$

and $|z_\delta(x) - z(x)| < \epsilon, \tau_1 \leq x \leq \tau_2$. Since $y''(x_1) > z''(x_1)$, it follows that, for ϵ sufficiently small, there exist $\tau_1 < p_1 < x_1 < p_2 < x_2$ such that

$$z_\delta(p_1) = y(p_1), \ z_\delta(p_2) = y(p_2), \ z_\delta(x_2) = y(x_2).$$

By condition (C), it follows that $z_\delta(x) \equiv y(x)$ on (a, b). But $z_\delta(x_1) = y(x_1) + \delta$, which is a contradiction.

Similarly, each $(1, 2)$ conjugate boundary value problem for (1.1.1) has at most one solution on (a, b). The proof is complete. □

For equation (1.1.1) of arbitrary order n, Jackson and Klaasen [80] dealt with the question of uniqueness of solutions of n-point conjugate boundary value problems implying the uniqueness of solutions of k-point conjugate boundary value problems, when solutions exist, for $2 \leq k < n$. They obtained partial results depending on the parity of n. Namely, if n is *odd*, then (A)–(C) imply the uniqueness of all $(n-1)$-point and all $(n-2)$-point conjugate boundary value problems on (a, b), when solutions exist, whereas, if n is *even*, the same assumptions imply the uniqueness of all $(n-1)$-point conjugate boundary value problems on (a, b), when solutions exist.

Later, Jackson [76] adapted the linear argument of Opial [96], by employing continuous dependence on boundary conditions (via application of the Brouwer Theorem on Invariance of Domain), and completed the earlier Jackson and Klaasen [80] uniqueness result. We present that result now.

Theorem 1.2.2. [[76], Jackson, Thm. 2] *Assume that with respect to (1.1.1), conditions (A)–(C) are satisfied. Then, each k-point conjugate boundary value problem for (1.1.1) on (a, b), $2 \leq k \leq n - 1$, has at most one solution.*

Proof. Assume that the conclusion to the theorem is false. Then, some k-point conjugate boundary value problem, $2 \leq k \leq n - 1$, for (1.1.1) has distinct solutions. Let h be the largest integer for which this is the case. Then $2 \leq h < n$, and all k-point conjugate boundary value problems, $h + 1 \leq k \leq n$, have at most one solution.

Let $a < x_1 < \cdots < x_h < b$ be such that, there exist distinct solutions, $y(x)$ and $z(x)$, of (1.1.1) with $y(x) - z(x)$ having a zero of order $r_j \geq 1$ at x_j, for $1 \leq j \leq h$, where $\sum_{j=1}^{h} r_j \geq n$. We assume that for each $1 \leq j \leq h$, the order of the zero at x_j is exactly r_j; that is, $y^{(i)}(x_j) - z^{(i)}(x_j) = 0, 0 \leq i \leq r_j - 1$, and $y^{(r_j)}(x_j) - z^{(r_j)}(x_j) \neq 0$. Since $h < n$, at least one of the zeros is a multiple zero.

Hereafter, let x_0 and x_{h+1}, with $a < x_0 < x_1$ and $x_h < x_{h+1} < b$, be fixed.

There are two cases.

Case (i): $h = n - 1$.

Suppose first, for some $1 \leq k \leq h$, r_k is even; that is, $y^{(i)}(x_k) - z^{(i)}(x_k) = 0, 0 \leq i \leq r_k - 1$, and $r_k = 2m, m \geq 1$, and we may assume $y^{(r_k)}(x_k) > z^{(r_k)}(x_k)$. By the uniqueness of solutions of $(h+1)$-point (that is, n-point), conjugate boundary value problems for (1.1.1), it follows from Theorem 1.1.6 that solutions of $(h + 1)$-point conjugate boundary value problems depend continuously on boundary conditions. Namely, given $\epsilon > 0$, there is a $\delta > 0$ such that there exists a solution, $z_\delta(x)$, of (1.1.1) satisfying,

$$z_\delta(x_k) = z(x_k) + \delta, \ z_\delta(x_j) = z(x_j), \ j \in \{1, \ldots, h+1\} \setminus \{k\},$$

and $|z_\delta^{(i-1)}(x) - z^{(i-1)}(x)| < \epsilon, 1 \leq i \leq n$, on $[x_0, x_{h+1}]$. It follows that $z_\delta(x_j) - y(x_j) = 0, j \in \{1, \ldots, h\} \setminus \{k\}$, and since r_k is even, for $\epsilon > 0$ sufficiently small, there exist $x_{k-1} < \tau_1 < x_k < \tau_2 < x_{k+1}$ such that $z_\delta(\tau_i) - y(\tau_i) = 0, i = 1, 2$. Thus, $z_\delta(x) - y(x)$ has at least $(h-1) + 2 = h + 1 = n$ distinct zeros, but this is a contradiction.

Suppose next that, for each $1 \leq j \leq h, r_j$ is odd. In particular, $y(x) - z(x)$ changes sign at x_j, $1 \leq j \leq h$. Furthermore, for some $1 \leq k \leq h$, $r_k > 1$. For $\epsilon > 0$, let $z_\epsilon(x)$ be the solution of the initial value problem for (1.1.1) satisfying

$$z_\epsilon^{(i)}(x_k) = z^{(i)}(x_k), \ 0 \leq i \leq n - 1, \ i \neq r_k - 1,$$
$$z_\epsilon^{(r_k-1)}(x_k) = z^{(r_k-1)}(x_k) + \epsilon.$$

By (B), solutions of initial value problems for (1.1.1) depend continuously on initial conditions. So, for $\epsilon > 0$ sufficiently small, $y(x) - z_\epsilon(x)$ has a zero in a neighborhood of x_j, $1 \leq j \leq h$, $j \neq k$, and a zero of multiplicity of $r_k - 1$ at x_k. And since $r_k > 1$ and odd, and $z_\epsilon^{(r_k-1)}(x_k) > y^{(r_k-1)}(x_k)$, it follows that $y(x) - z_\epsilon(x)$ has at least one zero in a deleted neighborhood of x_k. That is, $y(x) - z_\epsilon(x)$ has at least n distinct zeros; again, a contradiction.

Case (ii): $h < n - 1$.

The argument in the second part of Case (i) shows that, if $h - 1$ zeros of $y(x) - z(x)$ are simple, (hence $y(x) - z(x)$ would have a sign change at each of those zeros), then we can find another solution $z_\epsilon(x)$ of (1.1.1) such that $y(x) - z_\epsilon(x)$ has more than h distinct zeros of total multiplicity at least n. This contradicts the definition of h.

Thus, it remains to consider the case where at least two zeros, x_k and x_m say, of $y(x) - z(x)$ are not simple. Without loss of generality we may assume $y^{(r_k)}(x_k) - z^{(r_k)}(x_k) > 0$. Now choose integers $q_1, \ldots, q_j$ and $s \geq 1$ such that

$$1 \leq q_j \leq r_j, \ 1 \leq j \leq h, \ j \neq k,$$
$$0 \leq q_k = r_k - 2s,$$
$$\text{and } q := q_1 + \cdots + q_h = n - 2.$$

By the maximality of h, it follows from Remark 1.1.1 that solutions of $(h + 1)$-point conjugate boundary value problems for (1.1.1) depend continuously on boundary conditions. That is, given $\epsilon > 0$, there is a $\delta > 0$ such that there is a solution $z_\delta(x)$ of (1.1.1) satisfying

$$z_\delta^{(i)}(x_j) = z^{(i)}(x_j), \ 0 \leq i \leq q_j - 1, 1 \leq j \leq h, \ j \neq k,$$
$$z_\delta(x_{h+1}) = z(x_{h+1}),$$
$$z_\delta^{(i)}(x_k) = z^{(i)}(x_k), \ 0 \leq i \leq q_k - 1,$$
$$z_\delta^{(q_k)}(x_k) = z^{(q_k)}(x_k) + \delta,$$

and $|z^{(i)}(x) - z_\delta^{(i)}(x)| < \epsilon$ on $[x_0, x_{h+1}]$, $0 \leq i \leq n - 1$. It follows that $y(x) - z_\delta(x)$ has a zero at x_j of multiplicity $q_j, 1 \leq j \leq h, j \neq k$. If $q_k \geq 1$, $y(x) - z_\delta(x)$ has a zero of order q_k at x_k. Moreover, since $y^{(r_k)}(x_k) - z_\delta^{(r_k)}(x_k) > 0$, $q_k = r_k - 2s$, and $y^{(q_k)}(x_k) - z_\delta^{(q_k)}(x_k) = -\epsilon < 0$, it follows from an argument similar to the one used in the second part of Case (i) that $y(x) - z_\delta(x)$ has at least two zeros in $(x_{k-1}, x_{k+1}) \setminus \{x_k\}$. So, $y(x) - z_\delta(x)$ has at least $h + 1$ distinct zeros of total multiplicity at least n, which again contradicts the definition of h. The proof is complete. $\qquad\square$

For our final results of this section on uniqueness of solutions of m-point conjugate boundary value problems for (1.1.1) implying uniqueness of solutions of k-point conjugate boundary value problems for (1.1.1), for $2 \leq k < m$, we present two theorems by Henderson and Jackson [62]. Both results are for (1.1.1) of arbitrary order n. And the first of the results is fundamental for the second result.

Theorem 1.2.3. [[62], Henderson and Jackson, Thm. 1] *Assume that conditions* (A) *and* (B) *are satisfied with respect to* (1.1.1), *and assume there is an integer h such that $2 \leq h - 1 < h < n$, and such that all h-point and all $(h-1)$-point conjugate boundary value problems for* (1.1.1) *on (a, b) have at most one solution. Then, all k-point conjugate boundary value problems for* (1.1.1) *on (a, b) have at most one solution, for $2 \leq k \leq h$.*

Proof. The proof follows much along the path of the proof of Theorem 1.2.2. Assume the hypotheses of the theorem are satisfied, but that the conclusion of the theorem is false. Then h must satisfy $2 < h - 1 < h < n$ and there is a largest integer k, with $2 \leq k < h - 1$ such that some k-point conjugate boundary value for (1.1.1) on (a, b) has two distinct solutions. It follows from the maximality of k and the fact that $k < h - 1$ that all $(k + 1)$-point and all $(k + 2)$-point conjugate boundary value problems for (1.1.1) on (a, b) have at most one solution.

Let $y(x)$ and $z(x)$ be distinct solutions of (1.1.1) such that, for some $a < x_1 < \cdots < x_k < b$, $y(x) - z(x)$ has a zero at x_j of exact order p_j, for each $1 \leq j \leq k$, where each $p_j \geq 1$ and $\sum_{j=1}^{n} p_j \geq n$. Because of (B), $p_j < n$, for $1 \leq j \leq k$, and since $k < n$, at least one $p_j > 1$. Let $1 \leq r \leq k$ be chosen such that $p_r = \max\{p_j \,|\, 1 \leq j \leq k\}$.

Assume that $y^{(p_r)}(x_r) > z^{(p_r)}(x_r)$, and choose x_0 and x_{k+1} such that $a < x_0 < x_1 \cdots < x_k < x_{k+1} < b$. There are two cases to consider.

The first case is $p_j = 1$, for $j \in \{1, \ldots, k\} \setminus \{r\}$, and the second case is that, for some $j_0 \in \{1, \ldots, k\} \setminus \{r\}$, $p_{j_0} > 1$.

Case: $p_j = 1$, for $j \in \{1, \ldots, k\} \setminus \{r\}$: Let $\epsilon > 0$ be given, and let $y_\epsilon(x)$ be the solution of the initial value problem for (1.1.1) satisfying

$$y_\epsilon^{(i)}(x_r) = y^{(i)}(x_r), \ 0 \leq i \leq n - 1, \ i \neq p_r - 1,$$
$$y_\epsilon^{(p_r-1)}(x_r) = y^{(p_r-1)}(x_r) - \epsilon.$$

By (B) and the Kamke Convergence Theorem,

$$|y_\epsilon^{(i)}(x) - y^{(i)}(x)| \to 0 \text{ uniformly on } [x_0, x_{k+1}],$$

as $\epsilon \to 0$, for each $0 \leq i \leq n - 1$. And so, for $\epsilon > 0$ sufficiently small, $y_\epsilon(x) - z(x)$ has a zero in a neighborhood of x_j, $j \in \{1, \ldots, k\} \setminus \{r\}$, and $y_\epsilon(x) - z(x)$ has a zero of order $p_r - 1$ at x_r, and $y_\epsilon(x) - z(x)$ has a zero in (x_r, x_{r+1}). Thus, $y_\epsilon(x)$ and $z(x)$ are distinct solutions of a $(k + 1)$-point conjugate boundary value problem, which contradicts the uniqueness of solutions of such problems.

Case: For some $j_0 \in \{1, \ldots, k\} \setminus \{r\}$, $p_{j_0} > 1$: There are integers $q_j, 1 \leq j \leq k$, and an integer $s \geq 1$ such that

$$1 \leq q_j \leq p_j, \ 1 \leq j \leq k, \ j \neq r,$$
$$0 \leq q_r \leq p_r - 2s,$$
$$\text{and } \sum_{j=1}^{k} q_j = n - 2.$$

Now, from the uniqueness of solutions of $(k+1)$-point conjugate boundary value problems for (1.1.1), Remark 1.1.1 implies that solutions of such problems depend continuously on boundary conditions. So, if τ with $x_k < \tau < x_{k+1}$ is fixed, then for $\epsilon > 0$ sufficiently small, there is a solution $y_\epsilon(x)$ of (1.1.1) satisfying the $(k+1)$-point conjugate boundary conditions,

$$y_\epsilon^{(i)}(x_j) = y^{(i)}(x_j),\ 0 \le i \le q_j - 1,\ 1 \le j \le k,\ j \ne r,$$
$$y_\epsilon^{(i)}(x_r) = y^{(i)}(x_r),\ 0 \le i \le q_r - 1\ (\text{omitted if } q_r = 0),$$
$$y_\epsilon^{(q_r)}(x_r) = y^{(q_r)}(x_r) - \epsilon,$$
$$y_\epsilon(\tau) = y(\tau),$$

and in addition,

$$|y_\epsilon^{(i)}(x) - y^{(i)}(x)| \to 0 \text{ uniformly on } [x_0, x_{k+1}],$$

as $\epsilon \to 0$, for each $0 \le i \le n - 1$.

Thus, for $\epsilon > 0$ sufficiently small, $y_\epsilon(x) - z(x)$ has a zero of order q_j at x_j, $j \in \{1, \ldots, k\} \setminus \{r\}$, has a zero in each of the open intervals (x_{r-1}, x_r) and (x_r, x_{r+1}), and if $q_r > 0$, has a zero of order q_r at x_r. Namely, for sufficiently small $\epsilon > 0$, $y_\epsilon(x)$ and $z(x)$ are distinct solutions of the same $(k+1)$-point conjugate boundary value problem, when $q_r = 0$, and are distinct solutions of the same $(k+2)$-point conjugate boundary value problem, when $q_r > 0$. In either case, we have a contradiction, because of the maximality of k. This completes the proof. $\qquad\square$

As stated earlier in this section, Peterson [97] proved that, in the case when (1.1.1) is linear, uniqueness of solutions of $(n-1)$-point conjugate boundary value problems implies the uniqueness of solutions of k-point conjugate boundary value problems, for each $2 \le k < n-1$. Our last result of this section can be considered as an analog for the nonlinear equation (1.1.1). Moreover, this last result of the section provides a sequence for uniqueness results of the next section.

Theorem 1.2.4. [[62], Henderson and Jackson, Thm. 2] *Assume that conditions* (A) *and* (B) *are satisfied with respect to* (1.1.1). *If all* $(n-1)$-point *conjugate boundary value problems for* (1.1.1) *on* (a, b) *have at most one solution, then all k-point conjugate boundary value problems for* (1.1.1) *on* (a, b) *have at most one solution, for $2 \le k < n-1$.*

Proof. As a consequence of Theorem 1.2.3, it is sufficient to show that all $(n-2)$-point conjugate boundary value problems for (1.1.1) on (a, b) have at most one solution. We assume this is not the case and that $y(x)$ and $z(x)$ are distinct solutions of (1.1.1) such that $y(x) - z(x)$ has a zero of exact order $p_j \geq 1$ at x_j, for each $1 \leq j \leq n - 2$, where $a < x_1 < \cdots < x_{n-2} < b$ and $\sum_{j=1}^{n-2} p_j \geq n$.

Fix x_0 and x_{n-1} such that $a < x_0 < x_1$ and $x_{n-2} < x_{n-1} < b$. As before, there are again two cases to consider.

For the first case, we assume that either each p_j is an odd integer, or that one p_j is an even integer and all the other $p_i = 1$. For either of these choices, we let $1 \leq r \leq n-2$ be such that $p_r = \max\{p_j \,|\, 1 \leq j \leq n-2\}$. Then $p_r \geq 3$, and using solutions of initial value problems with initial conditions given at x_r, as was done in the proof of Theorem 1.2.3, we obtain a contradiction to the uniqueness of solutions of $(n - 1)$-point conjugate boundary value problems.

In the second case, which is the complement of the first case, there exists an r, with $1 \leq r \leq n - 2$, such that p_r is even and $\sum_{j=1, j \neq r}^{n-2} p_j \geq n - 2$. It follows that we can choose $1 \leq q_j \leq p_j, j \in \{1, \ldots, n - 2\} \setminus \{r\}$, such that $\sum_{j=1, j \neq r}^{n-2} q_j = n-2$. Assume $y^{(p_r)}(x_r) > z^{(p_r)}(x_r)$, and let τ be chosen such that $x_{n-2} < \tau < x_{n-1}$. Then, as in the second part of the proof of Theorem 1.2.3, for $\epsilon > 0$ sufficiently small, there is a solution $y_\epsilon(x)$ of (1.1.1) such that

$$y_\epsilon^{(i)}(x_j) = y^{(i)}(x_j),\ 0 \leq i \leq q_j - 1,\ 1 \leq j \leq n - 1,\ j \neq r,$$
$$y_\epsilon(x_r) = y(x_r) - \epsilon,$$
$$y_\epsilon(\tau) = y(\tau),$$

and

$$|y_\epsilon^{(i)}(x) - y^{(i)}(x)| \to 0 \text{ uniformly on } [x_0, x_{n-1}],$$

as $\epsilon \to 0$, for $0 \leq i \leq n - 1$. Then, for $\epsilon > 0$ sufficiently small, $y_\epsilon(x) - z(x)$ has a zero in each of the open intervals (x_{r-1}, x_r) and (τ_r, x_{r+1}), and $y_\epsilon(x) - z(x)$ has a zero of order q_j at x_j, for $j \in \{1, \ldots, n - 2\} \setminus \{r\}$. This again contradicts the uniqueness of solutions of $(n - 1)$-point conjugate boundary value problems for (1.1.1), and the theorem is proved. $\qquad \square$

1.3 Conjugate boundary value problems: for $m < k$, uniqueness of m-point implies uniqueness of k-point

This section is also devoted to uniqueness results for solutions of conjugate boundary value problems (1.1.1), (1.1.2) that are converses and partial converses of results obtained in Section 1.2. Namely, we are concerned with uniqueness of solutions of m-point conjugate boundary value problems for (1.1.1), this time for fixed $2 \leq m < n$, implying the uniqueness of solutions of k-point conjugate boundary value problems for (1.1.1), this time for $m < k \leq n$.

In 1960, in one of the first results, Azbelev and Tsalyuk [7] proved that, if $n = 3$ and (1.1.1) is linear, then uniqueness of solutions of 2-point conjugate boundary value problems implies the uniqueness of solutions of 3-point conjugate boundary value problems. In 1965, Sherman [103] proved, for (1.1.1) linear and of arbitrary order n, that uniqueness of solutions of 2-point conjugate boundary value problems implies the uniqueness of solutions of k-point conjugate boundary value problems, for all $2 < k \leq n$. And it follows from Muldowney's [94] work in 1984, for (1.1.1) linear and of arbitrary order n, that given $2 \leq m < n$, if solutions of m-point conjugate boundary value problems are unique, then solutions of k-point conjugate boundary value problems are unique, for all $2 \leq k \leq n$.

Again, our results will involve from Section 1.1 conditions (A), (B), (C) and "compactness condition" (CP), and continuous dependence from Theorem 1.1.6 and Remark 1.1.1, along with other conditions we will impose.

It was mentioned in Section 1.1 that the long running conjecture about the "compactness condition" (CP) was based, in part at that time, on a proof for the case of $n = 3$ by Jackson and Schrader [82]. Also in that paper, as we presented in Theorem 1.2.1, the authors established that uniqueness of solutions of 3-point conjugate problems implies uniqueness of solutions of 2-point conjugate problems. But, in their proof concerning the validity of (CP), for $n = 3$, they actually made use of conditions (A) and (B) and the uniqueness condition on 2-point conjugate boundary value problems:

(D) For $n = 3$ and for each $m_1, m_2 \in \mathbb{N}$ such that $m_1 + m_2 = 3$, there exists at most one solution of each (m_1, m_2) conjugate boundary value problem for (1.1.1) on (a, b).

Remark 1.3.1. *That is, for* (1.1.1), *when* $n = 3$, *conditions* (A), (B) *and* (D) *imply* (CP).

For (1.1.1) nonlinear and $n = 3$, Jackson [75] proved that the theorem of Azbelev and Tsalyuk is valid. In particular, he proved that conditions (A), (B) and (D) imply (C). That is the first result we present in this section.

Theorem 1.3.1. [[75], Jackson, Thm. 1] *Let $n = 3$ and assume that with respect to (1.1.1), conditions (A), (B) and (D) are satisfied. Then, each 3-point conjugate boundary value problem for (1.1.1) on (a, b) has at most one solution; that is, condition (C) is satisfied.*

Proof. We assume the hypotheses, but that the conclusion of the theorem is false. Then, there are points $a < x_1 < x_2 < x_3 < b$ and distinct solutions $y(x)$ and $z(x)$ of (1.1.1) such that $y(x_j) = z(x_j)$, for $j = 1, 2, 3$. It follows from (D) that $y'(x_j) \neq z'(x_j)$, for $j = 1, 2, 3$. Hence, without loss of generality, we can assume $y(x) > z(x)$ on (x_1, x_2) and $y(x) < z(x)$ on (x_2, x_3).

For each $n \geq 1$, let $y_n(x)$ be the solution on (a, b) of (1.1.1) satisfying the initial conditions,

$$y_n(x_1) = y(x_1), \quad y_n'(x_1) = y'(x_1), \quad \text{and} \quad y_n''(x_1) = y''(x_1) + n.$$

Again, by (D), it follows that, for each $n \geq 1$,

$$y_{n+1}(z) > y_n(x) > y(x) \text{ on } (a, b) \setminus \{x_1\}.$$

For each $n \geq 1$, let

$$E_n := \{x \mid x_2 \leq x \leq x_3 \text{ and } y_n(x) \leq z(x)\}.$$

It is not difficult to see, from continuous dependence of solutions on initial conditions, that each of the sets $E_n \neq \emptyset$. So, $E_{n+1} \subset E_n \subset (x_2, x_3)$, for each n, and each E_n is nonempty and compact. Therefore,

$$\bigcap_{n=1}^{\infty} E_n := E \neq \emptyset.$$

Next, we observe that $E = \{x_0\}$, for some $x_2 < x_0 < x_3$. Because, if there are points $t_1, t_2 \subset E$, where $x_2 < t_1 < t_2 < x_3$, then the same type of argument used to show each $E_n \neq \emptyset$ leads to the conclusion that the interval $[t_1, t_2] \subseteq E$. But then, the sequence $\{y_n(x)\}$ is uniformly bounded on $[t_1, t_2]$, and by the Remark 1.3.1, there is a subsequence $\{y_{n_j}(x)\}$ such that $\{y_{n_j}^{(i)}(x)\}$ converges uniformly on each compact subinterval of (a, b), for $i = 0, 1, 2$. But that is in contradiction to $y_{n_j}''(x_1) = y''(x_1) + n_j$. Therefore, $E = \{x_0\}$, with $x_2 < x_0 < x_3$ and $\lim_{n \to \infty} y_n(x_0) := y_0 \leq z(x_0)$.

We claim that this is not possible. There are two cases:

First, assume $y_0 = z(x_0)$. Then, for $\epsilon > 0$ sufficiently small, by continuous dependence of solutions of (1.1.1) on $(2, 1)$ conjugate boundary conditions, there is a solution $z(x, \epsilon)$ of (1.1.1) such that

$z(x_1, \epsilon) = z(x_1), z'(x_1, \epsilon) = z'(x_1), z(x_0, \epsilon) = z(x_0) - \epsilon$, and $z(x, \epsilon) \leq z(x)$ on $[x_1, x_3]$. Then, for ϵ chosen such that $z(x_0, \epsilon) = z(x_0) - \epsilon > y(x_0)$, the solution $z(x, \epsilon)$ can be used in place of $z(x)$ in defining the sequence $\{E_n\}$ with respect to the given sequence $\{y_n(x)\}$. Then, the previous arguments would yield that each of the new sets would be nonempty, which is impossible.

Second, assume $y(x_0) < y_0 < z(x_0)$. The line segment $\{(p_1, p_2, p_3) \,|\, p_i = \lambda y^{(i-1)}(x_1) + (1 - \lambda)z^{(i-1)}(x_1), \ i = 1, 2, 3, \ 0 \leq \lambda \leq 1\}$ is a connected subset of $\mathbb{R}^3$, and so there exists a $0 \leq \lambda_0 \leq 1$, and a solution $y(x, \lambda_0)$ of (1.1.1) satisfying

$$y^{(i)}(x_1, \lambda_0) = \lambda_0 y^{(i)}(x_1) + (1 - \lambda_0)z^{(i)}(x_1), \quad i = 0, 1, 2,$$

and such that

$$y(x_0, \lambda_0) = y_0.$$

And we also have

$$y(x_1, \lambda_0) = y(x_1) = z(x_1).$$

By continuity, there is an $\eta > 0$ such that $[x_0 - \eta, x_0 + \eta] \subset (x_2, x_3)$ and

$$y(x, \lambda_0) < z(x) \text{ on } [x_0 - \eta, x_0 + \eta].$$

Then, with $\{y_n(x)\}$ the same sequence as previously defined, we have

$$\lim_{n \to \infty} y_n(x) > z(x) > y(x, \lambda_0) \text{ on } [x_0 - \eta, x_0 + \eta] \setminus \{x_0\},$$

and

$$\lim_{n \to \infty} y_n(x_0) = y_0 = y(x_0, \lambda_0).$$

This leads to the same contradiction as above in the case $y_0 = z(x_0)$.

Hence, from the two contradictions, we conclude $y_0 \leq z(x_0)$ is impossible. Therefore, equation (1.1.1) must satisfy condition (C). □

When Theorem 1.2.1 is coupled with Theorem 1.3.1, we obtain the following uniqueness result for solutions of (1.1.1) when $n = 3$.

Theorem 1.3.2. *Let $n = 3$ and assume that with respect to (1.1.1), conditions (A) and (B) are satisfied. Then, condition (C) holds if, and only if, condition (D) holds.*

When $n = 4$, D. Peterson [101] obtained, in some sense, extensions of Theorem 1.3.1 for uniqueness of solutions of 2-point conjugate boundary value problems implying uniqueness of solutions for some 3-point and for all 4-point conjugate boundary value problems. It is important to note that Theorem 1.1.5 possibly does not hold for Peterson's results, and so his arguments require assumption of the "compactness condition" (CP). Peterson established his uniqueness results in a single theorem, but for clarity purposes, we choose to present his results in two theorems.

Theorem 1.3.3. [[101], D. Peterson, Thm. 3.1, p. 22] *For $n = 4$, assume that with respect to* (1.1.1), *conditions* (A), (B) *and* (CP) *are satisfied. If, for each $m_1, m_2 \in \mathbb{N}$ such that $m_1 + m_2 = 4$, there exists at most one solution of each (m_1, m_2) conjugate boundary value problem for* (1.1.1) *on (a, b), then solutions of all $(2, 1, 1)$ conjugate boundary value problems and all $(1, 1, 2)$ conjugate boundary value problems for* (1.1.1) *on (a, b) are unique, when they exist.*

Proof. Consider the $(2, 1, 1)$ conjugate boundary value problem. Suppose to the contrary that there exist distinct solutions $y(x)$ and $z(x)$ of (1.1.1) and points $a < x_1 < x_2 < x_3 < b$ such that $y(x_i) = z(x_i), i = 1, 2, 3$, and $y'(x_1) = z'(x_1)$. By the hypotheses, $y''(x_1) \neq z''(x_1), y'(x_2) \neq z'(x_2)$ and $y'(x_3) \neq z'(x_3)$. We may assume without loss of generality that $y''(x_1) > z''(x_1)$, $y(x) > z(x)$ on (x_1, x_2) and $y(x) < z(x)$ on (x_2, x_3).

For each $n \geq 1$, let $y_n(x)$ be the solution of the initial value problem for (1.1.1) satisfying

$$y_n^{(i)}(x_1) = y^{(i)}(x_1), \ i = 0, 1, 2, \text{ and } y_n'''(x_1) = y'''(x_1) + n.$$

It follows from uniqueness of solutions of $(3, 1)$ conjugate boundary value problems for (1.1.1) that $y_{n+1}(x) > y_n(x) > y(x)$ on (x_1, b), for each $n \geq 1$. Next, for each $n \geq 1$, let

$$E_n := \{x \,|\, x_2 \leq x \leq x_3 \text{ and } y_n(x) \leq z(x)\}.$$

By their uniqueness, solutions of $(3, 1)$ conjugate boundary value problems depend continuously on boundary conditions. Using that continuous dependence, as in the previous theorem, it is not difficult to argue that each $E_n \neq \emptyset$. Thus, for each $n \geq 1$, $\emptyset \neq E_{n+1} \subset E_n$, and each E_n is compact. Therefore,

$$\bigcap_{n=1}^{\infty} E_n := E \neq \emptyset, \text{ and } E \text{ is compact.}$$

As in the previous theorem, our claim is that $E = \{x_0\}$. For, if $t_1, t_2 \in E$ with $x_2 < t_1 < t_2 < x_3$, then invoking an argument similar to showing each $E_n \neq \emptyset$ leads to $[t_1, t_2] \subset E$. It then follows that $y(x) \leq y_n(x) \leq z(x)$ on $[t_1, t_2]$; that is, $\{y_n(x)\}$ is uniformly bounded on $[t_1, t_2]$. In view of assumed condition (CP), this is not possible since $y_n''(x_1) \to +\infty$, as $n \to \infty$. Hence $E = \{x_0\}$, where $x_2 < x_0 < x_3$, and

$$\lim_{n \to \infty} y_n(x) := y_0 \leq z(x_0).$$

By arguments much along the lines of those used in the previous theorem, we show that each of the cases, $y_0 = z(x_0)$ and $y_0 < z(x_0)$, is impossible.

First, assume $y_0 = z(x_0)$. By continuous dependence of $(3,1)$ conjugate boundary value problems, for $0 < \epsilon < z(x_0) - y(x_0)$ sufficiently small, there is a solution $z(x, \epsilon)$ of (1.1.1) satisfying the boundary conditions,

$$z^{(i)}(x_1, \epsilon) = z^{(i)}(x_1), \ i = 0, 1, 2, \ z(x_0, \epsilon) = z(x_0) - \epsilon,$$

and $z(x, \epsilon) < z(x)$ on $(x_1, x_3]$. For such an $\epsilon > 0$, $z(x_0, \epsilon) > y(x_0)$, and the solution $z(x, \epsilon)$ can be used in place of $z(x)$ in defining the sequence $\{E_n\}$ with respect to the same sequence of solutions $\{y_n(x)\}$. The previous arguments in this proof could be repeated to conclude that these new sets would each be nonempty, which is impossible.

Next, assume $y(x_0) < y_0 < z(x_0)$. For $0 \leq \lambda \leq 1$, let $z(x, \lambda)$ be the solution of (1.1.1) with

$$z^{(i)}(x_1, \lambda) = \lambda y^{(i)}(x_1) + (1 - \lambda)z^{(i)}(x_1), \ i = 0, 1, 2, 3.$$

We note that $z(x, 0) \equiv z(x)$ and $z(x, 1) \equiv y(x)$. Since the initial data is ranging over a connected set, there is a $0 < \lambda_0 < 1$ such that $z(x_0, \lambda_0) = y_0$. We also observe that $z^{(i)}(x_1, \lambda_0) = \lambda y^{(i)}(x_1), i = 0, 1$, and $z''(x_1) < z''(x_1, \lambda_0) < y''(x_1)$. Now, there is a $\delta > 0$ such that $[x_0 - \delta, x_0 + \delta] \subset [x_2, x_3]$ and such that $z(x, \lambda_0) < z(x)$ on $[x_0 - \delta, x_0 + \delta] \setminus \{x_0\}$. With $\{y_n(x)\}$ the originally defined sequence of solutions, we have

$$\lim_{n \to \infty} y_n(x) > z(x, \lambda_0) \text{ on } [x_0 - \delta, x_0 + \delta] \setminus \{x_0\},$$

and

$$\lim_{n \to \infty} y_n(x_0) = y_0 = z(x_0, \lambda_0).$$

Now the argument used in the first case can be repeated here; and so, the second case, $y_0 < z(x_0)$, is also impossible.

From this final contradiction, we conclude that solutions for $(2, 1, 1)$ conjugate boundary value problems for (1.1.1) on (a, b) are unique, when they exist.

The proof for the uniqueness of solutions of $(1, 1, 2)$ conjugate boundary value problems is completely analogous. The proof is complete. □

The next result is from the same theorem by Peterson [101], but which shows that uniqueness of solutions of 2-point conjugate boundary value problems yields uniqueness of solutions of 4-point conjugate boundary value problems and the remaining 3-point conjugate boundary value problem.

Theorem 1.3.4. [[101], D. Peterson, Thm. 3.1, p. 22] *For $n = 4$, assume that with respect to (1.1.1), conditions (A), (B) and (CP) are satisfied. If, for each $m_1, m_2 \in \mathbb{N}$ such that $m_1 + m_2 = 4$, there exists at most one solution of each (m_1, m_2) conjugate boundary value problem for (1.1.1) on (a, b), then solutions of all $(1, 1, 1, 1)$ conjugate boundary value problems and all $(1, 2, 1)$ conjugate boundary value problems for (1.1.1) on (a, b) are unique, when they exist.*

Proof. We consider the $(1, 1, 1, 1)$ conjugate boundary value problem. The proof very much reflects the pattern of proof in Theorem 1.3.3, but it does have a few steps unique to this problem. For the purpose of contradiction, assume there exist distinct solutions $y(x)$ and $z(x)$ of (1.1.1) and points $a < x_1 < x_2 < x_3 < x_4 < b$ with $y(x_j) = z(x_j), j = 1, 2, 3, 4$. From Theorem 1.3.3, $y'(x_j) \neq z'(x_j), j = 1, 2, 3, 4$, and so we may assume, without loss of generality, that $y'(x_i) < z'(x_i), i = 1, 3$, and $y'(x_j) > z'(x_j), j = 2, 4$, so that $y(x) < z(x)$ on $(x_1, x_2) \cup (x_3, x_4)$ and $y(x) > z(x)$ on (x_2, x_3).

As fits the pattern and routine, for each $n \geq 1$, let $y_n(x)$ be the solution of (1.1.1) satisfying the initial conditions,

$$y_n^{(i)}(x_1) = y^{(i)}(x_1), \ i = 0, 1, 2, \text{ and } y_n'''(x_1) = y'''(x_1) + n.$$

Then, it follows from uniqueness of solutions of $(3, 1)$ conjugate boundary value problems for (1.1.1) that $y_{n+1}(x) > y_n(x) > y(x)$ on (x_1, b), for each $n \geq 1$. Note that, by condition (CP), for each $n \geq 1$, there exists $\tau_n \in (x_1, x_2)$ such that $y_n(\tau_n) = z(\tau_n), y_n(x) < z(x)$ on (x_1, τ_n) and $y_n(x) > z(x)$ on $(\tau_n, x_3]$.

Next, for each $n \geq 1$, let

$$E_n := \{x \mid x_3 \leq x \leq x_4 \text{ and } y_n(x) \leq z(x)\}.$$

Each E_n is compact, and as before, using the continuous dependence of solutions of $(3, 1)$ conjugate boundary value problems, it follows that, for each $n \geq 1$, $\emptyset \neq E_{n+1} \subset E_n$, and then by invoking (CP) again, there exists $x_3 < x_0 < x_4$ such that

$$\bigcap_{n=1}^{\infty} E_n = \{x_0\} \text{ and } \lim_{n \to \infty} y_n(x_0) := y_0 \leq z(x_0).$$

If $y_0 = z(x_0)$, then repeating much of the previous theorem's arguments, for $\epsilon > 0$ sufficiently small, there is a solution $z(x, \epsilon)$ of (1.1.1) satisfying

$$z^{(i)}(x_1, \epsilon) = z^{(i)}(x_1), \ i = 0, 1, 2, \ z(x_0, \epsilon) = z(x_0) - \epsilon > y(x_0),$$

and $z(x, \epsilon) < z(x)$ on $(x_1, x_4]$. The solution, $z(x, \epsilon)$, can be used in place of $z(x)$ in defining the sequence $\{E_n\}$ with respect to the sequence $\{y_n(x)\}$, which then leads to the impossibility that each of the sets is nonempty.

For the other case that $y(x_0) < y_0 < z(x_0)$, let $z(x, \lambda)$ be the solution of (1.1.1) satisfying

$$z^{(i)}(x_1, \lambda) = \lambda y^{(i)}(x_1) + (1 - \lambda) z^{(i)}(x_1), \ i = 0, 1, 2, 3, \ 0 \leq \lambda \leq 1.$$

As before, $z(x, 0) \equiv z(x)$ and $z(x, 1) \equiv y(x)$. By continuity, there exists $0 < \lambda_0 < 1$ such that $z(x_0, \lambda_0) = y_0$, and we note that $z(x_1, \lambda_0) = y(x_1)$, and $y'(x_1) < z'(x_1, \lambda_0) \leq z'(x_1)$.

Choose an integer n_0 such that $y_{n_0}(x) - z(x, \lambda_0)$ has two zeros, say $t_1 < t_2$, in the interval (x_3, x_4). Let $w(x)$ be the solution of (1.1.1) satisfying $w^{(i)}(x_1) = z^{(i)}(x_1, \lambda_0), i = 0, 1, 2$, and $y_{n_0}(x) < w(x_0) < y_0$. Choose an integer n_1 such that $y_{n_1}(x) > w(x)$ on (t_1, t_2).

There are two cases to consider. Suppose first that $w(x) < z(x)$ on (x_1, x_2). Then there exists $\tau_1 \in (x_1, x_2)$ such that $w(\tau_1) = y_{n_1}(\tau_1)$. Now, we proceed as in the proof of the previous theorem, but in this context, to contradict the uniqueness of solutions of $(1, 1, 2)$ conjugate boundary value problems for (1.1.1).

The second possibility is that $w(x) > z(x)$ at some points in (x_1, x_2). Since, by (CP), the sequence $\{y_n(x)\}$ cannot be uniformly bounded on any subinterval, there exists $n_2 > n_1$ and $\tau \in (x_1, x_2)$ such that $y_{n_2}(\tau) > w(\tau)$. Let $w_1(x)$ be the solution of (1.1.1) with $w_1^{(i)}(x_1) = w^{(i)}(x_1), i = 0, 1, 2$, and $y_{n_2}(x_0) < w_1(x_0) < y_0$. Again, we proceed to contradict the uniqueness of solutions of $(1, 1, 2)$ conjugate boundary value problems for (1.1.1).

We conclude that solutions of 4-point conjugate boundary value problems for (1.1.1) on (a, b) are unique, when they exist.

Finally, once we have, when $n = 4$, uniqueness of solutions of 4-point conjugate boundary value problems, it follows from Theorem 1.2.2 that solutions $(1, 2, 1)$ conjugate boundary value problems are also unique, when they exist. And the proof is complete. ☐

Recall from Theorem 1.2.4, for arbitrary order n, that uniqueness of solutions of $(n - 1)$-point conjugate boundary value problems for (1.1.1) implies the uniqueness of solutions of all k-point conjugate boundary value problems for (1.1.1), for $2 \leq k < n - 1$. In the paper by Henderson and Jackson [62], a part of one of their results (Theorem 3, p. 378), established that, under (A), (B) and (CP), existence of solutions of all 2-point conjugate boundary value problems for (1.1.1), along with uniqueness of solutions of all $(n - 1)$-point conjugate boundary value problems, yields the existence of unique solutions for all k-point conjugate boundary value problems for (1.1.1) on (a, b), for all $2 \leq k \leq n - 2$.

From another part of the same theorem in the Henderson and Jackson paper, we extrapolate a uniqueness result that fits within the theme of this section. And in addition, this result will extend in some sense the previous two theorems, Theorem 1.3.3 and Theorem 1.3.4.

Theorem 1.3.5. [[62], Henderson and Jackson, part of Thm. 3, p. 378] *Assume, for arbitrary order n, that with respect to (1.1.1), conditions (A), (B) and (CP) are satisfied, that all $(n - 1)$-point conjugate boundary value problems for (1.1.1) on (a, b) have at most one solution, and that all 2-point conjugate boundary value problems for (1.1.1) on (a, b) have solutions. Then, all n-point conjugate boundary value problems for (1.1.1) on (a, b) have at most one solution.*

Proof. At the outset, it follows from the comments preceding the statement of this theorem that there exist unique solutions of all $(n - 2)$-point conjugate boundary value problems for (1.1.1) on (a, b).

Assume now to the contrary of the conclusion of the theorem, that $y(x)$ and $z(x)$ are distinct solutions of (1.1.1) and that there are points, $a < x_1 < \cdots < x_n < b$, such that $y(x_j) = z(x_j), 1 \leq j \leq n$. By uniqueness of solutions of $(n - 1)$-point problems, $y'(x_j) \neq z'(x_j), 1 \leq j \leq n$. We will assume that the x_j's are successive zeros of $y(x) - z(x)$, that $y(x) < z(x)$ on $(x_{n-3}, x_{n-2}) \cup (x_{n-1}, x_n)$, and that $y(x) > z(x)$ on (x_{n-2}, x_{n-1}).

For each $k \geq 1$, let $y_k(x)$ denote the solution of the $(1, m_2, \ldots, m_{n-2})$ conjugate boundary value problem for (1.1.1), where $m_j = 1$, for $2 \leq j \leq$

$n - 3$, $m_{n-2} = 3$, satisfying the boundary conditions,

$$y_k(x_j) = y(x_j), \ 1 \le j \le n - 2,$$
$$y_k'(x_{n-2}) = y'(x_{n-2}),$$
$$y_k''(x_{n-2}) = y''(x_{n-2}) + k.$$

By uniqueness of solutions of $(n - 1)$-point problems, it follows that, for each $k \ge 1$,

$$y_{k+1}(x) > y_k(x) > y(x) \text{ on } (x_{n-3}, b) \setminus \{x_{n-2}\}.$$

For each $k \ge 1$, let

$$E_k := \{x \mid x_{n-1} \le x \le x_n \text{ and } y_k(x) \le z(x)\}.$$

Solutions of $(n - 2)$-point conjugate boundary value problems depend continuously on the boundary conditions, and from this and the uniqueness of solutions of $(n-1)$-point problems, we conclude that each E_k is a nonempty compact set, and moreover, we have

$$E_{k+1} \subset E_k \subset (x_{n-1}, x_n),$$

so that

$$\bigcap_{k=1}^{\infty} E_k := E \text{ is a nonempty compact subset of } (x_{n-1}, x_n).$$

Similar arguments used to show each $E_k \ne \emptyset$ and conditions (B) and (CP), along with the fact that $y_k''(x_{n-2}) \to \infty$, as $k \to \infty$, yield that $E = \{x_0\}$, where $x_{n-1} < x_0 < x_n$, and $\lim_{k \to \infty} y_k(x_0) := y_0 \le z(x_0)$. As is typical of what is to follow, we will show that each of the cases, $y_0 = z(x_0)$ and $y_0 < z(x_0)$, is impossible. This will show the existence of the two distinct solutions of the same n-point boundary value problem is not possible.

First, assume that $y_0 = z(x_0)$. In this case, let $0 < \epsilon < z(x_0) - y(x_0)$. Then, there is an $\eta > 0$ such that, the solutions $z(x, \eta)$ of (1.1.1) satisfying the boundary conditions,

$$z(x_j, \eta) = z(x_j), \ 1 \le j \le n - 2,$$
$$z'(x_{n-2}, \eta) = z'(x_{n-2}),$$
$$z''(x_{n-2}, \eta) = z''(x_{n-2}) - \eta,$$

also satisfies

$$y(x_0) < z(x_0) - \epsilon < z(x_0, \eta) < z(x_0) = y_0.$$

The solutions $z(x, \eta)$ can be used in place of $z(x)$ in defining the sequence of sets $\{E_k\}$ in terms of the original sequence of solutions $\{y_k(x)\}$. It would follow as before that each E_k is nonempty and compact, which is not possible. So this case is eliminated.

Now, assume that $y(x_0) < y_0 < z(x_0)$. In this case, let $z(x, \lambda)$, for $0 \leq \lambda \leq 1$, be the solution of (1.1.1) satisfying the boundary conditions,

$$z(x_j, \lambda) = \lambda y(x_j) + (1 - \lambda)z(x_j),\ 1 \leq j \leq n - 2,$$
$$z'(x_{n-2}, \lambda) = \lambda y'(x_{n-2}) + (1 - \lambda)z'(x_{n-2}),$$
$$z''(x_{n-2}, \lambda) = \lambda y''(x_{n-2}) + (1 - \lambda)z''(x_{n-2}).$$

It follows, again by the continuity of solutions for this type of boundary value problem with respect to the boundary conditions, that there is a $0 < \lambda_0 < 1$ such that $z(x_0, \lambda_0) = y_0$. And there is a $\delta > 0$ such that $[x_0 - \delta, x_0 + \delta] \subset (x_{n-1}, x_n)$ and $z(x, \lambda_0) < z(x)$ on $[x_0 - \delta, x_0 + \delta]$. As in many of the preceding results, with $\{y_k(x)\}$ being the original sequence of solutions, we have

$$\lim_{k \to \infty} y_k(x) > z(x, \lambda_0) \text{ on } [x_0 - \delta, x_0 + \delta] \setminus \{x_0\},$$

and

$$\lim_{k \to \infty} y_k(x_0) = y_0 = z(x_0, \lambda_0).$$

Repeating an argument similar to the one used in the first case shows that this is impossible, and hence the second case, $y_0 < z(x_0)$, is also impossible. This completes the proof. $\qquad\square$

Subsequent sections of this chapter will deal with uniqueness quotiono of the type dealt with in each of Section 1.2 and Section 1.3, but for boundary value problems for (1.1.1) satisfying right focal boundary conditions. A later chapter will deal in part with similar uniqueness questions for boundary value problems satisfying nonlocal boundary conditions.

1.4   Right focal boundary value problems: for $m > r$, uniqueness of m-point implies uniqueness of r-point

In this section, we present uniqueness results for right focal boundary value problems (1.1.1), (1.1.3) that are somewhat in analogy to those of

Section 1.2 which were obtained for conjugate boundary value problems for (1.1.1). In particular, we are concerned with uniqueness results for solutions of (1.1.1), (1.1.3) that are along the lines of uniqueness results in Section 1.2 for (1.1.1), (1.1.2).

For local convenience, we recall from Section 1.1 what we mean by an r-point right focal boundary value problem (or an $(m_1, \ldots, m_r)$ right focal boundary value problem):

Given $2 \leq r \leq n$, $m_1, \ldots, m_r \in \mathbb{N}$ such that $\sum_{j=1}^{r} m_j = n$, $s_0 := 0$, $s_k := \sum_{j=1}^{k} m_j$, $1 \leq k \leq r$, points $a < x_1 < \cdots < x_r < b$, and $y_{ik} \in \mathbb{R}$, $s_{k-1} \leq i \leq s_k - 1, 1 \leq k \leq r$, a boundary value problem for (1.1.1) satisfying

$$y^{(i)}(x_k) = y_{ik}, \ s_{k-1} \leq i \leq s_k - 1, \ 1 \leq k \leq r,$$

is called either an r-*point right focal boundary value problem*, or an $(m_1, \ldots, m_r)$ *right focal boundary value problem*. (These boundary conditions are designated in Section 1.1 by (1.1.3).)

Primarily, our concern will be with uniqueness of solutions of n-point right focal boundary value problems implying uniqueness of solutions of r-point right focal boundary value problems for (1.1.1) on (a, b), for all $2 \leq r < n$. For the case when (1.1.1) is linear, Muldowney [93, Prop. 4] proved such a result.

Also, later in this section, for the case when $n = 4$, we will exhibit a result for uniqueness of solutions of 3-point right focal boundary value problems implying uniqueness of solutions of certain 2-point right focal boundary value problems for (1.1.1) on (a, b).

The results of this section will involve from Section 1.1 conditions (A), (B) and "compactness condition" (CP), and continuous dependence from Theorem 1.1.6 and Remark 1.1.1, along with other conditions we will impose.

In analogy to Jackson's result [76], we now present a result by Henderson [46] for right focal boundary value problems (1.1.1), (1.1.3).

Theorem 1.4.1. [[46], Henderson, The Theorem] *Assume that with respect to (1.1.1), conditions (A) and (B) are satisfied, and that each n-point right focal boundary value problem for (1.1.1) on (a, b) has at most one solution. Then, each r-point right focal boundary value problem for (1.1.1) on (a, b), $2 \leq r \leq n - 1$, has at most one solution.*

Proof. We assume the conclusion of the theorem is false. Then, for some $2 \leq k \leq n - 1$, some k-point right focal boundary value problem for (1.1.1) has distinct solutions; that is, for some $2 \leq k \leq n - 1$, some $(m_1, \ldots, m_r)$

right focal boundary value problem for (1.1.1) on (a, b) has at least two distinct solutions. Let

$$r = \max\{2 \le k \le n - 1 \,|\, \text{some } (m_1, \ldots, m_k) \text{ right focal boundary value}$$
$$\text{problem for (1.1.1) on } (a, b) \text{ has at least two distinct solutions}\}.$$

So, there are points $a < x_1 < \cdots < x_r < b$, positive integers $m_1, \ldots, m_r$ partitioning n, and distinct solutions $y(x)$ and $z(x)$ of (1.1.1) such that

$$y^{(i)}(x_k) = z^{(i)}(x_k), \quad s_{k-1} \le i \le s_k - 1, \ 1 \le k \le r.$$

Hereafter, let x_0 and x_{r+1} be fixed points satisfying $a < x_0 < x_1$ and $x_r < x_{r+1} < b$. Completion of the proof involves consideration of several cases.

Case 1: For the $(m_1, \ldots, m_r)$ right focal problem under consideration, we first consider the case for which there exists an integer, $1 \le j_0 \le r$ such that $m_{j_0} = 2k$, with $k \ge 1$, and $y^{(s_{j_0}-1)}(x) - z^{(s_{j_0}-1)}(x)$ does not change sign at x_{j_0}. We may assume that this difference is positive on $(x_{j_0-1}, x_{j_0+1}) \setminus \{x_{j_0}\}$. Let τ_1 and τ_2 be two points such that $x_{j_0-1} < \tau_1 < x_{j_0} < \tau_2 < x_{j_0+1}$, and for future reference, choose $\epsilon > 0$ such that $y^{(s_{j_0}-1)}(\tau_i) - z^{(s_{j_0}-1)}(\tau_i) > \epsilon$, $i = 1, 2$.

By the maximality of r and Remark 1.1.1, the uniqueness of solutions of $(r+1)$-point right focal boundary value problems for (1.1.1) yields that solutions depend continuously on $(r+1)$-point right focal boundary conditions. In particular, for $\delta > 0$ sufficiently small, there is a solution $u_\delta(x)$ of (1.1.1) satisfying

$$u_\delta^{(i)}(x_k) = z^{(i)}(x_k) \ s_{k-1} \le i \le s_k - 1, \ k \in \{1, \ldots, r\} \setminus \{j_0\},$$
$$u_\delta^{(i)}(x_{j_0}) = z^{(i)}(x_{j_0}) \ s_{j_0-1} \le i \le s_{j_0} - 3, \quad \text{(omitted if } m_{j_0} = 2),$$
$$u_\delta^{(s_{j_0}-2)}(x_{j_0}) = z^{(s_{j_0}-2)}(x_{j_0}) + \delta,$$
$$u_\delta^{(s_{j_0}-1)}(\tau_2) = z^{(s_{j_0}-1)}(\tau_2),$$

and

$$\lim_{\delta \to 0^+} u_\delta^{(i)}(x) = z^{(i)}(x)$$

uniformly on $[x_0, x_{r+1}]$, for $0 \le i \le n$. From the positivity of $y^{(s_{j_0}-1)}(x) - z^{(s_{j_0}-1)}(x)$ on $(x_{j_0-1}, x_{j_0+1}) \setminus \{x_{j_0}\}$ and the choice of τ_1, τ_2 and ϵ, it follows for $\delta > 0$ sufficiently small, that there are points ξ_0 and σ_0 with $\tau_1 < \xi_0 < x_{j_0} < \sigma_0 < \tau_2$ such that $u_\delta^{(s_{j_0}-1)}(\xi_0) = y^{(s_{j_0}-1)}(\xi_0)$ and $u_\delta^{(s_{j_0}-1)}(\sigma_0) = y^{(s_{j_0}-1)}(\sigma_0)$. We also have that $u_\delta^{(i)}(x_{j_0}) = y^{(i)}(x_{j_0}), s_{j_0-1} \le i \le s_{j_0} - 3$, and so, by repeated applications of Rolle's Theorem, there exist points $\xi_1, \ldots, \xi_{m_{j_0}-2}, \sigma_1, \ldots, \sigma_{m_{j_0}-2}$ with $\xi_0 < \xi_1 < \cdots < \xi_{m_{j_0}-2} < x_{j_0} < \sigma_{m_{j_0}-2} < \cdots < \sigma_1 < \sigma_0$ such that

$$u_\delta^{(s_{j_0-1}+k)}(\xi_k) = y^{(s_{j_0-1}+k)}(\xi_k), \text{ and}$$
$$u_\delta^{(s_{j_0-1}+k)}(\sigma_k) = y^{(s_{j_0-1}+k)}(\sigma_k), \ 0 \le k \le m_{j_0} - 2.$$

An additional application of Rolle's Theorem yields a point $\xi_{m_{j_0}-2} < \xi_{m_{j_0}-1} < \sigma_{m_{j_0}-2}$ such that

$$u_\delta^{(s_{j_0-1}+m_{j_0}-1)}(\xi_{m_{j_0}-1}) = y^{(s_{j_0-1}+m_{j_0}-1)}(\xi_{m_{j_0}-1}).$$

That is, for the points, $\xi_0 < \xi_1 < \cdots < \xi_{m_{j_0}-1} < \sigma_0$,

$$u_\delta^{(s_{j_0-1}+k)}(\xi_k) = y^{(s_{j_0-1}+k)}(\xi_k), \quad 0 \le k \le m_{j_0} - 1,$$

(note here that $s_{j_0-1} + m_{j_0} - 1 = s_{j_0} - 1$), and also

$$u_\delta^{(i)}(x_k) = y^{(i)}(x_k), \quad s_{k-1} \le i \le s_k - 1, \ k \in \{1, \ldots, r\} \setminus \{j_0\}.$$

Thus, $u_\delta(x)$ and $y(x)$ are distinct solutions of an $(i_1, \ldots, i_{r-1+m_{j_0}})$ right focal boundary value problem for (1.1.1), which contradicts the maximality of r. Thus, Case 1 is not possible.

Case 2: In analogy to Case 1, for the $(m_1, \ldots, m_r)$ right focal problem being considered, we deal with the case for which there is an index $1 \le j_0 \le r$ such that $m_{j_0} = 2k+1$, with $k \ge 1$, and $y^{(s_{j_0-1})}(x) - z^{(s_{j_0-1})}(x)$ changes sign at x_{j_0}. We may assume that this difference is negative on (x_{j_0-1}, x_{j_0}) and is positive on (x_{j_0}, x_{j_0+1}). The argument is very much along the lines used for Case 1. Let τ_1 and τ_2 be two points such that $x_{j_0-1} < \tau_1 < x_{j_0} < \tau_2 < x_{j_0+1}$, and choose $\epsilon > 0$ such that $|y^{(s_{j_0-1})}(\tau_i) - z^{(s_{j_0-1})}(\tau_i)| > \epsilon$, $i = 1, 2$.

Again, solutions of $(r+1)$-point right focal boundary value problems depend continuously on boundary conditions, and so, for $\delta > 0$ sufficiently small, there is a solution $u_\delta(x)$ of (1.1.1) satisfying

$$u_\delta^{(i)}(x_k) = z^{(i)}(x_k) \ s_{k-1} \le i \le s_k - 1, \ k \in \{1, \ldots, r\} \setminus \{j_0\},$$
$$u_\delta^{(i)}(x_{j_0}) = z^{(i)}(x_{j_0}) \ s_{j_0-1} \le i \le s_{j_0} - 3, \quad \text{(one term if } m_{j_0} = 3),$$
$$u_\delta^{(s_{j_0}-2)}(x_{j_0}) = z^{(s_{j_0}-2)}(x_{j_0}) + \delta,$$
$$u_\delta^{(s_{j_0}-1)}(\tau_2) = z^{(s_{j_0}-1)}(\tau_2),$$

and

$$\lim_{\delta \to 0^+} u_\delta^{(i)}(x) = z^{(i)}(x)$$

uniformly on $[x_0, x_{r+1}]$, for $0 \le i \le n$. Then, by proceeding exactly as in Case 1, the same contradiction is reached. Hence, Case 2 is not possible.

Case 3: For the $(m_1, \ldots, m_r)$ right focal problem of interest, we deal with the case in which some $m_{j_0} = 2k$, with $k \ge 1$, $j_0 \ne r$, and $y^{(s_{j_0}-1)}(x) - z^{(s_{j_0}-1)}(x)$ changes sign at x_{j_0}.

This implies that $y^{(i)}(x_{j_0}) = z^{(i)}(x_{j_0})$, $s_{j_0-1} \le i \le s_{j_0} - 1 + 1 = s_{j_0}$, and furthermore that the zero of $y^{(s_{j_0}-1)}(x) - z^{(s_{j_0}-1)}(x)$ at x_{j_0} must be of odd multiplicity. This differs from Case 1; namely in Case 1, $y^{(s_{j_0})}(x) = z^{(s_{j_0})}(x)$ was possible, yet it had to be the case that the zero of $y^{(s_{j_0}-1)}(x) - z^{(s_{j_0}-1)}(x)$ at x_{j_0} was of even multiplicity.

Let us assume now that $y^{(s_{j_0}-1)}(x) - z^{(s_{j_0}-1)}(x)$ is negative on (x_{j_0-1}, x_{j_0}) and is positive on (x_{j_0}, x_{j_0+1}). Again, we choose $x_{j_0-1} < \tau_1 < x_{j_0} < \tau_2 < x_{j_0+1}$, and $\epsilon > 0$ such that $|y^{(s_{j_0}-1)}(\tau_i) - z^{(s_{j_0}-1)}(\tau_i)| > \epsilon$, $i = 1, 2$.

To resolve this case, we employ the uniqueness of solutions of a type of boundary value problem, not of the right focal type, but whose uniqueness follows from Rolle's Theorem and uniqueness of solutions of each $(i_1, \ldots, i_s)$ right focal problem, where $s = r + m_{j_0}$ or $s = r + m_{j_0} - 1$; in particular, given $a < t_1 < \cdots < t_{j_0-1} < \tau < t_{j_0} < t_{j_0+1} < \cdots < t_r < b$ and $y_{ik} \in \mathbb{R}$, $s_{k-1} \le i \le s_k - 1$, $1 \le k \le r$, the boundary value problem for (1.1.1) satisfying

$$u^{(i)}(t_k) = y_{ik}, \ s_{k-1} \le i \le s_k - 1, \ k \in \{1, \ldots, r\} \setminus \{j_0, j_0 + 1\},$$
$$u^{(s_{j_0}-1)}(\tau) = y_{s_{j_0}-1, j_0},$$
$$u^{(i)}(t_{j_0}) = y_{i+1, j_0}, \ s_{j_0-1} \le i \le s_{j_0} - 2,$$
$$u^{(s_{j_0}-1)}(t_{j_0}) = y_{s_{j_0}, j_0+1},$$
$$u^{(i)}(t_{j_0+1}) = y_{i, i_0+1}, \ s_{j_0} + 1 \le i \le s_{j_0+1} - 1, \quad \text{(omitted if } m_{j_0+1} = 1),$$

has at most one solution on (a, b).

And so solutions of (1.1.1) depend continuously on boundary conditions of the type in the just above described boundary value problem. Hence, for sufficiently small $\delta > 0$ there is a solution $u_\delta(x)$ of (1.1.1) satisfying

$$u_\delta^{(i)}(x_k) = z^{(i)}(x_k), \ s_{k-1} \le i \le s_k - 1, \ k \in \{1, \ldots, r\} \setminus \{j_0, j_0 + 1\},$$
$$u_\delta^{(s_{j_0}-1)}(\tau_1) = z^{(s_{j_0}-1)}(\tau_1),$$
$$u_\delta^{(i)}(x_{j_0}) = z^{(i)}(x_{j_0}), \ s_{j_0-1} \le i \le s_{j_0} - 2,$$
$$u_\delta^{(s_{j_0}-1)}(x_{j_0}) = z^{(s_{j_0}-1)}(x_{j_0}) + \delta,$$
$$u_\delta^{(i)}(x_{j_0+1}) = z^{(i)}(x_{j_0+1}), \ s_{j_0} + 1 \le i \le s_{j_0+1} - 1, \quad \text{(omitted if } m_{j_0+1} = 1),$$

and

$$\lim_{\delta \to 0^+} u_\delta^{(i)}(x) = z^{(i)}(x)$$

uniformly on $[x_0, x_{r+1}]$, for $0 \le i \le n$. By the choice or ϵ and the negativity on (x_{j_0-1}, x_{j_0}) and the positivity on (x_{j_0}, x_{j_0+1}) of $y^{(s_{j_0}-1)}(x) - z^{(s_{j_0}-1)}(x)$, it follows, for δ sufficiently small, that there are points $\tau_1 < \rho_0 < x_{j_0} < \sigma_0 < \tau_2$ such that

$$u_\delta^{(s_{j_0}-1)}(\rho_0) = y^{(s_{j_0}-1)}(\rho_0) \text{ and } u_\delta^{(s_{j_0}-1)}(\sigma_0) = y^{(s_{j_0}-1)}(\sigma_0).$$

Since we also have $u_\delta^{(i)}(x_{j_0}) = y^{(i)}(x_{j_0}), s_{j_0-1} \le i \le s_{j_0} - 2$, by repeated applications of Rolle's Theorem, there are points $\rho_0 < \rho_1 < \cdots < \rho_{m_{j_0}} < \sigma_0$ such that

$$u_\delta^{(s_{j_0}-1+k)}(\rho_k) = y^{(s_{j_0}-1+k)}(\rho_k), \ 0 \le k \le m_{j_0} \text{ (note } s_{j_0-1} + m_{j_0} = s_{j_0}).$$

Furthermore,

$$u_\delta^{(i)}(x_k) = y^{(i)}(x_k), \ s_{k-1} \le i \le s_k - 1, \ k \in \{1, \ldots, r\} \setminus \{j_0, j_0 + 1\},$$
$$u_\delta^{(i)}(x_{j_0+1}) = y^{(i)}(x_{j_0+1}), \ s_{j_0} + 1 \le i \le s_{j_0+1} - 1.$$

That is, $u_\delta(x)$ and $y(x)$ are distinct solutions of an $(i_1, \ldots, i_s)$ right focal boundary value problem for (1.1.1), where $s = r + m_{j_0} - 1$ if $m_{j_0+1} = 1$, and $s = r + m_{j_0}$ if $m_{j_0+1} > 1$. In each case, $s > r$, which contradicts the maximality of r. Hence, Case 3 is not possible.

Case 4: This case is an exact analogue of Case 3, in that, for the $(m_1, \ldots, m_r)$ right focal boundary value problem in question, we consider when some $m_{j_0} = 2k + 1$, $j_0 \ne r$, and $y^{(s_{j_0}-1)}(x) - z^{(s_{j_0}-1)}(x)$ does not change sign at x_{j_0}. Assuming that $y^{(s_{j_0}-1)}(x) - z^{(s_{j_0}-1)}(x)$ is positive on $(x_{j_0-1}, x_{j_0+1}) \setminus \{x_{j_0}\}$, one can follow essentially the same steps as in Case 3, and arrive at the same contradiction. Thus, Case 4 also is not possible.

Case 5: For the $(m_1, \ldots, m_r)$ right focal boundary value problem in question, we consider, in our last case, various values of m_r.

(a) $m_r = 1$: Then, there exists $1 \le j_0 \le r - 1$ such that $m_{j_0} \ge 2$. The impossibility of this subcase has been exhibited in Cases 1–4.

(b) $m_r > 1$: There are two further subcases to deal with here. Either, (i) there exists $1 \le j_0 \le r - 1$ such that $m_{j_0} \ge 2$, or (ii) $m_j = 1$, for all $1 \le j \le r$.

As with (a), subcase (i) is not possible due to Cases 1–4. We shall now exhibit the impossibility of subcase (ii). We observe in subcase (ii), that the problem in question is a $(1, \ldots, 1, n - (r-1))$ right focal boundary value problem for (1.1.1), where $n - (r - 1) \geq 2$. Resolution of subcase (ii) now involves the two additional possibilities:

(ii.1) $y^{(j_0-1)}(x) - z^{(j_0-1)}(x)$ does not change sign at x_{j_0}, for some $1 \leq j_0 \leq r - 1$, or

(ii.2) $y^{(j-1)}(x) - z^{(j-1)}(x)$ changes sign at x_j, for each $1 \leq j \leq r - 1$.

If (ii.1) holds, it follows that, for some $1 \leq j_0 \leq r - 1$, $y^{(i)}(x_{j_0}) = z^{(i)}(x_{j_0}), i = j_0 - 1, j_0$. Since the problem is of the $(1, \ldots, 1, n - (r - 1))$ right focal type, $y^{(i-1)}(x_i) = z^{(i-1)}(x_i), 1 \leq i \leq r - 1$, and $y^{(i)}(x_r) = z^{(i)}(x_r), r - 1 \leq i \leq n - 1$, and so by Rolle's Theorem, $y(x)$ and $z(x)$ are distinct solutions of a $(1, \ldots, 1, m_{j_0} + 1, 1, \ldots, 1, m_s)$ right focal boundary value problem for (1.1.1), where $m_{j_0} = 1, m_s = 1$, and $s \geq r$. The situation when $s = r$ was ruled out in Case 1, as well as in (a) just above, and if $s > r$, the maximality of r is contradicted. Hence (ii.1) is not possible.

If (ii.2) holds, choose points τ_j and τ_j', $1 \leq j \leq r - 1$, such that $x_0 < \tau_1 < x_1 < \tau_1' < \tau_2 < x_2 < \tau_2' < \cdots < \tau_{r-1} < x_{r-1} < \tau_{r-1}' < x_r$. By Rolle's Theorem and the maximality of r, we have that $|y^{(j-1)}(x) - z^{(j-1)}(x)| > 0$ on $(x_{j-1}, x_{j+1}) \setminus \{x_j\}$, for $1 \leq j \leq r - 1$. Moreover, by the maximality of r, we may assume, without loss of generality, that $y^{(n-1)}(x) > z^{(n-1)}(x)$ on (x_r, x_{r+1}). Next, we fix τ with $x_r < \tau < x_{r+1}$ and choose $\epsilon > 0$ such that $|y^{(j-1)}(\tau_j) - z^{(j-1)}(\tau_j)| > \epsilon$, $|y^{(j-1)}(\tau_j') - z^{(j-1)}(\tau_j')| > \epsilon$, $1 \leq j \leq r - 1$, and $|y^{(n-1)}(\tau) - z^{(n-1)}(\tau)| > \epsilon$. Now, by condition (B), solutions of initial value problems for (1.1.1) depend continuously on initial conditions. That is, there exists a $\delta(\epsilon) > 0$ such that, if $u(x)$ is the solution of (1.1.1) satisfying the initial conditions,

$$u^{(i)}(x_r) = y^{(i)}(x_r), \quad 0 \leq i \leq n - 2,$$
$$u^{(n-1)}(x_r) = y^{(n-1)}(x_r) - \delta,$$

then $|u^{(i-1)}(x) - y^{(i-1)}(x)| < \epsilon$ on $[x_0, x_{r+1}]$, $1 \leq i \leq n$. It follows that, for each $1 \leq j \leq r - 1$, there exists $\tau_j < \sigma_j < \tau_j'$ such that $u^{(j-1)}(\sigma_j) = z^{(j-1)}(\sigma_j)$. Moreover, $u^{(i)}(x_r) = z^{(i)}(x_r)$, $r - 1 \leq i \leq n - 2$. Also, $u^{(n-1)}(x_r) < z^{(n-1)}(x_r)$, whereas by our choice of ϵ, $u^{(n-1)}(\tau) > z^{(n-1)}(\tau)$. So, there is a point $x_r < \sigma < \tau$ such that $u^{(n-1)}(\sigma) = z^{(n-1)}(\sigma)$. And so, $u(x)$ and $z(x)$ are distinct solutions of a $(1, \ldots, 1, i_r, i_{r+1})$ right focal boundary value problem for (1.1.1) on (a, b), where $i_r = n - r$ and $i_{r+1} = 1$, which again contradicts the maximality of r. Thus, (ii.2) is not possible, which completes the impossibility of subcase (b), and ultimately the impossibility of Case 5.

Therefore, based on the assumption that $y(x)$ and $z(x)$ are distinct solutions of some $(m_1, \ldots, m_r)$ right focal boundary value problem for (1.1.1), each case to be considered produced a contradiction. We conclude that assumption is false, and the proof is complete. $\qquad\square$

For $n = 4$, Henderson [49] obtained a partial result concerning uniqueness of solutions of 3-point right focal boundary value problems for (1.1.1) implying the uniqueness of solutions of 2-point right focal boundary value problems for (1.1.1). This result constitutes a partial analogue for $n = 4$ of Theorem 1.2.4 from Section 1.2. For convenience of reference, we introduce the condition:

(D$_3$) There exists at most one solution of each (m_1, m_2, m_3) right focal boundary value problem for (1.1.1) on (a, b).

Theorem 1.4.2. [[49], Henderson, Thm. 3.1] *Let $n = 4$, and assume that with respect to (1.1.1), conditions (A), (B) and (D$_3$) are satisfied. Then, each $(1, 3)$ right focal boundary value problem for (1.1.1) on (a, b) has at most one solution, and each $(2, 2)$ right focal boundary value problem for (1.1.1) on (a, b) has at most one solution.*

Proof. First, we consider the $(1, 3)$ right focal boundary value problem. We assume in this case that the conclusion is false. Then, there exist distinct solutions $y(x)$ and $z(x)$ of (1.1.1) and points $a < x_1 < x_2 < b$ such that $z(x_1) = y(x_1)$ and $z^{(i)}(x_2) = y^{(i)}(x_2), i = 1, 2, 3$. Let x_0 and x_3 be arbitrary, but fixed points with $a < x_0 < x_1$ and $x_2 < x_3 < b$. From (D$_3$) and Rolle's Theorem, $z(x) - y(x)$ has a simple zero at x_1, and in addition, we may assume $z'''(x) > y'''(x)$ on (x_2, b).

Next, given $\delta > 0$, let $u_\delta(x)$ be the solution of (1.1.1) satisfying the initial conditions, $u_\delta^{(i)}(x_2) = z^{(i)}(x_2), i = 0, 1, 2$, and $u_\delta'''(x_2) = z'''(x_2) - \delta$. By continuous dependence on initial conditions, for sufficiently small δ, there exist $x_0 < t_1 < x_2 < t_2 < x_3$ such that $u_\delta(t_1) = y(t_1)$ and $u_\delta'''(t_2) = y'''(t_2)$. Yet, we also have that $y_\delta^{(i)}(x_2) = y^{(i)}(x_2), i = 1, 2$, which contradicts the uniqueness of solutions of $(1, 2, 1)$ right focal boundary value problems for (1.1.1). This disposes of the case concerning uniqueness of solutions of each $(1, 3)$ right focal boundary value problem.

As we next deal with the $(2, 2)$ right focal boundary value problem for (1.1.1), we assume the conclusion of the theorem is false. So there are distinct solutions $y(x)$ and $z(x)$ of (1.1.1) and points $a < x_1 < x_2 < b$ such that $z^{(i)}(x_1) = y^{(i)}(x_1), i = 0, 1$, and $z^{(j)}(x_2) = y^{(j)}(x_2), j = 2, 3$. Again, fix x_0 and x_3, with $a < x_0 < x_1$ and $x_2 < x_3 < b$. There are two additional cases to consider.

Case 1: $y(x) - z(x)$ has a zero of order 2 at $x = x_1$. From (D3), we may assume $y''(x) > z''(x)$ on $[x_1, x_2)$. Then, $y^{(i)}(x) > z^{(i)}(x)$ on $(x_1, x_2]$, for $i = 0, 1$. This time, given $\delta > 0$, let $u_\delta(x)$ be the solution of (1.1.1) satisfying the initial conditions, $u_\delta(x_2) = y(x_2) - \delta$, and $u_\delta^{(i)}(x_2) = y^{(i)}(x_2), i = 1, 2, 3$. Uniqueness of solutions of $(1, 3)$ right focal boundary value problems implies $u_\delta(x) < y(x)$ on $(a, x_2]$. In addition, by continuous dependence of solutions on initial conditions, for sufficiently small δ, there exist $x_0 < t_1 < x_1 < t_2 < x_2$ such that $u_\delta(t_i) = z(t_i), i = 1, 2$. Also, $u_\delta^{(i)}(x_2) = z^{(i)}(x_2), i = 2, 3$. So, by Rolle's Theorem, there exists $t_1 < \tau < t_2$, such that $u_\delta(t_1) = z(t_1), u_\delta'(\tau) = z'(\tau)$ and $u_\delta^{(i)}(x_2) = z^{(i)}(x_2), i = 2, 3$. But this contradicts the uniqueness of solutions of $(1, 1, 2)$ right focal boundary value problems.

Case 2: $y(x) - z(x)$ has a zero of order 3 at $x = x_1$, (that is, $y^{(i)}(x_1) = z^{(i)}(x_2), i = 0, 1, 2$). Assume $y'''(x_1) > z'''(x_1)$. By (D3), we may assume $y''(x) > z''(x)$. There are two further subcases:

Subcase 2a: Assume $y'''(x) < z'''(x)$ on (x_2, b). Then $y''(x) < z''(x)$ on (x_2, b), and hence $y''(x) - z''(x)$ changes sign at x_2. For $\delta > 0$, let $u_\delta(x)$ be the solution of (1.1.1) satisfying the initial conditions, $u_\delta^{(i)}(x_1) = y^{(i)}(x_1), i = 0, 1, 3$, and $u_\delta''(x_1) = y''(x_1) - \delta$. By continuous dependence, for $\delta > 0$ sufficiently small, there exist $x_1 < t_1 < t_2 < x_3$ such that $u_\delta(t_1) = z(t_1)$ and $u_\delta''(t_2) = z''(t_2)$.) We also have $u_\delta^{(i)}(x_1) = z^{(i)}(x_1), i = 0, 1$. By repeated applications of Rolle's Theorem, there are points $x_1 < \tau_1 < \tau_2 < t_2$ such that $u_\delta^{(i)}(x_1) = z^{(i)}(x_1), i = 0, 1, u_\delta''(\tau_1) = z''(\tau_1)$, and $u_\delta'''(\tau_2) = z'''(\tau_2)$, which contradicts the uniqueness of solutions of $(2, 1, 1)$ right focal boundary value problems for (1.1.1).

Subcase 2b: Assume $y'''(x) > z'''(x)$ on (x_2, b). Then $y''(x) > z''(x)$ on (x_2, b), and so $y''(x) - z''(x)$ does not change sign at x_2. Consequently, $y''(x) - z''(x)$ attains a positive maximum at some point $\tau_0 \in (x_1, x_2)$, and so there exist points $\tau_0 < t_1 < x_2 < t_2 < x_3$ such that $y'''(t_1) < z'''(t_1)$ and $y'''(t_2) > z'''(t_2)$ (recall $y'''(x) > z'''(x)$ on (x_2, b)). Next, for $\delta > 0$, let $u_\delta(x)$ be the solution of the initial value problem for (1.1.1) satisfying $u_\delta^{(i)}(x_1) = y^{(i)}(x_1), i = 0, 1, 3$, and $u_\delta''(x_1) = y''(x_1) - \delta$. By continuous dependence of solutions on initial conditions, for sufficiently small δ, there are points $x_1 < \tau_1 < t_1 < \tau_2 < t_2$ such that $u_\delta(\tau_1) = z(\tau_1)$

and $u_\delta'''(\tau_2) = z'''(\tau_2)$. Also, $u_\delta^{(i)}(x_1) = z^{(i)}(x_1), i = 0, 1$, and so repeated applications of Rolle's Theorem contradicts the uniqueness of solutions of $(2, 1, 1)$ right focal boundary value problems for (1.1.1) on (a, b).

And so, Case 2 is also impossible, and thus the conclusion concerning uniqueness of solutions of $(2, 2)$ right focal boundary value problems for (1.1.1) is valid.   □

1.5   Right focal boundary value problems: for $m < r$, uniqueness of m-point implies uniqueness of r-point

In this section, converses and partial converses of results of the previous section, Section 1.4, are presented. And our uniqueness results for right focal boundary value problems (1.1.1), (1.1.3) are somewhat in analogy to those of Section 1.3 which were obtained for conjugate boundary value problems for (1.1.1). In particular, we are concerned with uniqueness results for solutions of (1.1.1), (1.1.3) that are along the lines of uniqueness results in Section 1.3 for (1.1.1), (1.1.2).

As in the previous section, Section 1.4, for local convenience, we recall from Section 1.1 what we mean by an r-point right focal boundary value problem (or an $(m_1, \ldots, m_r)$ right focal boundary value problem): Given $2 \leq r \leq n$, $m_1, \ldots, m_r \in \mathbb{N}$ such that $\sum_{j=1}^{r} m_j = n$, $s_0 := 0$, $s_k := \sum_{j=1}^{k} m_j$, $1 \leq k \leq r$, points $a < x_1 < \cdots < x_r < b$, and $y_{ik} \in \mathbb{R}$, $s_{k-1} \leq i \leq s_k - 1$, $1 \leq k \leq r$, a boundary value problem for (1.1.1) satisfying

$$y^{(i)}(x_k) = y_{ik}, \ s_{k-1} \leq i \leq s_k - 1, \ 1 \leq k \leq r,$$

is called either an r-*point right focal boundary value problem*, or an $(m_1, \ldots, m_r)$ *right focal boundary value problem*. (These boundary conditions are designated in Section 1.1 by (1.1.3).)

The results of this section will involve from Section 1.1 conditions (A), (B) and "compactness condition" (CP), and continuous dependence from Theorem 1.1.6 and Remark 1.1.1, along with other conditions we will impose.

Primarily, for fixed $2 \leq m < n$, our concern will be with uniqueness of solutions of m-point right focal boundary value problems implying uniqueness of solutions of r-point right focal boundary value problems for (1.1.1) on (a, b), for all $m < r \leq n$. Converse results for right focal boundary value problems are more rare than those for the conjugate problems obtained in Section 1.3.

In 1979, Muldowney [93, Prop. 1(b)] proved, for (1.1.1) linear, that uniqueness of solutions of 2-point right focal boundary value problems implies the uniqueness of solutions of each r-point right focal point boundary value problem, $2 < r \leq n$. Also, in 1979, Peterson [98] established a result analogous to the Muldowney result for *focal-type* (more general than right focal) boundary value problems for linear (1.1.1).

The first result of this section, which deals with (1.1.1) for $n = 3$, is actually an analogue of Theorem 1.3.1, but of more interest to this section, it is a converse of Theorem 1.4.1 (that is, we show that uniqueness of solutions of 2-point right focal boundary value problems implies uniqueness of solutions of 3-point right focal boundary value problems). For convenience of reference, we provide a uniqueness statement as a condition.

(D$_2$) There exists at most one solution of each (m_1, m_2) right focal boundary value problem for (1.1.1) on (a, b).

Theorem 1.5.1. [[49], Henderson, Thm. 2.1] *Let $n = 3$ and assume that with respect to (1.1.1), conditions (A), (B) and (D$_2$) are satisfied. Then, each 3-point right focal boundary value problem for (1.1.1) on (a, b) has at most one solution; that is, each $(1, 1, 1)$ right focal boundary value problem for (1.1.1) has at most one solution.*

Proof. Assume the hypotheses of the theorem, but assume that the conclusion is false. Then, there are points $a < x_1 < x_2 < x_3 < b$ and distinct solutions $y(x)$ and $z(x)$ of (1.1.1) satisfying $y^{(i-1)}(x_i) = z^{(i-1)}(x_i), i = 1, 2, 3$. As a consequence (D$_2$), we may assume $y'(x_1) > z'(x_1)$; moreover, we may also assume that $y^{(i)}(x) - z^{(i)}(x) \neq 0$ on (x_i, x_{i+1}), for $i = 1, 2$. It follows that $y'(x) > z'(x)$ on $[x_1, x_2)$ and $y''(x) < z''(x)$ on $(x_2, x_3]$.

Next, for $\epsilon > 0$, let $y_\epsilon(x)$ be the solution of the initial value problem for (1.1.1) satisfying $y_\epsilon^{(i)}(x_1) = y^{(i)}(x_1), i = 0, 1$, and $y_\epsilon''(x_1) = y''(x_1) + \epsilon$. One can then show from (D$_2$) and continuous dependence of solutions on initial conditions that, for each $\epsilon > 0$, there exists an interval $[x_2(\epsilon), x_3(\epsilon)] \subset (x_2, x_3)$ such that $y_\epsilon'(x) > z'(x)$ on $[x_1, x_2(\epsilon))$, $y_\epsilon'(x_2(\epsilon)) = z'(x_2(\epsilon))$, $y'(x) < y_\epsilon'(x) < z'(x)$ on $(x_2(\epsilon), x_3(\epsilon)]$, $y_\epsilon''(x_3(\epsilon)) = z''(x_3(\epsilon))$, and $y''(x) < y_\epsilon''(x) < z''(x)$ on $[x_2(\epsilon), x_3(\epsilon))$. Moreover, for $0 < \epsilon_1 < \epsilon_2$, the intervals are nested, in that $[x_2(\epsilon_2), x_3(\epsilon_2)] \subset (x_2(\epsilon_1), x_3(\epsilon_1))$. These intervals are obtained by choosing $x_2(\epsilon)$ to be the first zero of $y_\epsilon'(x) - z'(x)$ in (x_1, x_3) and then choosing $x_3(\epsilon)$ to be the first zero of $y_\epsilon''(x) - z''(x)$ in $(x_3(\epsilon), x_3)$.

Then, there exists a point $x_0 \in \bigcap_{k=1}^{\infty} [x_2(k), x_3(k)]$, and for each $k \geq 1$, $y^{(i)}(x_0) < y_k^{(i)}(x_0) < z^{(i)}(x_0)$, for $i = 2, 3$. In particular, each of the sequences $\{y'_k(x_0)\}_{k=1}^{\infty}$ and $\{y''_k(x_0)\}_{k=1}^{\infty}$ is bounded. Moreover, from (D_2), for each $k \geq 1$, $y''(x) < y''_k(x)$ on $[x_1, b)$, and so, $y^{(i)}(x) < y_k^{(i)}(x)$ on (x_1, b), $i = 0, 1$, and also, both $y'_k(x) - y'(x)$ and $y_k(x) - y(x)$ are increasing on $[x_1, b)$. Since $\{y'_k(x_0)\}_{k=1}^{\infty}$ is a bounded sequence, it follows that there is an $M > 0$ such that $|y'_k(x)| \leq M$, on $[x_1, x_0]$ and for all $k \geq 1$. Also, for all $k \geq 1$, $y_k(x_1) = y(x_1)$, and so

$$|y_k(x_0)| \leq \int_{x_1}^{x_0} |y'_k(s)| ds + |y(x_1)| \leq M(b-a) + |y(x_1)|.$$

Hence, each of the sequences $\{y_k^{(i)}(x_0)\}_{k=1}^{\infty}$, $i = 0, 1, 2$, is as bounded sequence. It follows from the Kamke Convergence Theorem that there is a subsequence $\{y_{k_j}(x)\}$ such that $\{y_{k_j}^{(i)}(x)\}$ converges uniformly on each compact subinterval of (a, b), for each $i = 0, 1, 2$. Yet, this contradicts the fact that $y''_{k_j}(x_1) = y''(x_1) + k_j \to \infty$.

We conclude that our assumptions concerning the existence of such distinct solutions $y(x)$ and $z(x)$ is false, and the proof is complete. $\square$

When Theorems 1.4.1 and 1.5.1 are combined, we obtain the following uniqueness characterization for solutions of 2-point and 3-point right focal boundary value problems when (1.1.1) is of third order.

Theorem 1.5.2. *Let $n = 3$ and assume that with respect to (1.1.1), conditions (A) and (B) are satisfied. Then, each 3-point right focal boundary value problem for (1.1.1) on (a, b) has at most one solution if, and only if, each 2-point right focal boundary value problem for (1.1.1) on (a, b) has at most one solution.*

For the case, when (1.1.1) is a fourth order equation, Henderson [49] obtained a uniqueness result for right focal boundary value problems which is an analogue of the Henderson and Jackson [62] result presented in Theorem 1.3.5 for conjugate boundary value problems. That result fits within the theme of this section.

Theorem 1.5.3. *[[49], Henderson, Thm. 3.2] Let $n = 4$, and assume that with respect to (1.1.1), conditions (A) and (B) are satisfied, that all 3-point right focal boundary value problems for (1.1.1) on (a, b) have at most one solution, and that all 2-point right focal boundary value problems for (1.1.1) on (a, b) have solutions. Then, all 4-point right focal boundary value problems for (1.1.1) on (a, b) have at most one solution.*

Proof. As a consequence of Theorem 1.4.2, all $(1,3)$ right focal and $(2,2)$ right focal boundary value problems for $(1.1.1)$ on (a,b) *have* unique solutions.

Now, assume the conclusion of the theorem is false. Then there are distinct solutions $y(x)$ and $z(x)$ of $(1.1.1)$ such that, for some $a < x_1 < x_2 < x_3 < x_4 < b$, $y^{(i-1)}(x_i) = z^{(i-1)}(x_i), i = 1,2,3,4$. By the uniqueness of solutions of 3-point right focal boundary value problems, x_i is a simple zero of $y^{(i-1)}(x) - z^{(i-1)}(x), i = 1,2,3$. And we may assume, without loss of generality, that $y^{(i)}(x) - z^{(i)}(x) \neq 0$ in (x_i, x_{i+1}), for $i = 1,2,3$. Also, we may assume the case where $y'(x_1) < z'(x_1)$. It then follows that $y''(x_2) > z''(x_2), y'''(x_3) < z'''(x_3)$, that $y'(x) < z'(x)$ on $[x_1, x_2)$, $y''(x) > z''(x)$ on $[x_2, x_3)$, and $y'''(x) < z'''(x)$ on $[x_3, x_4)$. The last inequality implies $y''(x) < z''(x)$ on $(x_3, x_4]$.

Hereafter, our proof proceeds much like the proof for Theorem 1.5.1. For $\epsilon > 0$, let $y_\epsilon(x)$ be the solution of $(1.1.1)$ satisfying the $(1,3)$ right focal boundary conditions,

$$y_\epsilon(x_1) = y(x_1), \ y_\epsilon^{(i)}(x_2) = y^{(i)}(x_2), \ i = 1,2, \text{ and } y_\epsilon'''(x_2) = y'''(x_2) + \epsilon.$$

By Remark 1.1.1, it follows from (B) and uniqueness of solutions of $(1,3)$ right focal boundary value problems that solutions depend continuously on such boundary conditions. From this and the fact that each $(1,1,2)$ right focal boundary value problems has at most one solution, it follows that, for each $\epsilon > 0$, there is a subinterval $[x_3(\epsilon), x_4(\epsilon)] \subset (x_3, x_4)$ such that $y_\epsilon''(x) > z''(x)$ on $[x_2, x_3(\epsilon))$, $y_\epsilon''(x_3(\epsilon)) = z''(x_3(\epsilon))$, $y''(x) < y_\epsilon''(x) < z''(x)$ on $(x_3(\epsilon), x_4(\epsilon)]$, $y_\epsilon'''(x_4(\epsilon)) = z'''(x_4(\epsilon))$, and $y'''(x) < y_\epsilon'''(x) < z'''(x)$ on $[x_3(\epsilon), x_4(\epsilon))$. Moreover, for $0 < \epsilon_1 < \epsilon_2$, the intervals are nested, in that $[x_3(\epsilon_2), x_4(\epsilon_2)] \subset (x_3(\epsilon_1), x_4(\epsilon_1))$.

So, there exists a point $x_0 \in \bigcap_{k=1}^\infty [x_3(k), x_4(k)]$, and the sequences $\{y_k''(x_0)\}_{k=1}^\infty$ and $\{y_k'''(x_0)\}_{k=1}^\infty$ are bounded. The same argument as in the proof of Theorem 1.5.1 yields $\{y_k'(x_0)\}_{k=1}^\infty$ is also bounded. So there exists a subsequence $\{y_{k_j}(x)\}$ and $\alpha_1, \alpha_2, \alpha_3 \in \mathbb{R}$ such that $y_{k_j}^{(i)}(x_0) \to \alpha_i, i = 1,2,3$.

Next, let $u(x)$ be the solution of the $(1,3)$ right focal boundary value problem for $(1.1.1)$ satisfying

$$u(x_1) = y(x_1) \text{ and } u^{(i)}(x_0) = \alpha_i, \ i = 1,2,3.$$

By continuous dependence on $(1,3)$ right focal boundary conditions, it follows that $\{y_{k_j}^{(i)}(x)\}$ converges to $u^{(i)}(x)$ uniformly on each compact subinterval of (a,b), for $i = 0,1,2,3$. But this is in contradiction to $y_{k_j}'''(x_2) \to +\infty$.

So, our assumptions concerning distinct solutions $y(x)$ and $z(x)$ of $(1.1.1)$ is false. The conclusion of the theorem follows. $\square$

For completeness of this section, we include the statement of a result concerning the validity of the "compactness condition" (CP), when $n = 4$ and under the hypotheses of Theorem 1.5.3.

Theorem 1.5.4. [[49], Henderson, Thm. 3.4] *Let $n = 4$, and assume the hypotheses of Theorem 1.5.3. Then, the "compactness condition" (CP) is satisfied.*

Our final result of this section involves a generalization of Theorem 1.5.3, which is also somewhat in analogy to Theorem 1.3.5. Its proof involves only slight modifications of the proof of Theorem 1.5.3.

Theorem 1.5.5. [[49], Henderson, Thm. 4.1] *Assume that with respect to (1.1.1), conditions (A) and (B) are satisfied, that all $(n-1)$-point right focal boundary value problems for (1.1.1) on (a, b) have at most one solution, and that all $(n-2)$-point right focal boundary value problems for (1.1.1) on (a, b) have solutions. Then, all $(1, 1, \ldots, 1)$ right focal boundary value problems for (1.1.1) on (a, b) have at most one solution.*

Chapter 2

Uniqueness Implies Existence

This chapter is devoted primarily to existence of solutions for boundary value problems for ordinary differential equations. In particular, focus will be on when uniqueness of solutions for one type of boundary value problem implies existence of solutions for the same type and other types of boundary value problems. Again it is pointed out that in the case of linear boundary value problems for linear ordinary differential equations, uniqueness of solutions is equivalent to their existence.

The "uniqueness implies existence" arguments that are produced in this chapter will be applied to conjugate boundary value problems for second, third and nth order differential equations, respectively, and right focal boundary value problems for nth order differential equations.

2.1 Conjugate boundary value problems: for $n = 2$, $k = 2$, uniqueness of 2-point implies existence of 2-point

This section is devoted to existence results of conjugate boundary value problems, (1.1.1)-(1.1.2), in the case that $n = k = 2$. In a seminal paper [88], Lasota and Opial showed that global existence and uniqueness of solutions of initial value problems and uniqueness of solutions for 2-point conjugate problems for $y'' = f(x, y, y')$, $a < x < b$, implies existence of solutions of 2-point conjugate problems. Theorem 2.1.1 contains the original work of Lasota and Opial.

Consider the 2-point conjugate boundary value problem for the second order ordinary differential equation,

$$y'' = f(x, y, y'), \quad a < x < b, \tag{2.1.1}$$

$$y(x_1) = y_{11}, \quad y(x_2) = y_{12}, \tag{2.1.2}$$

where $a < x_1 < x_2 < b$ and $y_{11}, y_{12} \in \mathbb{R}$.

Theorem 2.1.1. [[88], Lasota and Opial, Thm. 1] *Assume that with respect to (2.1.1), conditions (A)–(C) are satisfied. Then for any $a < x_1 < x_2 < b$, and any $y_{11}, y_{12} \in \mathbb{R}$, the boundary value problem, (2.1.1)-(2.1.2), has a solution.*

Proof. Let $a < x_1 < x_2 < b$, and $y_{11}, y_{12} \in \mathbb{R}$, be given. Let $y(x; \mu)$ denote the solution of (2.1.1) that satisfies the initial conditions, $y(x_1) = y_{11}, y'(x_1) = \mu$. Let

$$S = \{y(x_2; \mu) \,|\, \mu \in \mathbb{R}\}.$$

It is shown that $S = \mathbb{R}$ and then Theorem 2.1.1 is proved. The proof given in [77] is reproduced here.

It follows from Continuous Dependence, Theorem 1.1.6, that $y(x_2; \mu)$ is continuous as a function of μ and so, S is an interval. Thus, the theorem is proved once it is shown that S is unbounded. Assume, for the sake of contradiction, that S is bounded above and let $s_0 > \sup S$. Denote by $u(x)$ the solution of the initial value problem, (2.1.1), with the initial conditions,

$$u(x_2) = s_0, \quad u'(x_2) = 0.$$

Since, $s_0 \notin S$, then $u(x_1) \neq y_{11}$. Assume first that $u(x_1) > y_{11}$. By the uniqueness hypothesis, condition (C), if $n_1 < n_2$, $n_1, n_2 \in \mathbb{N}$, then

$$y(x; n_1) < y(x; n_2), \quad x_1 < x < b.$$

Moreover, $u(x_1) > y(x_1; n)$ and $u(x_2) > y(x_2; n)$; so, again condition (C) implies

$$y(x; n_1) \leq y(x; n_2) \leq u(x), \quad x_1 \leq x \leq x_2,$$

for $n_1 < n_2$, $n_1, n_2 \in \mathbb{N}$. Thus, there exists $M > 0$ such that $|y(x; n)| \leq M$ on $[x_1, x_2]$. For each n, there exists $t_n \in (r_1, r_2)$ such that

$$|y'(t_n; n)|(x_2 - x_1) = |y(x_2; n) - y(x_1; n)| \leq 2M.$$

Relabeling if necessary, select a sequence $y(x; n)$ such that each of the sequences, $\{t_n\}, \{y(t_n; n)\}, \{y'(t_n; n)\}$, converges. Then by Kamke's Theorem, Theorem 1.1.4, and by condition (A) there exists a subsequence (relabeling again) such that $\{y(x; n)\}$ and $\{y'(x; n)\}$ each converge uniformly on $[x_1, x_2]$. This however, contradicts the original construction of $\{y(x; n)\}$ since $y'(x_1; n) = n$. Thus, the assumption $u(x_1) > y(x_1; n)$ is false, and it is the case that $u(x_1) < y(x_1; n)$.

Let $x_3 \in (x_2, b)$ be given. Since

$$u(x_1) < y(x_1; n), \quad u(x_2) > y(x_2; n),$$

it is the case that

$$y(x; n_1) \leq y(x; n_2) \leq u(x), \quad x_2 \leq x \leq x_3,$$

if $n_1 < n_2$, $n_1, n_2 \in \mathbb{N}$. The Kamke Theorem, Theorem 1.1.4, and condition (A) again imply there is a subsequence $\{y'(x; n)\}$ which converges uniformly on $[x_1, x_3]$, leading to the same contradiction.

Thus, S is not bounded above. A similar argument gives that S is not bounded below and so $S = \mathbb{R}$. ☐

2.2 Conjugate boundary value problems: for $n = 3$, uniqueness of 3-point implies existence of 2-point and 3-point

Jackson and Schrader [82] extended the Lasota and Opial result, Theorem 2.1.1, to $n = 3$. A major contribution of the work in [82] was the proof of the "compactness condition" (CP) in the case $n = 3$; although the method of proof did not lend itself to the proof of Theorem 1.1.5, we reproduce the details for the proof of "compactness condition" (CP) in the case $n = 3$ here.

Consider the conjugate boundary value problem for the third order ordinary differential equation,

$$y''' = f(x, y, y', y''), \quad a < x < b, \tag{2.2.1}$$

and consider each of the conjugate boundary conditions

$$y(x_1) = y_{11}, \quad y'(x_1) = y_{21}, \quad y(x_2) = y_{12}, \tag{2.2.2}$$

$$y(x_1) = y_{11}, \quad y(x_2) = y_{12}, \quad y'(x_2) = y_{22}, \tag{2.2.3}$$

where $a < x_1 < x_2 < b$ and $y_{11}, y_{21}, y_{12}, y_{22} \in \mathbb{R}$, and

$$y(x_1) = y_{11}, \quad y(x_2) = y_{12}, \quad y(x_3) = y_{13}, \tag{2.2.4}$$

where $a < x_1 < x_2 < x_3 < b$ and $y_{11}, y_{12}, y_{13} \in \mathbb{R}$.

Theorem 2.2.1. [[82], Jackson and Schrader, Lem. 2] *Assume that with respect to (2.2.1), conditions (A)–(C) are satisfied. Then given a compact interval $[c, d] \subset (a, b)$ and given $M > 0$, there exists $\delta(M) > 0$ such that for any $[x_1, x_2] \subset [c, d]$ and $x_2 - x_1 \leq \delta(M)$, and any real α with $|\alpha| \leq M$, each of the 2-point boundary value problems, (2.2.1)-(2.2.2) and (2.2.1)-(2.2.3) has a solution in the cases that $\alpha = y_{11} = y_{12}$, and $0 = y_{21} = y_{22}$. Moreover, any such solution satisfies $|y'(x)| \leq 1$ and $|y''(x)| \leq 1$ for $x \in [x_1, x_2]$.*

Proof. The proof of Theorem 2.2.1 is a straightforward application of the Schauder-Tychonoff fixed point theorem. However, a similar argument is not provided in this account, and so, we outline the details with the boundary value problem, (2.2.1)-(2.2.2). Let $G(x,s)$ denote the Green's function for the boundary value problem, (2.2.1)-(2.2.2). It can be shown that [3, p. 145]

$$\max_{x_1 \le x \le x_2} \int_{x_1}^{x_2} |G(x,s)| ds = \frac{2}{81}(x_2 - x_1)^3,$$

$$\max_{x_1 \le x \le x_2} \int_{x_1}^{x_2} |G_x(x,s)| ds = \frac{1}{6}(x_2 - x_1)^2,$$

$$\max_{x_1 \le x \le x_2} \int_{x_1}^{x_2} |G_{xx}(x,s)| ds = \frac{2}{3}(x_2 - x_1).$$

Define the fixed point operator T on the Banach space $C^2[x_1, x_2]$ by

$$Ty(x) = \alpha + \int_{x_1}^{x_2} G(x,s) f(s, x(s), x'(s), x''(s)) ds.$$

Define the closed convex subset of $C^2[x_1, x_2]$ by

$$K = \{y \in C^2[x_1, x_2] : ||y|| \le 2M, ||y'|| \le 2, ||y''|| \le 2\}$$

where $||u|| = \max_{x_1 \le x \le x_2} |u(x)|$. Let

$$Q = \max\{|f(x, u_1, u_2, u_3)| : x \in [x_1, x_2], |u_1| \le 2M, |u_2| \le 2, |u_3| \le 2\}.$$

If

$$\delta = \min\left\{ \sqrt[3]{\frac{81M}{2Q}}, \sqrt{\frac{6}{Q}}, \frac{3}{2} \right\}$$

and $x_2 - x_1 < \delta$, then T maps K into K and the Schauder fixed point theorem applies.   □

The next result is a technical result which is used quite specifically in the proof of the "compactness condition" (CP), for $n = 3$.

Theorem 2.2.2. [[82], Jackson and Schrader, Lem. 3] *Let $a < b$, and let $y \in C^2[a, b]$ such that $y(x) \le M$ on $[a, b]$. Then there exists a constant $K = K(M, b - a) > 0$ such that if*

$$\max\{|y'(x)|, |y''(x)|\} > K, \text{ for all } a \le x \le b,$$

then there exist $x_0 \in (a, b)$ such that $y'(x_0) = 0$.

Proof. Jackson and Schrader [82] left the proof of Theorem 2.2.2 to the reader. We present a proof provided by Lloyd Jackson through a private communication.

Let $[a, b]$ and $M > 0$ be given. Assume the conclusion of the theorem is false; that is, assume $y'(x) \neq 0$ for all $a < x < b$.

Assume $y \in C^2[a, b]$ such that $y(x) \leq M$ on $[a, b]$ and assume

$$|y'(x)| + |y''(x)| > K_1, \text{ for all } a \leq x \leq b, \tag{2.2.5}$$

where $K_1 = K_2 + \frac{2M}{b-a} + 1$. The argument is to exhibit $K_2 > 0$ sufficiently large such that if y satisfies (2.2.5) then the assumption $y'(x) \neq 0$ for all $a < x < b$ is false. Assume without loss of generality that $y'(x) > 0$ for all $a < x < b$.

Note, there exists $x_1 \in (a, b)$ such that

$$|y'(x_1)| = y'(x_1) = \left| \frac{y(b) - y(a)}{b - a} \right| \leq \frac{2M}{b - a}.$$

There are two cases to consider, $a < x_1 \leq \frac{a+b}{2}$ or $\frac{a+b}{2} \leq x_1 < b$. We show the details for the case $a < x_1 \leq \frac{a+b}{2}$. Note, we may assume $y(x_1) < M$; if $y(x_1) = M$, and $0 < y'(x_1)$, then the hypotheses $y(x) \leq M$ on $[a, b]$ is contradicted.

Define the parameter $\eta = \frac{b-a}{8}$.

Now address each possibility, $y''(x_1) \leq 0$ or $y''(x_1) > 0$. First, assume $y''(x_1) \leq 0$. By (2.2.5), $y''(x_1) < -K_2$, and y' is decreasing at x_1. Again by (2.2.5), $y''(x) < -K_2$ on a right neighborhood of x_1. Arguing by contradiction, it follows that for $x_1 \leq x \leq b$, $0 < y'(x) \leq \frac{2M}{b-a}$ and so, $y''(x) \leq -K_2$, if $x_1 \leq x \leq b$ by (2.2.5). Apply Taylor's theorem

$$y(b) = y(x_1) + y'(x_1)(b - x_1) + y''(c)\frac{(b - x_1)^2}{2}$$

for some $c \in (x_1, b)$ and the assumption, $x_1 \leq \frac{a+b}{2}$, which implies $b - x_1 \geq \frac{b-a}{2}$. Then

$$y(b) < M + \frac{2M(b - a)}{b - a} - K_2\frac{(b - a)^2}{8} \leq -M$$

if $K_2 \geq 32\frac{M}{(b-a)^2} = \frac{M}{2\eta^2}$.

Thus, the assumption $y''(x_1) \leq 0$ is false. Now assume $y''(x_1) > 0$ and in fact, $y''(x_1) > K_2$ by (2.2.5). Assume for the sake of contradiction that

$$y''(x) > \frac{K_2}{2}, \quad x_1 \leq x \leq x_1 + \eta.$$

Then there exists $c \in (x_1, x_1 + \eta)$ such that

$$y(x_1 + \eta) = y(x_1) + y'(x_1)\eta + \frac{y''(c)}{2}\eta^2 > -M + \frac{K_2}{4}\eta^2.$$

If $K_2 \geq \frac{8M}{\eta^2}$, then $y(x_1 + \eta) > M$ which is a contradiction. Thus, there exists, $x_2 \in (x_1, x_1 + \eta)$ such that

$$y''(x_2) = \frac{K_2}{2} \text{ and } y''(x) > \frac{K_2}{2}, \text{ if } x_1 \leq x < x_2.$$

By (2.2.5),

$$y'(x) > \frac{2M}{b-a} + \frac{K_2}{2}, \quad x_1 \leq x < x_2.$$

Now assume (again, for the sake of contradiction) that $y'(x) > \frac{2M}{b-a} + \frac{K_2}{2}$ on $[x_2, x_2 + \eta)$. Then there exists $c \in (x_2, x_2 + \eta)$ such that

$$y(x_2 + \eta) = y(x_2) + y'(c)\eta > -M + \frac{K_2}{2}\eta + \frac{2M}{b-a}\eta = -M + \frac{K_2}{2}\eta + \frac{M}{4}.$$

If $K_2 \geq \frac{7M}{2\eta}$, then $y(x_2 + \eta) > M$, which is a contradiction. Thus, there exists $x_3 \in (x_2, x_2 + \eta)$ such that

$$y'(x_3) = \frac{K_2}{2}\eta + \frac{2M}{b-a} \text{ and } y'(x) > \frac{K_2}{2}\eta + \frac{2M}{b-a} \text{ for } x_2 \leq x < x_3.$$

This implies that $y''(x_3) < 0$ and it follows by (2.2.5) that $y''(x_3) < -\frac{K_2}{2}$. Thus, in a right neighborhood of x_3, $y''(x) < -\frac{K_2}{2}$ and y' is decreasing. Continued applications of (2.2.5) imply that $y''(x) < -\frac{K_2}{2}$ and y' is decreasing on (x_3, b).

From the original assumption that $y'(x) \neq 0$ for all $a < x < b$, it is now the case that

$$0 < y'(x) \leq \frac{K_2}{2}\eta + \frac{2M}{b-a}, \quad x_3 \leq x < b. \tag{2.2.6}$$

Then there exists $c \in (x_3, b)$ such that

$$y(x_3) = y(b) + y'(b)(x_3 - b) + \frac{y''(c)}{2}(x_3 - b)^2 < M - \frac{K_2}{4}(b-a)^2.$$

So, if $K_2 \geq \frac{M}{8\eta^2}$, $y(x_3) < -M$ which is a contradiction. Thus, the assumption, (2.2.6) is false.

We conclude that if $y'(x_1) = \frac{2M}{b-a}$ and $x_1 \leq \frac{a+b}{2}$, and if

$$K_2 \geq \max\left\{ \frac{8M}{\eta^2}, \frac{7M}{2\eta} \right\},$$

there exists $x_0 \in (x_1, b)$ such that $y'(x_0) = 0$. The argument is analogous if $\frac{a+b}{2} \leq x_1$ and the proof is complete. $\qquad\square$

The next theorem is the "compactness condition" (CP) in the case $n = 3$.

Theorem 2.2.3. [[82], Jackson and Schrader, Thm. 1] *Assume that with respect to* (2.2.1), *conditions* (A)–(C) *are satisfied. Let* $[c, d]$ *be a compact subinterval of* (a, b) *and let* $\{y_n(x)\}$ *denote a sequence of solutions of* (2.2.1) *such the* $|y_n(x)| \leq M$ *on* $[c, d]$ *for some* $M > 0$ *and all* $n \geq 1$. *Then there exists a subsequence* $\{y_{n_j}(x)\}$ *of* $\{y_n(x)\}$ *such that* $\{y_{n_j}^{(i-1)}(x)\}$ *converges uniformly on* $[c, d]$, $i = 1, 2, 3$. *Moreover,* $\{y_{n_j}(x)\}$ *converges to a solution of* (2.2.1) *on* $[c, d]$.

Proof. Assume for the sake of contradiction that there is no subsequence of $\{y_n(x)\}$ converging in $C^2[c, d]$. Then by the Kamke Theorem, Theorem 1.1.4, it is the case that

$$|y_n'(x)| + |y_n''(x)| \to \infty$$

uniformly on $[c, d]$. Let $c \leq x_1 < x_2 < x_3 < x_4 \leq d$ be such that $x_4 - x_1 < \delta(M)$ where $\delta(M)$ is given by Theorem 2.2.1.

Apply Theorem 2.2.2 and find $K > 1$ such that if

$$\max\{|y_n'(x)||, |y_n''(x)|\} > K$$

for each $x \in [c, d]$, then y_n' vanishes in each subinterval, (x_1, x_2), (x_2, x_3), and (x_3, x_4). Since $|y_n'(x)| + |y_n''(x)| \to \infty$, uniformly on $[c, d]$ there exists n_0 such that

$$\max\left\{ \max_{c \leq x \leq d} |y_{n_0}'(x)|, \max_{c \leq x \leq d} |y_{n_0}''(x)| \right\} > K.$$

Now let

$$c \leq x_1 < t_1 < x_2 < t_2 < x_3 < t_3 < x_4 \leq d$$

be such that

$$y_{n_0}'(t_i) = 0, \quad i = 1, 2, 3,$$

and recall $|y_{n_0}''(t_i)| > K > 1$, for each i.

First, assume that $y_{n_0}(t_i) = y_{n_0}(t_j) = \alpha$, for $t_i < t_j$. Then y_{n_0} is a solution of a boundary value problem, (2.1.1), (2.1.2), with $\alpha = y_{11} = y_{12}$, and $0 = y_{21}$. Since $t_j - t_i < \delta(M)$, apply Theorem 2.2.1 and it follows that $|y_{n_0}''(x)| \leq 1$ on $[t_i, t_j]$. This contradicts the construction that $|y_{n_0}''(t_i)| > K > 1$ and so, it is the case that $y_{n_0}(t_i) \neq y_{n_0}(t_j)$ if $i \neq j$.

So now assume $y_{n_0}(t_i) \neq y_{n_0}(t_j)$ if $i \neq j$; again, there are cases, so first assume,

$$y_{n_0}(t_1) < y_{n_0}(t_2) < y_{n_0}(t_3).$$

If $y_{n_0}''(t_2) > K$, then there exists $\tau_1 \in (t_1, t_2)$ such that $y_{n_0}(\tau_1) = y_{n_0}(t_2)$; apply Theorem 2.2.1 with $x_1 = \tau_1$, $x_2 = t_2$, $y_{n_0}(t_2) = y_{11} = y_{12}$, and $0 = y_{22}$ to obtain a contradiction. If $y_{n_0}''(t_2) < -K$, then there exists $\tau_2 \in (t_2, t_3)$ such that $y_{n_0}(t_2) = y_{n_0}(\tau_2)$; again, Theorem 2.2.1 produces a contradiction.

If

$$y_{n_0}(t_1) > y_{n_0}(t_2) > y_{n_0}(t_3)$$

the argument is completely analogous. If

$$y_{n_0}(t_1) < y_{n_0}(t_2), \quad y_{n_0}(t_2) > y_{n_0}(t_3),$$

assume that $y_{n_0}''(t_2) < K$. If $y_{n_0}(t_1) < y_{n_0}(t_3)$, there exists $\tau_1 \in (t_1, t_2)$ such that $y_{n_0}(\tau_1) = y_{n_0}(t_3)$. Apply Theorem 2.2.1 with $x_1 = \tau_1$, $x_2 = t_3$, $y_{n_0}(t_3) = y_{11} = y_{12}$, and $0 = y_{22}$ to obtain a contradiction. If $y_{n_0}(t_1) > y_{n_0}(t_3)$, there exists $\tau_1 \in (t_2, t_3)$ such that $y_{n_0}(t_1) = y_{n_0}(\tau_1)$. Apply Theorem 2.2.1 with $x_1 = t_1$, $x_2 = \tau_1$, $y_{n_0}(t_1) = y_{11} = y_{12}$, and $0 = y_{21}$ to obtain a contradiction. All other cases can now be handled similarly. $\square$

Theorem 2.2.4. [[82], Jackson and Schrader, Lem. 6] *Assume that with respect to (2.2.1), conditions (A)–(C) are satisfied. Let $a < x_1 < x_2 < x_3 < b$ and let u and v be solutions of (1.1.1) on (a, b) with*

$$u(x) > v(x), \quad x_1 \le x < x_2 \text{ and } u(x) < v(x), \quad x_2 < x \le x_3. \tag{2.2.7}$$

Then there exist solutions, y_1, y_2 of (1.1.1) on (a, b) such that

$$y_1(x_1) = u(x_1), \quad y_1(x_2) = u(x_2) = v(x_2), \quad y_1(x_3) = v(x_3),$$

and

$$y_2(x_1) = v(x_1), \quad y_2(x_2) = v(x_2) = u(x_2), \quad y_2(x_3) = u(x_3).$$

Furthermore, for $i = 1, 2$,

$$v(x) \le y_i(x) \le u(x), \quad x_1 \le x \le x_2,$$
$$u(x) \le y_i(x) \le v(x), \quad x_2 \le x \le x_3.$$

Proof. The existence of y_1 is shown; details to prove existence of y_2 are analogous.

First, assume $u'(x_2) = v'(x_2)$. It follows from (2.2.7) that $u''(x_2) = v''(x_2)$ and u and v satisfy the same initial conditions at x_2. Thus,

$$y_1(x) = \begin{cases} u(x), & x_1 \le x \le x_2, \\ \\ v(x), & x_2 \le x \le x_3, \end{cases}$$

and y_1 exists.

Second, assume $u'(x_2) < v'(x_2)$ and let μ be a fixed real number satisfying $u'(x_2) \leq \mu < v'(x_2)$. It will be shown that for sufficiently large ν_1, the solution, $z(x; \nu_1)$, of (2.2.1) on (a, b) satisfying the initial conditions

$$z(x_2; \nu_1) = u(x_2), \quad z'(x_2; \nu_1) = \mu, \quad z''(x_2; \nu_1) = \nu_1,$$

satisfies $z(x_3; \nu_1) > v(x_3)$. It will also be shown that for sufficiently small ν_2, the solution, $z(x; \nu_2)$ of (2.2.1) on (a, b) satisfying the initial conditions

$$z(x_2; \nu_2) = u(x_2), \quad z'(x_2; \nu_2) = \mu, \quad z''(x_2; \nu_2) = \nu_2,$$

satisfies $z(x_3; \nu_2) \leq v(x_3)$.

Since $u'(x_2) \leq \mu < v'(x_2)$, if follows from condition (B) and Theorem 1.2.1 that for $\nu \geq 1$,

$$z(x; 1) \leq z(x; \nu) \leq v(x), \quad x_2 \leq x \leq x_3.$$

This contradicts the "compactness condition" (CP), Theorem 1.1.5 or Theorem 2.2.3, since $z''(x_2; \nu) = \nu$. Hence, there exists ν_1 such that $z(x_3; \nu_1) > v(x_3)$. If $u'(x_2) < \mu$, a similar argument implies there exist $\nu_2 < \nu_1$ such that $z(x_3; \nu_2) < u(x_3) < v(x_3)$. Finally, if $u'(x_2) = \mu$, and if $\nu_2 < u''(x_2)$, then $z(x_3; \nu_2) < u(x_3) < v(x_3)$. It now follows by continuous dependence, Theorem 1.1.6 and the intermediate value theorem that if $u'(x_2) \leq \mu < v'(x_2)$, there is a solution, $y(x; \mu)$ of (2.2.1) on (a, b) satisfying

$$y(x_2) = u(x_2), \quad y'(x_2) = \mu, \quad y(x_3) = v(x_3).$$

If $\mu = v'(x_2)$, then v serves that purpose and we conclude for all $u'(x_2) \leq \mu \leq v'(x_2)$, there is a solution, $y(x; \mu)$ of (2.2.1) on (a, b) satisfying

$$y(x_2) = u(x_2), \quad y'(x_2) = \mu, \quad y(x_3) = v(x_3).$$

Moreover, by uniqueness of solutions, Theorem 1.2.1, it follows that if

$$u'(x_2) \leq \mu_1 \leq \mu_2 \leq v'(x_2),$$

then

$$u(x) \leq y(x; \mu_1) \leq y(x; \mu_2) \leq v(x), \quad x_2 \leq x \leq x_3.$$

A similar argument shows that for each $u'(x_2) \leq \mu \leq v'(x_2)$, there is a solution, $w(x; \mu)$ of (2.2.1) on (a, b) satisfying

$$w(x_1 = u(x_1), \quad w(x_2) = u(x_2), \quad w'(x_2) = \mu$$

and if

$$u'(x_2) \leq \mu_1 \leq \mu_2 \leq v'(x_2),$$

then

$$v(x) \leq w(x; \mu_2) \leq w(x; \mu_1) \leq u(x), \quad x_1 \leq x \leq x_2.$$

Note that this implies

$$y''(x_2; v'(x_2)) = v''(x_2) \leq w''(x_2; v'(x_2)),$$
$$y''(x_2; u'(x_2)) \geq u''(x_2) \leq w''(x_2; u'(x_2)).$$

Set

$$\mu_0 = \inf\{\mu \in [u'(x_2, v'(x_2)] : y''(x_2; r) \leq w''(x_2; r), \ \mu \leq v'(x_2)\}.$$

It follows from the "compactness condition" (CP), Theorem 1.1.5 or Theorem 2.2.3, that

$$y''(x_2; \mu_0) = w''(x_2; \mu_0).$$

Define

$$y_1(x) = \begin{cases} w(x; \mu_0), & x_1 \leq x \leq x_2, \\ \\ w(x; \mu_0), & x_2 \leq x \leq x_3. \end{cases}$$

Then y_1 is the desired solution of (2.2.1) on (a, b) satisfying

$$y_1(x_1) = u(x_1), \quad y_1(x_2) = u(x_2) = v(x_2), \quad y_1(x_3) = v(x_3). \qquad \square$$

Theorem 2.2.5. [[82], Jackson and Schrader, Thm. 2] *Assume that with respect to (2.2.1), conditions (A)–(C) are satisfied. Then for any $a < x_1 < x_2 < b$, $y_{11}, y_{12}, y_{13} \in \mathbb{R}$, each boundary value problem, (2.2.1)-(2.2.2) and (2.2.1)-(2.2.3), has a solution.*

Proof. The details are given for the boundary value problem, (2.2.1)-(2.2.3); the details for the boundary value problem, (2.2.1)-(2.2.4) are completely analogous. Let $y(x; \mu)$ denote the solution of the initial value problem, (2.2.1), with initial conditions

$$y(x_1) = y_{11}, \quad y'(x_1) = y_{21}, \quad y''(x_1) = \mu.$$

Set

$$S = \{y(x_2; \mu) | \mu \in \mathbb{R}\}.$$

It follows by condition (B) that $S \neq \emptyset$ and again, it follows from Continuous Dependence, Theorem 1.1.6, that S is an interval. Thus, the argument is to show that S is not bounded above or below. The details contradicting an upper bound are given here.

Assume for the sake of contradiction that S is bounded above and let $s_0 > \sup S$. Denote by $u(x)$ the solution of the initial value problem, (2.1.1), with the initial conditions,

$$u(x_2) = s_0, \quad u'(x_2) = 0, \quad u''(x_2) = 0.$$

Assume first that $u(x_1) \geq y_{11}$. Assume further (the first of two subcases) that $y(x; \mu) \leq u(x)$ on $[x_1, x_2]$ for all $\mu \in \mathbb{R}$. Note then that for each $n \in \mathbb{N}$, $y(x; 1) \leq y(x; n) \leq u(x)$ on $[x_1, x_2]$. This contradicts the "compactness condition" (CP), Theorem 1.1.5 or Theorem 2.2.3, since $y''(x_1; n) = n$. So, now assume the second subcase, $y(x; \hat{\mu}) > u(x)$ at some points in (x_1, x_2) for some $\hat{\mu} \in \mathbb{R}$. Since $y(x; \mu) \geq y(x; \hat{\mu})$ on $[x_1, x_2]$ if $m > \hat{\mu}$, then $y(x; \mu) > u(x)$ at some points in (x_1, x_2) if $\mu > \hat{\mu}$. Since $y(x_1; \mu) = y_{11} \leq u(x_1), y(x_2; \mu) < s_0 = u(x_2)$, it follows by the uniqueness condition (C), $y(x; \mu) < u(x)$ on $[x_2, b]$ if $m > \hat{\mu}$. Again apply the "compactness condition" (CP), Theorem 1.1.5 or Theorem 2.2.3, and the Kamke Theorem, Theorem 1.1.4, to obtain a contradiction. Thus, the assumption $u(x_1) \geq y_{11}$ is false.

Now assume the case $u(x_1) < y_{11}$. Note that $y(x_1; 1) > u(x_1)$ and $y(x_2; 1) < u(x_2)$. Applying the uniqueness condition (C), there exists $x_3 \in (x_1, x_2)$ such that

$$y(x; 1) > u(x), \quad x_1 \leq x < x_3, \quad y(x; 1) < u(x) \quad x_3 < x \leq x_2.$$

Theorem 2.2.4 applies and there exists $w(x)$ a solution of (2.2.1) satisfying

$$w(x_1) = y(x_1; 1), \quad w(x_3) = y(x_3; 1) = u(x_3), \quad w(x_2) = u(x_2) = s_0.$$

Moreover, $w(x) \leq y(x; 1)$, on $[x_1, x_2]$. Thus, if $n > 1$, $y(x_1; n) = y_{11} = w(x_1)$, $y(x; n) > w(x)$, on $(x_1, x_2]$, and $y(x_2; n) < s_0 = w(x_2)$. Hence for each n, there exists $\tau_n \in (x_3, x_2)$ such that $y(\tau_n; n) = u(\tau_n)$. The uniqueness condition (C) then implies that $y(x; n) < w(x)$ on $[x_2, b]$ which produces the same contradiction as before.

Hence, S is not bounded above. A similar argument shows that S is not bounded below and the proof is complete. $\qquad \square$

The proof of existence of solutions of the 3-point boundary value problem, (2.2.1)-(2.2.4), is constructive in the sense that solutions of 2-point boundary value problems, whose existences are given by Theorem 2.2.5 are matched to give a solution of the boundary value problem, (2.2.1)-(2.2.4).

Theorem 2.2.6. [[82], Jackson and Schrader, Thm. 3] *Assume that with respect to* (2.2.1), *conditions* (A)–(C) *are satisfied. Then for any* $a < x_1 < x_2 < x_3 < b$, $y_1, y_2, y_3 \in \mathbb{R}$, *the boundary value problem,* (2.2.1)-(2.2.4), *has a solution.*

Proof. By Theorem 2.2.5, the boundary value problem, (2.2.1), with boundary conditions

$$y(x_1) = y_{11}, \quad y'(x_1) = 0, \quad y(x_3) = y_{13},$$

has a solution, $z(x)$. If $z(x_2) = y_2$, the z is a solution of (2.2.1)-(2.2.4) and the proof is complete. So assume $z(x_2) \neq y_2$. Assume first that $z(x_2) > y_2$. Let $s \in \mathbb{R}$. Again, apply Theorem 2.2.5 to obtain a solution, $u(x; s)$, of a boundary value problem, (2.2.1), with boundary conditions

$$y(x_1) = y_{11}, \quad y(x_2) = y_{12}, \quad y'(x_2) = s.$$

The "compactness condition" (CP), Theorem 1.1.5 or Theorem 2.2.3, implies there exists s_0 such that $u(x_3; s_0) > y_3$.

Similarly, there exists $s_1 \in \mathbb{R}$ and a solution $v(x, s_1)$ of the boundary value problem, (2.2.1), with boundary conditions

$$y(x_2) = y_{12}, \quad y'(x_2) = s_1, \quad y(x_3) = y_{13}$$

such that $v(x_1; s_1) > y_{11}$. By the uniqueness condition (C) and Theorem 1.2.1 it follows that either

$$v(x; s_1) > u(x : s_0), \quad x_1 \leq x < x_2,$$
$$v(x; s_1) < u(x : s_0), \quad x_2 < x \leq x_3,$$
$$v'(x_2; s_1) < u'(x_2 : s_0),$$

or

$$v'(x_2; s_1) = u'(x_2 : s_0), \quad v''(x_2; s_1) = u''(x_2 : s_0).$$

In the first case, (2.2.1)-(2.2.4) has a solution by Theorem 2.2.4. In the second case, define y by

$$y(x) = \begin{cases} u(x; s_0), & x_1 \leq x \leq x_2, \\ v(x; s_1), & x_2 \leq x \leq x_3. \end{cases}$$

Then y is a solution of the boundary value problem, (2.2.1)-(2.2.4). $\square$

2.3 Conjugate boundary value problems: nth order

An extension of Theorem 2.1.1 to the nth order k-point conjugate problem was done independently by Hartman [43] and Klaasen [85]. Initially, in 1958, Hartman [41] proved that, if conditions (A) and (B) are satisfied, then all n-point conjugate boundary value problems for (1.1.1) have unique solutions, if, and only if, all k-point conjugate boundary value problems, $2 \leq k \leq n-1$, for (1.1.1) have unique solutions. Then, in 1971, Hartman [43] proved that if conditions (A)–(C) are satisfied, then all n-point conjugate boundary value problems for (1.1.1) have unique solutions. Klaasen [85] obtained the existence of solutions of n-point conjugate boundary value problems using an alternative argument.

In this section, we shall modify Hartman's, [41] and [43], approach to extend the original Lasota and Opial result, Theorem 2.1.1, to the general k-point conjugate boundary value problem for the nth order equation. We shall also comment on Klassen's approach [85] in a remark at the end of this section.

Theorem 2.3.1. [[43], Hartman, Thm. II 1.1] *Assume that with respect to* (1.1.1), *conditions* (A)–(C) *are satisfied. Then each n-point conjugate boundary value problem for* (1.1.1) *on* (a,b) *has a solution.*

Proof. This proof is an adaptation of Hartman's [43] argument.

Let $a < x_1 < x_2 < \cdots < x_n < b$ and $y_i \in \mathbb{R}$, $i = 1, \ldots, n$, be given and consider the boundary value problem, (1.1.1), with the n-point conjugate conditions

$$y(x_i) = y_i, \quad i = 1, \ldots, n. \tag{2.3.1}$$

The proof is by induction on i and so, to begin, let z_0 denote a solution of (1.1.1) on (a,b). The argument is to construct a solution, z_1, of (1.1.1) on (a,b) satisfying

$$z_1(x_1) = y_1, \quad z_1(x_i) = z_0(x_i), \quad i = 2, \ldots, n.$$

Define the set S_1 by

$$S_1 = \{y(x_1) : y(x) \text{ is a solution of } (1.1.1) \text{ and } y(x_i) = z_0(x_i), i = 2, \ldots, n\}.$$

As in the proofs of Theorems 2.1.1 and 2.2.5 the method is to show $S_1 = \mathbb{R}$. This argument is modified and it is shown that S_1 is nonempty, open and closed. Since $z_0(x_1) \in S_1$, $S_1 \neq \emptyset$. That S_1 is open is an immediate consequence of Continuous Dependence, Theorem 1.1.6. It remains to show that S_1 is closed.

Assume for the sake of contradiction that S_1 is not closed and let $s_0 \in \overline{S_1} \setminus S_1$. Let $\{s_q\}_{q=1}^{\infty} \subset S_1$ be such that $\lim_{q \to \infty} s_q = s_0$ and assume, without loss of generality that the convergence is strictly monotone increasing. Let $\{y_q\}_{q=1}^{\infty}$ denote the corresponding solution of (1.1.1) satisfying

$$y_q(x_1) = s_q, \quad y_q(x_i) = z_0(x_i), \quad i = 2, \dots, n.$$

Note that $y_q - z_0$ has $n - 1$ zeros at x_i, $i = 2, \dots, n$, and so it follows by condition (C) that if $x \neq x_i$, $i = 2, \dots, n$, then $y_q(x) \neq z_0(x)$; moreover, by the uniqueness arguments in Chapter 1.2, Theorem 1.2.2, it is also the case that, $y_q'(x_i) \neq z_0'(x_i)$, $i = 2, \dots, n$, and $y_q - z_0$ changes sign at each x_i, $i = 2, \dots, n$. Thus, $\{y_q\}_{q=1}^{\infty}$ is strictly increasing on intervals (a, x_2) and (x_{2i+1}, x_{2i+2}) (and (x_n, b) if n is odd) and $\{y_q\}_{q=1}^{\infty}$ is strictly decreasing on each (x_{2i}, x_{2i+2}) (and (x_n, b) if n is even).

If $\{y_q\}_{q=1}^{\infty}$ is bounded on any compact subinterval of (a, b), the "compactness condition" (CP), Theorem 1.1.5, and the Kamke Theorem, Theorem 1.1.4, imply there is a solution z and a subsequence $\{y_{q_j}\}$ of $\{y_q\}$ which converges uniformly to z on each compact subinterval of (a, b). But that implies that $s_0 \in S_1$. So, $\{y_q\}_{q=1}^{\infty}$ is not bounded on any compact subinterval of (a, b). Let $u(x)$ denote the solution of an initial value problem, (1.1.1), with initial conditions

$$u(x_1) = s_0, \quad u^{(i-1)}(x_1) = 0, \quad i = 2, \dots, n.$$

Due to the monotonicity and unboundedness properties of $\{y_q\}$, eventually $y_q - u$ vanishes in disjoint neighborhoods of x_i, $i = 1, \dots, n$. This violates the uniqueness assumption (C). Thus, $\overline{S_1} \setminus S_1 = \emptyset$ and $S_1 = \mathbb{R}$.

Let z_1 denote the solution of the boundary value problem, (1.1.1), satisfying the n boundary conditions

$$z_1(x_1) = y_1, \quad z_1(x_i) = z_0(x_i), \quad i = 2, \dots, n.$$

To continue the induction, let z_1 play the role of z_0 and define

$$S_2 = \{y(x_2) : y(x) \text{ is a solution of } (1.1.1) \text{ and } y(x_i) = z_1(x_i), i \neq 2\}.$$

The argument to show $S_2 = \mathbb{R}$ is completely analogous; the theorem is proved using a straightforward induction on i. $\qquad\square$

The next result modifies Hartman's [41] Lemma 3, a fundamental result and technique of proof produced in [41].

Theorem 2.3.2. [[41], Hartman, Lem. 3] *Assume that with respect to* (1.1.1), *conditions* (A)–(C) *are satisfied. Let* $k \in \{2, \ldots, n\}$ *and* $l \in \{1, \ldots, k\}$ *be given. Then each* $(m_1, \ldots, m_k)$ *conjugate boundary value problem for* (1.1.1) *on* (a, b) *has a solution in the case that* $m_j = 1$ *if* $j \neq l$ *and* $m_l = n - (k-1)$.

Proof. The proof is by induction on m_l and for $m_l = 1$, the result has been proved in Theorem 2.3.1. So, assume $m_l > 1$, and assume the assertion of the theorem for each $1 \le \mu_l < m_l = n - (k-1)$. Let $a < x_1 < \cdots < x_k < b$, $y_{il} \in \mathbb{R}$, $i = 1, \ldots, m_l$, and $y_{1j} \in \mathbb{R}$, $j = 1, \ldots, k$, $j \neq l$ be given. We show the existence of a solution, y, of (1.1.1) on (a, b) satisfying the $(m_1, \ldots, m_k)$ conjugate conditions

$$y(x_j) = y_{1j}, \quad j = 1, \ldots, k, j \neq l,$$
$$y^{(i-1)}(x_l) = y_{il}, \quad i = 1, \ldots, m_l.$$

Let $x_l < \hat{x} < x_{l+1}$ (or $x_l < \hat{x} < b$, if $l = k$) and let z denote the solution of the $(i_1, \ldots, i_k, i_{k+1})$ conjugate boundary value problem for (1.1.1) on (a, b) where $i_j = 1$ if $j \neq l$ and $i_l = m_l - 1$, satisfying the conditions

$$z(x_j) = y_{1j}, \quad j = 1, \ldots, k, j \neq l, \quad z(\hat{x}) = 0,$$
$$z^{(i-1)}(x_l) = y_{il}, \quad i = 1, \ldots, m_l - 1.$$

z exists by the induction hypothesis on m_l.

Let $y(x; \alpha)$ denote the solution of the $(i_1, \ldots, i_k, i_{k+1})$ conjugate boundary value problem for (1.1.1) on (a, b) where $i_j = 1$ if $j \neq l$ and $i_l = m_l - 1$, satisfying the conditions

$$y(x_j) = z(x_j) = y_{1j}, \quad j = 1, \ldots, k, j \neq l, \quad y(\hat{x}) = \alpha,$$
$$y^{(i-1)}(x_l) = z^{(i-1)}(x_l) = y_{il}, \quad i = 1, \ldots, m_l - 1.$$

Let

$$S = \{y^{(m_l-1)}(x_l; \alpha) \,|\, \alpha \in \mathbb{R}\}.$$

We show $S = \mathbb{R}$ and obtain the existence of solutions for the $(m_1, \ldots, m_k)$ conjugate boundary value problem for (1.1.1) on (a, b) in the case that $m_i = 1$ if $i \neq l$ and $m_l = n - (k-1)$.

S is nonempty since $z^{(m_l-1)}(x_l) \in S$ and S is open by continuous dependence; see Theorem 1.1.6. It remains to show that S is closed. Assume for the sake of contradiction that S is not closed and assume $s_0 \in \overline{S} \setminus S$.

Without loss of generality assume $\{s_q\} \subset S$, $s_q \to s_0$ as $q \to \infty$ and assume $\{s_q\}$ is monotone increasing. Let y_q denote the solution of (1.1.1), satisfying the boundary conditions,

$$y(x_j) = y_{1j}, \quad j = 1, \ldots, k, j \neq l,$$
$$y^{(i-1)}(x_l) = y_{il}, \quad i = 1, \ldots, m_l - 1, \quad y^{(m_l-1)}(x_l) = s_q.$$

It follows by the "compactness condition" (CP), Theorem 1.1.5, that $\{y_q\}$ is not uniformly bounded on any compact subinterval of (a, x_2). Moreover, since y_q and y_{q+1} agree at $n - 1$ conjugate boundary conditions, and $s_q < s_0$, it follows that the sequence $\{y_q\}$ is monotone increasing without bound on a neighborhood to the right of x_l, $\{(-1)^{m_l-1}y_q\}$ is monotone increasing without bound on a neighborhood to the left of x_l and the direction of monotonicity of $\{y_q\}$ changes at each x_j, $j \neq l$.

Let u be the solution of (1.1.1) on (a, b) satisfying the initial conditions

$$u^{(i-1)}(x_l) = y_{il}, i = 1, \ldots, m_l - 1, \quad u^{(m_l-1)}(x_l) = s_0,$$
$$u^{(i-1)}(x_l) = 0, \quad i = m_l + 1, \ldots, n.$$

Note that there exists $\epsilon > 0$ such that $u(x) > y_q(x)$ on $(x_l, x_l + \epsilon)$ and $(-1)^{m_l-1}u(x) > (-1)^{m_l-1}y_q(x)$ on $(x_l - \epsilon, x_l)$. It now follows by the unboundedness and monotone properties of $\{y_q\}$ that there exist q, sufficiently large, and points,

$$a < \tau_1 < \cdots \tau_l < x_l < \tau_{l+1} < \cdots < \tau_{n-m_l+1} < b,$$

such that

$$y_q(\tau_j) = u(\tau_j), \quad j = 1, \ldots, n - m_l + 1,$$
$$y_q^{(i-1)}(x_l) = u^{(i-1)}(x_l), \quad 1 \leq i \leq m_l - 1.$$

Thus, u and y_q are distinct solutions of an $(i_1, \ldots, i_k, i_{k+1})$ conjugate boundary value problem for (1.1.1) on (a, b) where $i_j = 1$ if $j \neq l$ and $i_l = m_l - 1$. This contradicts uniqueness of solutions of conjugate boundary value problems of (1.1.1) on (a, b), proved in Theorem 1.2.2. In particular, S is closed and the theorem is proved. □

Hartman's [41] approach proceeds with an induction on the number of boundary points with assigned multiplicities exceeding 1. We modify Hartman's [41] approach to prove the following theorem.

Theorem 2.3.3. [[41], Hartman, Lem. 5] *Assume that with respect to* (1.1.1), *conditions* (A)–(C) *are satisfied. Let* $k \in \{2, \ldots, n\}$ *and* $l_1, l_2 \in \{1, \ldots, k\}$ *be given. Let* $m_{l_i} \geq 1$, $i = 1, 2$. *Then each* $(m_1, \ldots, m_k)$ *conjugate boundary value problem for* (1.1.1) *on* (a, b) *has a solution in the case that* $m_j = 1$ *if* $j \neq l_1$ *or* $j \neq l_2$, *and* $m_{l_1} + m_{l_2} = n - (k - 2)$.

Proof. First assume $m_{l_2} = 2$. The proof proceeds by an induction on m_{l_1}. Let $a < x_1 < \cdots < x_k < b$, $y_{il_1} \in \mathbb{R}$, $1 \leq i \leq m_{l_1}$, $y_{il_2} \in \mathbb{R}$, $i = 1, 2$, and $y_{1j} \in \mathbb{R}$, $j = 1, \ldots, k$, $j \neq l_1$ or $j \neq l_2$ be given. We show the existence of a solution, y of (1.1.1) on (a, b) satisfying the $(m_1, \ldots, m_k)$ conjugate conditions

$$y(x_j) = y_{1j}, \quad j = 1, \ldots, k, j \neq l_1, \text{ or } j \neq l_2,$$
$$y^{(i-1)}(x_{l_1}) = y_{il_1}, \quad 1 \leq i \leq m_{l_1},$$
$$y^{(i-1)}(x_{l_2}) = y_{il_2}, \quad i = 1, 2.$$

Note that for this problem, $k + m_{l_1} = n$.

If $m_{l_1} = 1$ the assertion of the theorem is true by Theorem 2.3.2 and so, for the induction hypothesis, assume the assertion of the theorem is true for all $a < x_1 < \cdots < x_k < b$, $y_{il_1} \in \mathbb{R}$, $1 \leq i \leq \mu_{l_1}$, $1 \leq \mu_{l_1} < m_{l_1}$, $y_{il_2} \in \mathbb{R}$, $i = 1, 2$, and $y_{1j} \in \mathbb{R}$, $j = 1, \ldots, k$, $j \neq l_1$ or $j \neq l_2$, such that $k - 2 + \mu_{l_1} + 2 = n$. Let $x_{l_1} < \hat{x} < x_{l_1+1}$ (or $x_{l_1} < \hat{x} < b$, if $l_1 = k$) and let z denote the solution of an $(i_1, \ldots, i_k, i_{k+1})$ conjugate boundary value problem for (1.1.1) on (a, b) satisfying the conditions

$$z(x_j) = y_{1j}, \quad j = 1, \ldots, k, \ j \neq l_1 \text{ or } j \neq l_2, \quad z(\hat{x}) = 0,$$
$$z^{(i-1)}(x_{l_1}) = y_{il_1}, \quad 1 \leq i \leq m_{l_1} - 1,$$
$$z^{(i-1)}(x_{l_2}) = y_{il_2}, \quad i = 1, 2.$$

z exists by the induction hypothesis on m_{l_1}.

Let $y(x; \alpha)$ denote the solution of the $(i_1, \ldots, i_k, i_{k+1})$ conjugate boundary value problem for (1.1.1) on (a, b) satisfying the conditions

$$y(x_j) = y_{1j}, \quad j = 1, \ldots, k, \ j \neq l_1, \text{ or } j \neq l_2, \quad y(\hat{x}) = \alpha,$$
$$y^{(i-1)}(x_{l_1}) = y_{il_1}, \quad 1 \leq i \leq m_{l_1} - 1,$$
$$y^{(i-1)}(x_{l_2}) = y_{il_2}, \quad i = 1, 2.$$

Let

$$S = \{ y^{(m_{l_1}-1)}(x_{l_1}; \alpha) \,|\, \alpha \in \mathbb{R} \}.$$

We show $S = \mathbb{R}$ and complete the proof of the theorem.

S is nonempty since $z^{(m_{l_1}-1)}(x_{l_1}) \in S$ and S is open by continuous dependence; see Theorem 1.1.6. It remains to show that S is closed. Assume for the sake of contradiction that S is not closed and assume $s_0 \in \overline{S} \setminus S$. Without loss of generality assume $\{s_q\} \subset S$, $s_q \to s_0$ as $q \to \infty$ and assume $\{s_q\}$ is monotone increasing. Let y_q denote the solution, (1.1.1), satisfying the boundary conditions,

$$y(x_j) = y_{1j}, \quad j = 1, \ldots, k, \; j \neq l_1, \text{ or } j \neq l_2,$$
$$y^{(i-1)}(x_{l_1}) = y_{il_1}, \quad 1 \leq i \leq m_{l_1} - 1, \quad y^{(m_{l_1}-1)}(x_{l_1}) = s_q,$$
$$y^{(i-1)}(x_{l_2}) = y_{il_2}, \quad i = 1, 2.$$

Due to the "compactness condition" (CP), Theorem 1.1.5, $\{y_q\}$ is monotonically unbounded on each compact subinterval of (a, x_1), (x_i, x_{i+1}), (x_k, b); $\{y_q\}$ is monotone increasing on $(x_{l_1}, x_{l_1}+1)$, $\{(-1)^{m_{l_1}-1}y_q\}$ is monotone increasing on $(x_{l_1} - 1, x_{l_1})$. Moreover, since $m_{l_2} = 2$ is even, $\{y_q\}$ is either monotone increasing on each of (x_{l_2-1}, x_{l_2}) and $(x_{l_2}, x_{l_2}+1)$ or monotone decreasing on each of (x_{l_2-1}, x_{l_2}) and $(x_{l_2}, x_{l_2}+1)$. Assume without loss of generality that $\{y_q\}$ is monotone increasing on each of (x_{l_2-1}, x_{l_2}) and $(x_{l_2}, x_{l_2}+1)$.

Let $\epsilon > 0$ and let u denote a solution of (1.1.1) satisfying the conjugate boundary conditions

$$y(x_j) = y_{1j}, \quad j = 1, \ldots, k, \; j \neq l_1, \text{ or } j \neq l_2,$$
$$y^{(i-1)}(x_{l_1}) = y_{il_1}, \quad 1 \leq i \leq m_{l_1} - 1,$$
$$y^{(m_{l_1}-1)}(x_{l_1}) = s_0, \quad y^{(m_{l_1})}(x_{l_1}) = 0$$
$$y(x_{l_2}) = y_{1l_2} + \epsilon.$$

Existence of u is guaranteed by Theorem 2.3.2. We also point out here that if it is the case $\{y_q\}$ is monotone decreasing on each of (x_{l_2-1}, x_{l_2}) and $(x_{l_2}, x_{l_2}+1)$, assume $\epsilon < 0$. Hence, it is the case that it is assumed that $\{y_q\}$ is monotone increasing on each of (x_{l_2-1}, x_{l_2}) and $(x_{l_2}, x_{l_2}+1)$ without loss of generality.

Since $s_q < s_0$, $(-1)^{m_{l_1}-1}y_q < u$ in a neighborhood to the left of x_{l_1} and $y_q < u$ in a neighborhood to the right of x_{l_1}; since $y_q(x_{l_2}) < u(x_{l_2})$, $y_q < u$ in a neighborhood to the left of x_{l_1} and $y_q < u$ in a neighborhood to the right of x_{l_1}. Apply the monotonic unboundedness of $\{y_q\}$ and there exist q, sufficiently large, and points $x_{l_1-1} < \tau_1 < x_{l_1} < \tau_2 < x_{l_1+1}$ and $x_{l_2-1} < \tau_3 < x_{l_2} < \tau_4 < x_{l_2+1}$ such that

$$y_q(\tau_i) = u(\tau_i), \quad 1 \leq i \leq 4.$$

In particular, y_q and u are distinct solutions of (1.1.1) satisfying

$$y_q(x_j) = u(x_j), \quad j = 1, \ldots, k, \; j \neq l_1, \text{ or } j \neq l_2,$$
$$y_q(\tau_i) = u(\tau_i), \quad 1 \leq i \leq 3,$$
$$y_q^{(i-1)}(x_{l_1}) = u^{(i-1)}(x_{l_1}), \quad i = 1, \ldots, m_{l_1} - 1.$$

Since $k + m_{l_1} = n$, this contradicts uniqueness of solutions of conjugate boundary value problems. In particular, S is closed and the assertion of the theorem is proved in the case that $m_{l_1} > 1$ and $m_{l_2} = 2$.

Now assume $m_{l_2} > 2$ and assume the assertion of the theorem is true for all $1 \leq \mu_{l_2} < m_{l_2}$. Let $a < x_1 < \cdots < x_k < b$, $y_{il_1} \in \mathbb{R}$, $1 \leq i \leq m_{l_1}$, $y_{il_2} \in \mathbb{R}$, $i = 1, 2$, and $y_{1j} \in \mathbb{R}$, $j = 1, \ldots, k$, $j \neq l_1$ or $j \neq l_2$ be given. If $m_{l_1} = 1$, the assertion of the theorem is true by Theorem 2.3.2. Proceed by induction on m_{l_1} precisely as in the case when $m_{l_2} = 2$ and the theorem is proved. ◻

The general theorem will now be stated and proved with an induction on the number of boundary points with assigned multiplicities exceeding 1.

Theorem 2.3.4. [[41], Hartman, Lem. 7] *Assume that with respect to (1.1.1), conditions (A)–(C) are satisfied. Then each $(m_1, \ldots, m_k)$ conjugate boundary value problem for (1.1.1) on (a, b), $2 \leq k \leq n$, has a unique solution.*

Proof. The technique of proof has been illustrated in the proof of Theorem 2.3.3 and so the details are briefly outlined; in particular, we shall show the existence of appropriate functions, z and u in each construction. Let $a < x_1 < \cdots < x_k < b$, $y_{ij} \in \mathbb{R}$, $1 \leq i \leq m_j$, $1 \leq j \leq k$ be given. We show the existence of a solution, y of (1.1.1) on (a, b) satisfying the $(m_1, \ldots, m_k)$ conjugate conditions

$$y^{(i-1)}(x_j) = y_{ij}, \quad 1 \leq i \leq m_j, \; 1 \leq j \leq k.$$

Assume that there exist $l_1, \ldots, l_q \in \{1, \ldots, k\}$, $m_{l_i} > 1$, $1 \leq i \leq q$, and $m_j = 1$, if $j \in \{1, \ldots, k\} \setminus \{l_1, \ldots, l_q\}$. Also assume for simplicity $l_1 < \cdots < l_q$. For the induction hypothesis, assume the assertion of the theorem is valid for all $(m_1, \ldots, m_k)$ satisfying the existence of $l_1, \ldots, l_{\hat{q}} \in \{1, \ldots, k\}$, $m_{l_i} > 1$, $1 \leq i \leq \hat{q}$, $2 \leq \hat{q} < q$, and $m_j = 1$, if $j \in \{1, \ldots, k\} \setminus \{l_1, \ldots, l_{\hat{q}}\}$.

The proof employs a double induction on m_{l_1} and m_{l_q}. So, first assume $m_{l_q} = 2$. The assertion of the theorem is valid for $m_{l_1} = 1$ by the induction hypothesis on l_q. So, now induct on m_{l_1} and assume in addition the assertion of the theorem for each $1 \leq \mu_{l_1} < m_{l_1}$.

Let $x_{l_1} < \hat{x} < x_{l_1+1}$ and let z denote the solution of the $(i_1, \ldots, i_k, i_{k+1})$ conjugate boundary value problem for (1.1.1) on (a, b) where $i_j = 1$ if $j \neq l_1$ or $j \neq l_2$, $i_{l_1} = m_{l_1} - 1$, and $i_{l_2+1} = m_{l_2}$, satisfying the conditions

$$z(x_j) = y_{1j}, \quad j \in \{1, \ldots, k\}, \ j \neq l_i, \ 1 \leq i \leq q, \quad z(\hat{x}) = 0,$$
$$z^{(i-1)}(x_{l_1}) = y_{il_1}, \quad 1 \leq m_{l_{j_1}} - 1,$$
$$z^{(i-1)}(x_{l_j}) = y_{il_j}, \quad 2 \leq j \leq q.$$

z exists by the induction hypothesis on m_{l_1}.

Let $\epsilon > 0$ and let u denote a solution of (1.1.1) satisfying the conjugate boundary conditions

$$y(x_j) = y_{1j}, \quad j \in \{1, \ldots, k\}, \ j \neq l_i, \ 1 \leq i \leq q,$$
$$y^{(i-1)}(x_{l_1}) = y_{il_1}, \quad i = 1, \ldots, m_{l_1} - 1,$$
$$y^{(m_{l_1}-1)}(x_{l_1}) = s_0, \ y^{(m_{l_1})}(x_{l_1}) = 0,$$
$$y^{(i-1)}(x_{l_j}) = y_{il_j}, \quad 2 \leq j \leq q - 1,$$
$$y(x_{l_q}) = y_{il_q} + \epsilon.$$

Existence of u is also guaranteed by the induction hypothesis on q.

So, the assertion of the theorem can be proved in the case $m_{l_q} = 2$. One proceeds by induction on m_2 precisely as in the proof of Theorem 2.3.3. □

Remark 2.3.1. *In an alternative approach to prove Theorem 2.3.4, Klaasen [85] also first proves Theorem 2.3.1. Klaasen uses a shooting method beginning with solutions of initial value problems and obtains existence of solutions for 2-point, $(n-1, 1)$, conjugate boundary value problems and then proceeds by induction to prove the following theorem.*

Theorem 2.3.5. [[85], Klaasen, Thm. 9] *Assume that with respect to (1.1.1), conditions (A)–(C) are satisfied. Let $k \in \{1, \ldots, n-1\}$ be given. Then each $(n-k, m_2, \ldots, m_{k+1})$ conjugate boundary value problem for (1.1.1) on (a, b) has a solution in the case that $m_j = 1$ for each $j = 2, \ldots, k+1$.*

Thus, in the case $k = n-1$, Theorem 2.3.1 follows as a corollary to Theorem 2.3.5.

2.4 Right focal boundary value problems: 2-point, uniqueness implies existence

In this section we present the arguments and analogous results for uniqueness implies existence for the 2-point right focal boundary value problem. These results are due to Henderson [45]. Assume $r = 2$, $1 \leq k \leq n-1$, and consider the right focal boundary value problem, (1.1.1)-(1.1.3), where $a < x_1 < x_2 < b$, $k \in \{1, \ldots, n-1\}$, $y_{i1}, 0 \leq i \leq k-1$, $y_{i2}, k \leq i \leq n-1$, are given. To establish the results for the right focal problem, first consider a family of in between boundary conditions,

$$y^{(i)}(x_1) = y_{i1}, \quad i = 0, \ldots, k-1, \tag{2.4.1}$$

$$y^{(j)}(x_2) = y_{j2}, \quad j = l, \ldots, l+n-k+1,$$

where $a < x_1 < x_2 < b$, $y_{i1} \in \mathbb{R}$, $i = 0, \ldots, k-1$, $y_{j2} \in \mathbb{R}, j = l, \ldots, l+n-k+1$, are given and $1 \leq l \leq k$. Note, if $l = 0$, (2.4.1) represents 2-point conjugate boundary conditions and if $l = k$, (2.4.1) represents 2-point right focal boundary conditions.

Assume throughout the remainder of this chapter that that each n-point right focal boundary value problem for (1.1.1) on (a, b) has at most one solution. Then it follows easily from Rolle's Theorem that condition (C) is continued to be assumed and the "compactness condition" (CP), Theorem 1.1.5 continues to apply.

This first result is labeled "Lemma $m.k$" in [45], but in the context of our presentation, the index l plays the role of the original index m.

Theorem 2.4.1. [[45], Henderson, Lem. *l.k*] *Assume that with respect to* (1.1.1), *conditions* (A)–(B) *are satisfied and assume that each n-point right focal boundary value problem for* (1.1.1) *on* (a, b) *has at most one solution. Let $k \in \{1, \ldots, n-1\}$. Then given any $1 \leq l \leq k$, $a < x_1 < x_2 < b$, $y_{i1} \in \mathbb{R}$, $i = 0, \ldots, k-1$, $y_{j2} \in \mathbb{R}, j = l, \ldots, l+n-k+1$, the boundary value problem,* (1.1.1)-(2.4.1), *has a unique solution.*

Proof. The proof is by induction on k and on l. First assume $l = k = 1$; that is, consider boundary conditions,

$$y(x_1) = y_{11}, \quad y^{(j)}(x_2) = y_{j2}, \quad j = 1, \ldots, n-1.$$

Uniqueness has been established for the 2-point right focal boundary value problem in Theorem 1.4.1.

Let $y(x; \mu)$ denote the solution of the initial value problem for (1.1.1) satisfying

$$y(x_2) = \mu, \quad y^{(j)}(x_2) = y_{j2}, \quad j = 1, \ldots, n-1.$$

Let

$$S = \{y(x_1; \mu) \mid \mu \in \mathbb{R}\}.$$

We show $S = \mathbb{R}$ and existence is established for $l = k = 1$. S is nonempty by condition (B). S is open by continuous dependence; see Theorem 1.1.6, and Remark 1.1.1. Now we show S is closed. Assume for the sake of contradiction that S is not closed and assume $s_0 \in \overline{S} \setminus S$. Without loss of generality assume $\{s_q\} \subset S$, $s_q \to s_0$ as $q \to \infty$ and assume $\{s_q\}$ is monotone increasing. Let y_q denote the solution of (1.1.1), satisfying the boundary conditions,

$$y(x_1) = s_q, \quad y^{(j)}(x_2) = y_{j2}, \quad j = 1, \ldots, n-1.$$

By uniqueness of solutions for $l = k = 1$, it follows that $y_q(x) < y_{q+1}(x)$ on (a, x_2). It follows by the "compactness condition" (CP), Theorem 1.1.5, that $\{y_q\}$ is not uniformly bounded on any compact subinterval of (a, x_2). By the existence of solutions of 2-point conjugate boundary value problems, let u denote the solution of (1.1.1) satisfying

$$u(x_1) = s_0, \quad u(x_2) = 0, \quad u^{(j)}(x_2) = y_{j2}, \quad j = 1, \ldots, n-2.$$

There exist q sufficiently large and $a < \tau_1 < x_1 < \tau_2 < x_2$ such that $y_q(\tau_1) = u(\tau_1)$, $y_q(\tau_2) = u(\tau_2)$. Now, using the boundary conditions $y_q^{(j)}(x_2) = u^{(j)}(x_2) = y_{j2}$, $j = 1, \ldots, n-2$, and Rolle's Theorem, construct

$$\tau_1 < \xi_2 < \cdots < \xi_n < x_2$$

such that

$$y_q(\tau_1) = u(\tau_1), \quad y_q^{(j+1)}(\xi_j) = u^{(j+1)}(\xi_j), \quad 1 \le j \le n-1.$$

This contradicts the assumption that each n-point right focal boundary value problem for (1.1.1) on (a, b) has at most one solution. In particular, S is closed and existence is proved in the case $l = k = 1$.

Proceed by induction and assume the theorem is true for each $1 < k \le n-1$, for each $1 \le i < k$, and $1 \le l \le i$. We show the theorem is true for all $1 \le l \le k$ and do so by induction on both k and l. To begin, let $k \in \{2, \ldots, n-1\}$ and let $l = 1$. Consider the boundary value problem, (1.1.1), with boundary conditions

$$y^{(i)}(x_1) = y_{i1}, \quad i = 0, \ldots, k-1,$$

$$y^{(j)}(x_2) = y_{j2}, \quad j = 1, \ldots, n-k.$$

Uniqueness of solutions follows from uniqueness of n-point right focal boundary value problem for (1.1.1) on (a, b) by Theorem 1.4.1 and repeated application of Rolle's Theorem.

Let $z(x)$ denote the solution of the 2-point conjugate problem for (1.1.1) with boundary conditions

$$y^{(i)}(x_1) = y_{i1}, \quad i = 0, \ldots, k-2,$$
$$y(x_2) = 0, \quad y^{(j)}(x_2) = y_{j2}, \quad j = 1, \ldots, n-k.$$

Let $y(x; \mu)$ denote the solution of the 2-point conjugate problem for (1.1.1) satisfying

$$y^{(i)}(x_1) = y_{i1}, \quad i = 0, \ldots, k-2,$$
$$y(x_2) = \mu, \quad y^{(j)}(x_2) = y_{j2}, \quad j = 1, \ldots, n-k.$$

Let

$$S = \{y^{(k-1)}(x_1; \mu) : \mu \in \mathbb{R}\}.$$

$S \neq \emptyset$ since $z^{(k-1)}(x_1) \in S$ and S is open by continuous dependence. So it remains to show that S is closed. Again, assume $s_0 \in \bar{S} \setminus S$, assume $\{s_q\} \subset S$, $s_q \to s_0$ as $q \to \infty$, assume $\{s_q\}$ is monotone increasing, and let y_q denote the solution of the boundary value problem, (1.1.1),

$$y^{(i)}(x_1) = z^{(i)}(x_1), \quad i = 0, \ldots, k-2, \quad y^{(k-1)}(x_1) = s_q,$$
$$y^{(j)}(x_2) = z^{(j)}(x_2), \quad j = 1, \ldots, n-k.$$

By uniqueness of solutions, Rolle's Theorem, and the boundary conditions,

$$y_q^{(i)}(x_1) = y_{q+1}^{(i)}(x_1), \quad i = 0, \ldots, k-2,$$
$$y_q^{(k-1)}(x_1) = s_q < s_{q+1} = y_{q+1}^{(k-1)}(x_1),$$

it is the case that

(i) $y_q(x) > y_{q+1}(x)$ on (a, x_1) and $y_q(x) < y_{q+1}(x)$ on (x_1, x_2) if k is even,
(ii) $y_q(x) < y_{q+1}(x)$ on $(a, x_2) \setminus \{x_1\}$ if k is odd.

Since $s_0 \notin S$, if k is even, then $\{y_q\}$ is not uniformly bounded below on each compact subinterval of (a, x_1) and $\{y_q\}$ is not uniformly bounded above on each compact subinterval of (x_1, x_2). If k is odd, then $\{y_q\}$ is not uniformly bounded above on each compact subinterval of (a, x_2).

Now let u denote the solution of a 2-point conjugate problem for (1.1.1) satisfying the boundary conditions,

$$u^{(i)}(x_1) = y_{i1}, \quad i = 0, \ldots, k-2, \quad u^{(k-1)}(x_1) = s_0,$$
$$u(x_2) = 0, \quad u^{(j)}(x_2) = y_{j2}, \quad j = 1, \ldots, n-k-1.$$

It follows readily from the monotonicity and the unboundedness of $\{y_q\}$ that there exist q sufficiently large and $a < \tau_1 < x_1 < \tau_2 < x_2$ such that y_q and u are solutions of (1.1.1) satisfying boundary conditions

$$u(\tau_i) = y_q(\tau_i), \quad i = 1, 2,$$
$$u^{(i)}(x_1) = y_q^{(i)}(x_1), \quad i = 0, \ldots, k - 2,$$
$$u^{(j)}(x_2) = y_q^{(j)}(x_2), \quad j = 1, \ldots, n - k - 1.$$

Repeated applications of Rolle's Theorem imply u and y_q are solutions of the same n-point right focal boundary value problem contradicting the uniqueness of n-point right focal boundary value problems. Thus, S is closed and the theorem is proved for $k \in \{1, \ldots, n - 1\}$, $l = 1$.

To finish the induction and complete the proof, assume the theorem is true for $k \in \{1, \ldots, n - 1\}$, $1 \le l < k$. We show the theorem is valid for $k \in \{1, \ldots, n - 1\}$, $l + 1 \le k$. The technique is analogous to the technique employed in each of the cases, $k = 1$, $l = 1$, and $k \in \{1, \ldots, n - 1\}$, $l = 1$.

Let $z(x)$ denote the solution of the 2-point conjugate problem for (1.1.1) with boundary conditions

$$y^{(i)}(x_1) = y_{i1}, \quad i = 0, \ldots, k - 2,$$
$$y^{(l-1)}(x_2) = 0, \quad y^{(j)}(x_2) = y_{j2}, \quad j = l, \ldots, n - k + l - 1.$$

z exists by the induction hypotheses with k replaced by $k - 1$ and l replaced by $l - 1$. Let $y(x; \mu)$ denote the solution of the 2-point conjugate problem for (1.1.1) satisfying

$$y^{(i)}(x_1) = y_{i1}, \quad i = 0, \ldots, k - 2, \ y^{(k-1)}(x_1) = \mu,$$
$$y^{(j)}(x_2) = y_{j2}, \quad j = l, \ldots, n - k + l - 1.$$

Let

$$S = \{y^{(k-1)}(x_1; \mu) : \mu \in \mathbb{R}\}.$$

$S \ne \emptyset$ and open. To show that S is closed, assume $s_0 \in \overline{S} \setminus S$, assume $\{s_q\} \subset S$, $s_q \to s_0$ as $q \to \infty$, assume $\{s_q\}$ is monotone increasing, and let y_q denote the solution of the boundary value problem, (1.1.1),

$$y^{(i)}(x_1) = z^{(i)}(x_1), \quad i = 0, \ldots, k - 2, \quad y^{(k-1)}(x_1) = s_q,$$
$$y^{(j)}(x_2) = z^{(j)}(x_2), \quad j = l, \ldots, n - k + l - 1.$$

Condition (i) holds in the case k is even, and condition (ii) holds in the case k is odd, and so the sequence $\{y_q\}$ satisfies the monotonicity and unboundedness conditions employed above.

Now let u denote the solution of a 2-point problem for (1.1.1) satisfying the boundary conditions,

$$u^{(i)}(x_1) = y_{i1}, \quad i = 0, \ldots, k-2, \quad u^{(k-1)}(x_1) = s_0,$$
$$u(x_2) = 0, \quad u^{(j)}(x_2) = y_{j2}, \quad j = 1, \ldots, n-k-1.$$

u exists by the induction hypothesis with k and l replaced by k and $l-1$, respectively. It follows readily from the monotonicity and the unboundedness of $\{y_q\}$ that there exist q sufficiently large and points $a < \tau_1 < x_1 < \tau_2 < x_2$ such that y_q and u are solutions of (1.1.1) satisfying boundary conditions

$$u(\tau_i) = y_q(\tau_i), \quad i = 1, 2,$$
$$u^{(i)}(x_1) = y_q^{(i)}(x_1), \quad i = 0, \ldots, k-2,$$
$$u^{(j)}(x_2) = y_q^{(j)}(x_2), \quad j = l, \ldots, n-k+l-1.$$

Repeated applications of Rolle's Theorem imply u and y_q are solutions of the same n-point right focal boundary value problem contradicting the uniqueness of n-point right focal boundary value problems. Thus, S is closed and the theorem is proved. □

2.5 Right focal boundary value problems: r-point, uniqueness implies existence

In this section, we present uniqueness implies existence results for the r-point right focal boundary value problem, (1.1.1)-(1.1.3), (or an $(m_1, \ldots, m_r)$ right focal boundary value problem), restated here for convenience:

Given $2 \le r \le n$, $m_1, \ldots, m_r \in \mathbb{N}$ such that $\sum_{j=1}^{r} m_j = n$, $s_0 := 0$, $s_k := \sum_{j=1}^{k} m_j$, $1 \le k \le r$, points $a < x_1 < \cdots < x_r < b$, and $y_{ik} \in \mathbb{R}$, $s_{k-1} \le i \le s_k - 1$, $1 \le k \le r$, a boundary value problem for (1.1.1) satisfying

$$y^{(i)}(x_k) = y_{ik}, \quad s_{k-1} \le i \le s_k - 1, \quad 1 \le k \le r.$$

This analysis will be carried out by induction on r and the following is adapted from the original work due to Henderson [45].

Theorem 2.5.1. [[45], Henderson, Thm. 3] *Assume that with respect to (1.1.1), conditions (A) and (B) are satisfied, and that each n-point right focal boundary value problem for (1.1.1) on (a, b) has at most one solution. Then, each r-point right focal boundary value problem for (1.1.1) on (a, b), $2 \le r \le n$, has a unique solution.*

Proof. Uniqueness, originally established by Henderson [45], is obtained in Theorem 1.4.1. Existence for $r = 2$ is established in Theorem 2.4.1. Let $r \in \{3, \ldots, n\}$ be given. Assume that for each $2 \leq \rho < r$, each $(m_1, \ldots, m_\rho)$ right focal boundary value problem for (1.1.1) has a unique solution.

Let $m_1 = m_2 = 1$, and let m_i, $i = 3, \ldots, r$, be positive integers such that $\sum_{i=3}^{r} m_i = n - 2$. Let $a < x_1 < \cdots < x_r < b$, $y_{ik} \in \mathbb{R}$, $s_{k-1} \leq i \leq s_k - 1$, $1 \leq k \leq r$, be given where $s_k = \sum_{j=1}^{k} m_j$, $1 \leq k \leq r$. The initial argument employs an induction on m_2.

Let $z(x)$ denote the solution of the $(i_1, \ldots, i_{r-1})$ right focal boundary value problem for (1.1.1) with $i_1 = 2$, $i_j = m_{j+1}$, $j = 2, \ldots, r - 1$,

$$y(x_2) = 0, \quad y'(x_2) = y_{12},$$
$$y^{(i)}(x_j) = y_{ij}, \quad s_{j-1} \leq i \leq s_j - 1, \quad 3 \leq j \leq r.$$

z exists by the induction hypothesis on r.

Let $y_\mu(x)$ denote the solution of (1.1.1) satisfying

$$y(x_2) = \mu, \quad y'(x_2) = y_{12},$$
$$y^{(i)}(x_j) = y_{ij}, \quad s_{j-1} \leq i \leq s_j - 1, \quad 3 \leq j \leq r.$$

Let

$$S = \{y_\mu(x_1) : \mu \in \mathbb{R}\}.$$

Then $S \neq \emptyset$ $(z(x_1) \in S)$, S is open by continuous dependence, and so it remains to be shown that S is also closed. To the contrary, assume $\eta_0 \in \bar{S} \setminus S$, assume $\{\eta_q\} \subset S$, $\eta_q \to \eta_0$ as $q \to \infty$ and assume $\{\eta_q\}$ is monotone increasing.

Let y_q denote the solution of the boundary value problem, (1.1.1),

$$y(x_1) = \eta_q, \quad y'(x_2) = y_{12},$$
$$y^{(i)}(x_j) = y_{ij}, \quad s_{j-1} \leq i \leq s_j - 1, \quad 3 \leq j \leq r.$$

Due to the uniqueness of solutions of $(1, 1, m_3, \ldots, m_r)$ right focal problems and the "compactness condition" (CP), Theorem 1.1.5 $\{y_q\}$ is an increasing sequence for each $x \in (a, x_2)$ and is not uniformly bounded above on each compact subinterval of (a, x_2).

Let u denote the unique solution of the $(i_1, \ldots, i_{r-1})$ right focal boundary value problem for (1.1.1) with $i_1 = 1$, $i_2 = m_3 + 1$, $i_j = m_{j+1}$, $j = 3, \ldots, r - 1$, satisfying

$$u(x_1) = \eta_0,$$
$$u'(x_3) = 0, \quad u^{(i)}(x_j) = y_{ij}, \quad s_{j-1} \leq i \leq s_j - 1, \quad 3 \leq j \leq r.$$

Using the monotonicity and unboundedness properties of y_q, there exist q sufficiently large and $a < \tau_1 < x_1 < \tau_2 < x_2$ such that $y_q(\tau_1) = u(\tau_1)$, $y_q(\tau_2) = u(\tau_2)$. Apply Rolle's Theorem to obtain $\tau_1 < \xi_1 < \tau_2$ such that $y_q'(\xi_1) = u'(\xi_1)$. In particular, y_q and u are solutions of (1.1.1) satisfying the $(1, 1, m_3, \ldots, m_r)$ right focal boundary conditions

$$y(\tau_1) = u(\tau_1), \quad y'(\xi_1) = u'(\xi_1)$$

$$y^{(i)}(x_j) = y_{ij}, \quad s_{j-1} \le i \le s_j - 1, \quad 3 \le j \le r.$$

Uniqueness of solutions of $(1, 1, m_3, \ldots, m_r)$ right focal boundary value problems is established in Theorem 1.4.1 and so, the contradiction is obtained. Thus, S is closed and the assertion is true for $(1, 1, m_3, \ldots, m_r)$ right focal boundary value problems; in particular, the assertion is true for $m_2 = 1$.

The next step is to perform an induction on m_2. Assume $m_2 > 1$. Assume for all $1 \le k < m_2$, and for all $m_3, \ldots, m_r$ such that $1 + k + m_3 + \cdots + m_r = n$, each $(1, k, m_3, \ldots, m_r)$ right focal boundary value problem for (1.1.1) on (a, b) has a unique solution.

Let $m_3, \ldots, m_r$ be such that $1 + m_2 + m_3 + \cdots + m_r = n$, let $a < x_1 < \cdots < x_r < b$, let $y_{ik} \in \mathbb{R}$, $s_{k-1} \le i \le s_k - 1$, $1 \le k \le r$, be given. Let z denote the unique solution of the $(i_1, \ldots, i_{r-1})$ right focal point boundary value problem for (1.1.1), with $i_1 = m_2 + 1$, $i_j = m_{j+1}$, $2 \le j \le r - 1$, with boundary conditions

$$z(x_2) = 0, \quad z^{(i)}(x_k) = y_{ik}, \quad s_{k-1} \le i \le s_k - 1, \quad 2 \le k \le r.$$

Let y_α denote the solution of (1.1.1) satisfying the boundary conditions

$$y(x_2) = \alpha, \quad y^{(i)}(x_k) = y_{ik}, \quad s_{k-1} \le i \le s_k - 1, \quad 2 \le k \le r,$$

and define

$$S = \{y_\alpha(x_1) : \alpha \in \mathbb{R}\}.$$

S is nonempty and S is open by Continuous Dependence, Theorem 1.1.6, so it remains to show S is closed. Assume $\eta_0 \in \overline{S} \setminus S$, assume $\{\eta_q\} \subset S$, $\eta_q \to \eta_0$ as $q \to \infty$ and assume $\{\eta_q\}$ is monotone increasing. Let y_q denote the solution of the boundary value problem, (1.1.1),

$$y(x_1) = \eta_q, \quad y^{(i)}(x_k) = y_{ik}, \quad s_{k-1} \le i \le s_k - 1, \quad 2 \le k \le r.$$

Let u denote the unique solution of the $(1, m_2 - 1, m_3 + 1, \ldots, m_r)$ right focal boundary value problem for (1.1.1) satisfying

$$u(x_1) = \eta_0,$$

$$u^{(i)}(x_2) = y_{i2}, \quad 1 \le i \le m_2 - 1 = s_2 - 2,$$

$$u^{(s_2 - 1)}(x_j) = 0, u^{(i)}(x_j) = y_{ij}, \quad s_{j-1} \le i \le s_j - 1, \quad 3 \le j \le r.$$

By the induction hypothesis on m_2, u exists.

As before, $\{y_q\}$ is not uniformly bounded above on each compact subinterval of (a, x_2). Since $u(x_1) = \eta_0 > y_q(x_1)$ for all q, there exist q sufficiently large and $a < \tau_1 < x_1 < \tau_2 < x_2$ such that $y_q(\tau_1) = u(\tau_1)$, $y_q(\tau_2) = u(\tau_2)$. Apply Rolle's Theorem to obtain $\tau_1 < \xi_1 < \tau_2$ such that $y_q'(\xi_1) = u'(\xi_1)$. Also note that $y_q^{(i)}(x_2) = u^{(i)}(x_2)$, $1 \le i \le s_2 - 2$. Apply Rolle's theorem to obtain $\tau_1 < \xi_1 < \cdots < \xi_{s_2-1} < x_2$ such that $y_q^{(i)}(\xi_i) = u^{(i)}(\xi_i)$, $1 \le i \le s_s - 1$. Thus, y_q and u are distinct solutions of (1.1.1) satisfying $(1, \ldots, 1, i_{s_2+1}, \ldots, i_{s_2+r-2})$ right focal boundary conditions

$$y(\tau_1) = u(\tau_1),$$
$$y^{(i)}(\xi_i) = u^{(i)}(\xi_i), \quad 1 \le i \le s_2 - 2,$$
$$y^{(i)}(x_j) = u^{(i)}(x_j), \quad s_{j-1} \le i \le s_j - 1, \quad 3 \le j \le r.$$

This violates the uniqueness theorem, Theorem 1.4.1, proved in Chapter 1.4. In particular, we have proved that each $(1, m_2, m_3, \ldots, m_r)$ right focal boundary value problem for (1.1.1) has a unique solution on (a, b) for all positive integers $m_2, \ldots, m_r$ satisfying $1 + m_2 + \cdots + m_r = n$.

We complete the proof of Theorem 2.5.1 by inducting on m_1 and m_2. Let $1 < m_1 \le n - (r - 1)$ and assume that for all $1 \le h < m_1$ and for all positive integers $m_2, \ldots, m_r$ satisfying $h + m_2 + \cdots + m_r = n$, each $(h, m_2, m_3, \ldots, m_r)$ right focal boundary value problem for (1.1.1) has a unique solution on (a, b).

Set $m_2 = 1$ and assume $m_3, \ldots, m_r$ are positive integers satisfying $m_1 + 1 + m_3 + \cdots + m_r = n$. Let $a < x_1 < \cdots < x_r < b$, let $y_{ik} \in \mathbb{R}$, be given and let $z(x)$ denote the solution the $(m_1 - 1, 2, m_3, \ldots, m_r)$ right focal boundary value problem for (1.1.1) with boundary conditions

$$z^{(i)}(\tau_1) = y_{i1}, \quad 0 \le i \le m_1 - 2,$$
$$z^{(m_1-1)}(x_2) = 0, \quad z^{(m_1)}(x_2) = y_{m_1,2},$$
$$z^{(i)}(x_j) = y_{ij}, \quad s_{j-1} \le i \le s_j - 1, \quad 3 \le j \le r.$$

Let y_α denote the solution of (1.1.1) satisfying the boundary conditions

$$y^{(i)}(x_1) = z^{(i)}(x_1), \quad 0 \le i \le m_1 - 2,$$
$$y^{(m_1-1)}(x_2) = \alpha, \quad y^{(m_1)}(x_2) = z^{(m_1)}(x_2) = y_{m_1,2},$$
$$y^{(i)}(x_j) = z^{(i)}(x_j), \quad s_{j-1} \le i \le s_j - 1, \quad 2 \le j \le r,$$

and define

$$S = \{y_\alpha^{(m_1-1)}(x_1) : \alpha \in \mathbb{R}\}.$$

S is nonempty and open and so the details are given to show that S is closed.

Assume $\eta_0 \in \overline{S} \setminus S$, assume $\{\eta_q\} \subset S$, $\eta_q \to \eta_0$ as $q \to \infty$ and assume $\{\eta_q\}$ is monotone increasing. Let y_q denote the solution of the boundary value problem, (1.1.1),

$$y^{(i)}(x_1) = z^{(i)}(x_1), \quad 0 \leq i \leq m_1 - 2, \quad y^{(m_1-1)}(x_1) = \eta_q,$$
$$y^{(i)}(x_j) = z^{(i)}(x_j), \quad s_{j-1} \leq i \leq s_j - 1, \quad 2 \leq j \leq r.$$

By uniqueness of solutions, Rolle's Theorem, and the boundary conditions,

$$y_q^{(i)}(x_1) = y_{q+1}^{(i)}(x_1), \quad i = 0, \ldots, k - 2,$$
$$y_q^{(k-1)}(x_1) = s_q < s_{q+1} = y_{q+1}^{(k-1)}(x_1),$$

it is the case that

(i) $y_q(x) > y_{q+1}(x)$ on (a, x_1) and $y_q(x) < y_{q+1}(x)$ on (x_1, x_2) if k is even,
(ii) $y_q(x) < y_{q+1}(x)$ on $(a, x_2) \setminus \{x_1\}$ if k is odd.

Since $\eta_0 \notin S$, if k is even, then $\{y_q\}$ is not uniformly bounded below on each compact subinterval of (a, x_1) and $\{y_q\}$ is not uniformly bounded above on each compact subinterval of (x_1, x_2). If k is odd, then $\{y_q\}$ is not uniformly bounded above on each compact subinterval of (a, x_2).

Let u denote the unique solution of the $(i_1, \ldots, i_{r-1})$ right focal boundary value problem for (1.1.1) where $i_1 = m_1$, $i_2 = m_3 + 1$, $i_j = m_{j+1}$, $3 \leq j \leq r - 1$, satisfying

$$u^{(i)}(x_1) = z^{(i)}(x_1), \quad 0 \leq i \leq m_1 - 2, \quad u^{(m_1-1)}(x_1) = \eta_0,$$
$$u^{(m_1)}(x_3) = 0,$$
$$u^{(i)}(x_j) = z^{(i)}(x_j), \quad s_{j-1} \leq i \leq s_j - 1, \quad 3 \leq j \leq r,$$

By the induction hypothesis on m_2, u exists.

It follows readily from the monotonicity and the unboundedness of $\{y_q\}$ that there exist q sufficiently large and $a < \tau_1 < x_1 < \tau_2 < x_2$ such that

$$y_q(\tau_i) = u(\tau_i), \quad i = 1, 2.$$

Moreover, $y_q^{(i)}(x_1) = u^{(i)}(x_1)$, $0 \leq i \leq m_1 - 2$. So, apply Rolle's Theorem and note that y_q and u are distinct solution of an $(i_1, \ldots, i_{s_2}, i_{s_2+1}, \ldots, i_{s_2+r-2})$ right focal boundary value problem for (1.1.1) where $i_j = 1$, $j = 1, \ldots, s_2 = s_1 + 1$, $i_{s_2+j} = m_{j+2}$, $1 \leq j \leq r - 2$. This contradicts uniqueness of solutions, Theorem 1.4.1 and so S is closed. Thus, each $(h, 1, m_3, \ldots, m_r)$ right focal boundary value problem for (1.1.1) has a unique solution on (a, b) if $h + 1 + m_3 + \cdots + m_r = n$.

Let $1 < m_1 \leq n - (r-1)$ and assume that for all $1 \leq h < m_1$ and for all positive integers $m_2, \ldots, m_r$ satisfying $h + m_2 + \cdots + m_r = n$, each $(h, m_2, m_3, \ldots, m_r)$ right focal boundary value problem for (1.1.1) has a unique solution on (a,b).

Finally, assume $m_2 > 1$. Since the case, $m_1 = 1$ is completely handled, assume through the remainder of the proof that $m_1 > 1$. Assume that for all $1 \leq h < m_2$ and for all $m_1, m_3, \ldots, m_r$ such that $m_1 + 1 + m_3 + \cdots + m_r = n - h$ each $(m_1, h, m_3, \ldots, m_r)$ right focal boundary value problem for (1.1.1) has a unique solution on (a,b). Let $a < x_1 < \cdots < x_r < b$, let $y_{ik} \in \mathbb{R}$, be given and let $z(x)$ denote the solution the $(m_1 - 1, m_2 + 1, m_3, \ldots, m_r)$ right focal boundary value problem for (1.1.1) with boundary conditions

$$z^{(i)}(x_1) = y_{i1}, \quad 0 \leq i \leq m_1 - 2,$$
$$z^{(m_1-1)}(x_2) = 0, \quad z^{(i)}(x_2) = y_{i,2}, \quad s_1 \leq i \leq s_2 - 1,$$
$$z^{(i)}(x_j) = y_{ij}, \quad s_{j-1} \leq i \leq s_j - 1, \quad 3 \leq j \leq r.$$

z exists by the preceding induction on m_1. Let y_α denote the solution of (1.1.1) satisfying the boundary conditions

$$y^{(i)}(x_1) = z^{(i)}(x_1), \quad 0 \leq i \leq m_1 - 2,$$
$$y^{(m_1-1)}(x_2) = \alpha, \quad y^{(i)}(x_2) = z^{(i)}(x_2), \quad s_1 \leq i \leq s_2 - 1,$$
$$y^{(i)}(x_j) = z^{(i)}(x_j), \quad s_{j-1} \leq i \leq s_j - 1, \quad 3 \leq j \leq r,$$

and define

$$S = \{y_\alpha^{(m_1-1)}(x_1) : \alpha \in \mathbb{R}\}.$$

S is nonempty and open and so the standard details are briefly given to show that S is closed.

Assume $\eta_0 \in \overline{S} \setminus S$, assume $\{\eta_q\} \subset S$, $\eta_q \to \eta_0$ as $q \to \infty$ and assume $\{\eta_q\}$ is monotone increasing. Let y_q denote the solution of the boundary value problem, (1.1.1),

$$y^{(i)}(x_1) = z^{(i)}(x_1), \quad 0 \leq i \leq m_1 - 2, \quad y^{(m_1-1)}(x_1) = \eta_q,$$
$$y^{(i)}(x_j) = z^{(i)}(x_j), \quad s_{j-1} \leq i \leq s_j - 1, \quad 2 \leq j \leq r.$$

As in the argument with $m_2 = 1$, conditions (i) and (ii) hold, and by uniqueness of solutions, Rolle's Theorem, and the boundary conditions, $\{y_q\}$ is not uniformly bounded on each compact subinterval of (a, x_1) or (x_1, x_2).

Let u denote the unique solution of the $(m_1, m_2 - 1, m_3 + 1, m_4, \ldots, m_r)$ right focal boundary value problem for (1.1.1) where $i_1 = m_1$, $i_2 = m_3 + 1$, $i_j = m_{j+1}$, $3 \le j \le r - 1$, satisfying

$$u^{(i)}(x_1) = z^{(i)}(x_1), \ 0 \le i \le m_1 - 2, \ u^{(m_1 - 1)}(x_1) = \eta_0,$$
$$u^{(i)}(x_2) = z^{(i)}(x_2), \ s_1 \le i \le s_2 - 2,$$
$$u^{(s_2 - 1)}(x_3) = 0, \ u^{(i)}(x_j) = z^{(i)}(x_j), \ s_{j-1} \le i \le s_j - 1, \le j \le r.$$

By the induction hypothesis on m_2, u exists.

Apply the unboundedness of the sequence $\{y_q\}$ and Rolle's Theorem in a standard way and conclude there exists q sufficiently large such that y_q and u are distinct solutions of a right focal boundary value problem that has already been shown to have a unique solution. The proof is complete. $\qquad\square$

Chapter 3

Nonlocal Boundary Value Problems: Uniqueness and Existence

This chapter is devoted to uniqueness and existence of solutions of nonlocal boundary value problems for (1.1.1). Broadly speaking, a nonlocal linear boundary condition involves a linear condition providing data for a solution at more than one boundary point. Given $\alpha_1, \ldots, \alpha_p \in \mathbb{R}$, $a \leq x_1 < \cdots < x_p \leq b$, and $A \in \mathbb{R}$,

$$\sum_{i=1}^{p} \alpha_i y(x_i) = A$$

and

$$\int_a^b y(x)dx = A$$

are two examples of nonlocal linear boundary conditions. The first expression is also often called a *multipoint* boundary condition.

As is fitting with the themes of this book, some sections of this chapter will deal with uniqueness of solutions of a certain type of nonlocal boundary value problem for (1.1.1) implying uniqueness of solutions of related types of nonlocal boundary value problems for (1.1.1). The arguments for these "uniqueness implies uniqueness" results will take different paths, which will be developed in the first and third sections of this chapter.

Then, for each path taken, respective subsequent sections (that is, the second and fourth sections, respectively) will make use of the uniqueness results in establishing "uniqueness implies existence" results for the nonlocal boundary value problems for (1.1.1). The first path taken will be developed in the next section.

3.1 Nonlocal problems: uniqueness implies uniqueness, I

The results of this section are for nonlocal boundary value problems for (1.1.1), when the order $n = 2$ or $n = 3$.

For our first result, for $m \geq 1$, we are concerned with questions of uniqueness of solutions of certain $(m + 2)$-point nonlocal boundary value problems for the second order differential equation,

$$y'' = f(x, y, y'), \quad a < x < b, \tag{3.1.1}$$

satisfying

$$y(x_0) = y_1, \quad y(x_{m+1}) - \sum_{i=1}^{m} y(x_i) = y_2, \tag{3.1.2}$$

where $a < x_0 < x_1 < \cdots < x_{m+1} < b$, and $y_1, y_2 \in \mathbb{R}$. In particular, under the conditions (A) and (B) of Section 1.1 along with the uniqueness condition,

(C_{m+2}) Given points $a < x_0 < x_1 < \cdots < x_{m+1} < b$, if $y(x)$ and $z(x)$ are solutions of (3.1.1)-(3.1.2), then $y(x) = z(x)$, for all $a < x < b$.

Henderson [59] studied condition (C_{m+2}) implying the uniqueness of $(k+2)$-point nonlocal boundary value problems, for $1 \leq k \leq m$.

Theorem 3.1.1. [[59], Henderson, Thm. 2.1] *Assume that for* (3.1.1)*, conditions* (A)*,* (B) *and* (C_{m+2}) *hold. Then, given* $1 \leq k \leq m$*, condition* (C_{k+2}) *also holds.*

Proof. Of course from assumption (C_{m+2}), the condition is true when $k = m$. So we assume for the purpose of contradiction that, for some $1 \leq k < m$, there exist distinct solutions $y(x)$ and $z(x)$ of (3.1.1) and points $a < x_0 < \cdots < x_{k+1} < b$, so that

$$y(x_0) = z(x_0),$$

$$y(x_{k+1}) - \sum_{i=1}^{k} y(x_i) = z(x_{k+1}) - \sum_{i=1}^{k} z(x_i).$$

Set $w := y - z$. Then,

$$w(x_0) = 0, \quad w(x_{k+1}) - \sum_{i=1}^{k} w(x_i) = 0.$$

By (B), we may assume $w'(x_0) > 0$, and so there exists $x_0 < \tau < b$ so that $w'(x) > 0$ on $[x_0, \tau]$. Then there exist $x_0 < \tau_1 < \cdots < \tau_m < \tau$ such that $\sum_{i=1}^{m} w(\tau_i) < w(\tau)$. By the monotonicity of w on $[x_0, \tau]$, there exists $\tau_m < \tau_{m+1} < \tau$ such that

$$w(\tau_{m+1}) = \sum_{i=1}^{m} w(\tau_i).$$

Since $w(x_0) = 0$, we conclude

$$y(x_0) = z(x_0),$$

$$y(\tau_{m+1}) - \sum_{i=1}^{m} y(\tau_i) = z(\tau_{m+1}) - \sum_{i=1}^{m} z(\tau_i),$$

which is a contradiction to assumption (C_{m+2}). The proof is complete. $\square$

Jones [83] extended Theorem 3.1.1 for second order (3.1.1), by modifying condition (C_{m+2}) to be valid for *some* $m \geq 2$. And then, Gray [34] extended Theorem 3.1.1 to the third order case. In particular, for $m \geq 4$, for the third order differential equation,

$$y''' = f(x, y, y', y''), \quad a < x < b, \tag{3.1.3}$$

Gray was concerned with uniqueness of solutions satisfying m-point nonlocal boundary conditions,

$$y(x_1) = y_1, \quad y(x_2) = y_2, \quad y(x_m) - \sum_{i=3}^{m-1} y(x_i) = y_3, \tag{3.1.4}$$

where $a < x_1 < x_2 < \cdots < x_m < b$, and $y_1, y_2, y_3 \in \mathbb{R}$. Then, under the conditions (A) and (B) of Section 1.1 along with the uniqueness condition,

(Cm) Given points $a < x_1 < x_2 < \cdots < x_m < b$, if $y(x)$ and $z(x)$ are solutions of (3.1.3)-(3.1.4), then $y(x) = z(x)$, for all $a < x < b$.

Gray [34] studied the uniqueness of k-point nonlocal boundary value problems, for $3 \leq k \leq m$.

Theorem 3.1.2. [[34], Gray, Thm. 2.1] *Assume that for* (3.1.3), *conditions* (A), (B), *and* (Cm) *hold. Then, given* $3 \leq k \leq m$, *condition* (Ck) *also holds.*

Proof. The proof is by induction. We will begin by showing the theorem is true for the case $m = 4$, $k = 3$. Here, we are assuming conditions (A) and (B) hold, and that solutions for the differential equation (3.1.3) with boundary conditions (3.1.4), where (the nonlocal condition) is $y(x_4) - y(x_3) = y_3$, for $a < x_1 < x_2 < x_3 < x_4 < b$ and $y_1, y_2, y_3 \in \mathbb{R}$, are unique when they exist.

Now, suppose that $u(x)$ and $v(x)$ are distinct solutions of (3.1.3) with boundary conditions (3.1.4) and where (the nonlocal condition is actually) $y(x_3) = y_3$, for $a < x_1 < x_2 < x_3 < b$ and $y_1, y_2, y_3 \in \mathbb{R}$. We let $w(x) = u(x) - v(x)$, and we have

$$u(x_1) = v(x_1),$$
$$u(x_2) = v(x_2),$$
$$u(x_3) = v(x_3),$$

so that

$$w(x_1) = w(x_2) = w(x_3) = 0.$$

We have by condition (B) that either $w'(x_2) \neq 0$ or $w''(x_2) \neq 0$. We examine each case.

Case 1: $w'(x_2) \neq 0$. We may assume, without loss of generality, that $w'(x_2) > 0$. Therefore, since $w(x_2) = 0$, $w(x_3) = 0$, and $w(x) = u(x) - v(x)$ is continuous, $w(x)$ has a local maximum on (x_2, x_3), say at $x = \beta$. Then we must have $\alpha \in (x_2, \beta)$ and $\gamma \in (\beta, x_3)$ such that

$$w(\alpha) = w(\gamma),$$
$$u(\alpha) - v(\alpha) = u(\gamma) - v(\gamma),$$
$$u(\alpha) - u(\gamma) = v(\alpha) - v(\gamma).$$

We have that $u(x)$ and $v(x)$ satisfy

$$u(x_1) = v(x_1),$$
$$u(x_2) = v(x_2),$$
$$u(\alpha) - u(\gamma) = v(\alpha) - v(\gamma).$$

Therefore, $u(x) \equiv v(x)$ on (a, b), by condition (C4). This contradicts our assumption that $u(x)$ and $v(x)$ are distinct.

Case 2: $w'(x_2) = 0$ *but* $w''(x_2) \neq 0$. We may assume, without loss of generality, that $w''(x_2) > 0$. Then there exists a $\delta > 0$ such that $x_2 < x_2 + \delta < x_3$ and $w''(x) > 0$ on $[x_2, x_2 + \delta]$. Thus $w'(x)$ is strictly increasing on $[x_2, x_2 + \delta]$. Since $w'(x_2) = 0$, it is the case that $w'(x)$ is positive on

$(x_2, x_2 + \delta]$; and so $w(x)$ is increasing on $[x_2,\ x_2 + \delta]$, and therefore positive on $(x_2, x_2 + \delta]$, since $w(x_2) = 0$. We see that $w(x)$ must in fact be positive on all of $(x_2,\ x_3)$ by condition (A) and the fact that the next zero of $w(x)$ occurs at x_3, so we may now repeat the argument of Case 1.

In both cases we reach a contradiction to our assumption that $u(x)$ and $v(x)$ are distinct solutions of the selected boundary value problem. We conclude that the theorem holds for $m = 4$ and $k = 3$. To complete the proof, we now show that the theorem holds for an arbitrary positive integer $m > 4$ and the case $k = m - 1$.

Assume conditions (A), (B), and (Cm) hold for a positive integer $m > 4$. Let $u(x)$ and $v(x)$ be distinct solutions of the differential equation (3.1.3) with boundary conditions (3.1.4) and

$$y(x_k) - \sum_{i=3}^{k-1} y(x_i) = y_3, \qquad (3.1.5)$$

for $k = m - 1$, any $a < x_1 < x_2 < \cdots < x_k < b$, and any $y_1, y_2, y_3 \in \mathbb{R}$. We let $w(x) = u(x) - v(x)$, and we have

$$w(x_1) = w(x_2) = 0,$$

$$w(x_k) - \sum_{i=3}^{k-1} w(x_i) = 0.$$

By condition (B), either $w'(x_2) \neq 0$ or $w''(x_2) \neq 0$. We will examine each of the cases.

Case 1: $w'(x_2) \neq 0$. Assume without loss of generality that $w'(x_2) > 0$. Then there exists $\alpha > 0$ such that $x_2 < x_2 + \alpha < x_3$ and $w(x)$ is strictly increasing on $(x_2, x_2 + \alpha)$. Observe that $w(x)$ is also positive on $(x_2, x_2 + \alpha)$ since $w(x_2) = 0$. Therefore we may choose $x_2 < t_1 < t_2 < \cdots < t_{k-1} \leq x_2 + \alpha$ such that

$$w(t_{k-1}) - \sum_{i=1}^{k-2} w(x_i) = 0.$$

That is, we have

$$u(x_1) = v(x_1),$$

$$u(x_2) = v(x_2),$$

$$u(t_{k-1}) - \sum_{i=1}^{k-2} u(x_i) = v(t_{k-1}) - \sum_{i=1}^{k-2} v(x_i).$$

Therefore, $u(x) \equiv v(x)$ on (a, b), by condition (Cm). This contradicts our assumption that $u(x)$ and $v(x)$ are distinct.

Case 2: $w'(x_2) = 0$ *but* $w''(x_2) \neq 0$. We may assume, without loss of generality, that $w''(x_2)$ is positive. Then there exists a $\delta > 0$ such that $x_2 < x_2 + \delta < x_3$ and $w'(x)$ is strictly increasing on $[x_2, x_2 + \delta]$. Since $w'(x_2) = 0$, it is the case that $w'(x)$ is positive on $(x_2, x_2 + \delta]$, and so $w(x)$ is increasing on $[x_2, x_2 + \delta]$, and hence is positive on $(x_2, x_2 + \delta]$ since $w(x_2) = 0$. We may now choose appropriate values of t_i, $i = 1, 2, \ldots, k - 1$, from $(x_2, x_2 + \delta]$ to repeat the argument of Case 1.

We conclude that the result holds for the case $k = m - 1$. This completes the proof of the theorem. $\qquad \square$

Later, Gray [33,35] generalized Theorem 3.1.2 to a nonlocal problem for (3.1.3) satisfying two expressions of nonlocal boundary conditions. To wit, for $m \geq 1$ and $l > 1$, Gray was concerned with uniqueness of solutions of (3.1.3) satisfying $(l + m)$-point nonlocal boundary conditions,

$$y(x_1) - \sum_{i=2}^{l-1} y(x_i) = y_1, \ y(x_l) = y_2, \ y(x_{m+l}) - \sum_{j=l+1}^{m+l-1} y(x_j) = y_3, \quad (3.1.6)$$

where $a < x_1 < x_2 < \cdots < x_{m+l} < b$, and $y_1, y_2, y_3 \in \mathbb{R}$. Then, under the conditions (A) and (B) of Section 1.1 along with the uniqueness condition,

(C_{l+m}) Given points $a < x_1 < x_2 < \cdots < x_{m+l} < b$, if $y(x)$ and $z(x)$ are solutions of (3.1.3)-(3.1.6), then $y(x) = z(x)$, for all $a < x < b$.

Gray [33, 35] studied the uniqueness of $(q + p)$-point nonlocal boundary value problems, for $1 < q \leq l$ and $1 \leq p \leq m$. The proof of his result involves only slight modifications of the proof of Theorem 3.1.2.

Theorem 3.1.3. [[33], Gray, Thm. 4.1], [[35], Gray] *Assume that for (3.1.3), conditions (A), (B), and (C_{l+m}) hold. Then, given $1 < p \leq l$ and $1 \leq q \leq m$, condition (C_{p+q}) also holds.*

3.2   Nonlocal problems: uniqueness implies existence, I

The uniqueness implies existence results of this section for nonlocal boundary value problems for (1.1.1) continue to be for second or third order equations. Uniqueness implies existence results for (3.1.1) for 3-point boundary value problems were first obtained by Henderson, Karna and Tisell [64] and Henderson [58]. The first result models the work produced in [58]; the hypotheses in [58] are weaker that than those employed in [64] where it was assumed that solutions of right focal boundary value problems were unique.

Lemma 3.2.1. *Assume that for* (3.1.1), *condition* (C_3) *holds. Then solutions of 2-point conjugate boundary value problems exist and are unique.*

Proof. Assume there exist $a < x_1 < x_4 < b$ and distinct solutions y, z of (3.1.1) satisfying $(y-z)(x_1) = (y-z)(x_4)$. Assume without loss of generality that $y - z$ has a positive maximum value at $c \in (x_1, x_4)$. Then there exist $x_2 \in (x_1, c)$ and $x_3 \in (c, x_4)$ such that $(y - z)(x_2) = \frac{(y-z)(c)}{2} = (y - z)(x_3)$. This contradicts condition (C_3) at $a < x_1 < x_2 < x_3 < b$. Thus, condition (C) holds for (3.1.1) and solutions of conjugate boundary value problems exist by Theorem 2.1.1 or 2.3.1. $\qquad\square$

As a consequence of Lemma 3.2.1, the "compactness condition" (CP) applies.

Theorem 3.2.1. [[58], Henderson, Thm. 3] *Assume that for* (3.1.1), *conditions* (A), (B) *and* (C_3) *hold. Then, given points* $a < x_1 < x_2 < x_3 < b$, *and* $y_1, y_2 \in \mathbb{R}$, *there exists a unique solution of* (3.1.1), (3.1.2) *on* (a, b) *in the case that* $m + 2 = 3$.

Proof. Let $a < x_1 < x_2 < x_3 < b$ and $y_1, y_2 \in \mathbb{R}$ be given. Let $z(x)$ denote the solution of the 2-point conjugate boundary value problem for (3.1.1) satisfying the boundary conditions at x_2 and x_3,

$$z(x_2) = y_2, \quad z(x_3) = 0.$$

z exists by Lemma 3.2.1.
Define the set

$$S = \{y(x_1) \mid y(x) \text{ is a solution (3.1.1) satisfying}$$
$$y(x_2) - y(x_3) = z(x_2) - z(x_3)\}.$$

We observe first that S is nonempty, since $z(x_1) \in S$. In addition, it follows from Continuous Dependence, Theorem 1.1.6, that S is an open subset of $\mathbb{R}$.

The remainder of the argument is devoted to showing that S is also a closed subset of $\mathbb{R}$. To that end, we assume for the purpose of contradiction that S is not closed. Then there exists an $s_0 \in \overline{S} \backslash S$ and a strictly monotone sequence $\{s_q\} \subset S$ such that $\lim_{q \to \infty} s_q = s_0$.

We may assume, without loss of generality, that $s_q \uparrow s_0$. By the definition of S, we denote, for each $q \in \mathbb{N}$, by $y_q(x)$ the solution of (3.1.1) satisfying

$$y_q(x_1) = s_q, \text{ and } y_q(x_2) - y_q(x_3) = z(x_2) - z(x_3).$$

Since $s_{q+1} > s_q$, it follows from the uniqueness condition, (C_3), that

$$y_q(x) < y_{q+1}(x) \text{ on } (a, x_2).$$

Consequently, it follows from the "compactness condition" (CP), Theorem 1.1.5, and the fact that $s_0 \notin S$, that $\{y_q(x)\}$ is not uniformly bounded above on each compact subinterval of each of (a, x_1) and (x_1, x_2).

Let $u(x)$ denote the solution of the initial value problem for (3.1.1) satisfying the initial conditions at x_1,

$$u(x_1) = s_0, \quad u'(x_1) = 0.$$

It follows that there exists q sufficiently large and

$$a < \tau_1 < x_1 < \tau_2 < x_2,$$

such that

$$y_q(\tau_1) = u(\tau_1), \qquad y_q(\tau_2) = u(\tau_2),$$

which is a contradiction to the uniqueness of solutions of 2-point conjugate boundary value problems, Lemma 3.2.1. Thus, S is also a closed subset of $\mathbb{R}$. In particular, $S = \mathbb{R}$ and the theorem is proved. □

Building inductively on the number of boundary points, Henderson [57] extended Theorem 3.2.1 to the 4-point boundary value problem, (3.1.1), (3,1,2), in the case $m - 2$. The induction step is primarily given in Theorem 3.1.1.

Theorem 3.2.2. [[57], Henderson, Thm. 3] *Assume that for (3.1.1), conditions (A), (B) and (C_4) hold. Then, given points $a < x_1 < x_2 < x_3 < x_4 < b$, and $y_1, y_2 \in \mathbb{R}$, there exists a unique solution of (3.1.1), (3.1.2) on (a, b) in the case that $m + 2 = 4$.*

Proof. Let $a < x_1 < x_2 < x_3 < x_4 < b$ and $y_1, y_2 \in \mathbb{R}$ be given. Let $z(x)$ denote the solution of the 3-point nonlocal boundary value problem for (3.1.1) satisfying the boundary conditions at x_2, x_3 and x_4,

$$z(x_2) = 0, \quad z(x_4) - z(x_3) = y_2.$$

Theorem 3.1.1 and Theorem 3.2.1 imply the existence of z. Define the set

$$S = \{y(x_1) \mid y(x) \text{ is a solution } (3.1.1) \text{ and}$$
$$y(x_4) - y(x_2) - y(x_3) = z(x_4) - z(x_2) - z(x_3)\}.$$

We observe first that S is nonempty, since $z(x_1) \in S$ and S is open.

To show that S is closed, assume $s_0 \in \overline{S} \backslash S$ and assume without loss of generality that there exists a strictly monotone increasing sequence $\{s_q\} \subset S$ such that $\lim_{q \to \infty} s_q = s_0$. By the definition of S, we denote, for each $q \in \mathbb{N}$, by $y_q(x)$ the solution of (3.1.1) satisfying

$$y_q(x_1) = s_q, \text{ and } y_q(x_4) - y_q(x_2) - y_q(x_3) = z(x_4) - z(x_2) - z(x_3).$$

By (C_3) and since $s_{q+1} > s_q$, we have

$$y_q(x) < y_{q+1}(x) \text{ on } (a, x_2).$$

Thus, $\{y_q(x)\}$ is not uniformly bounded above on each compact subinterval of each of (a, x_1) or (x_1, x_2).

Now, let $u(x)$ be the solution of the initial value problem for (3.1.1) satisfying the initial conditions at x_1,

$$u(x_1) = s_0, \quad u'(x_1) = 0.$$

It follows that, there exists q sufficiently large, and points,

$$a < \tau_1 < x_1 < \tau_2 < x_2,$$

such that

$$y_q(\tau_1) = u(\tau_1), \qquad y_q(\tau_2) = u(\tau_2).$$

Thus, y_q and u are distinct solutions of a 2-point conjugate boundary value problem for (3.1.1). Apply first Lemma 3.1.1 and then Lemma 3.2.1 to see that 2-point conjugate boundary value problems have unique solution. Thus, a contradiction is obtained and S is also a closed subset of $\mathbb{R}$. □

In [59], Henderson obtained the existence result for the nonlocal $(m+2)$-point boundary conditions, (3.1.2), for (3.1.1) on (a, b). The fundamental uniqueness assumption employed for the general nonlocal $(m + 2)$-point problem is condition (C_{m+2}), which then implies condition (C_{k+2}), $1 \leq k \leq m-1$, by Theorem 3.1.1. The proof of the following theorem has been motivated above, or we refer the reader to [59].

Theorem 3.2.3. [[59], Henderson, Thm. 3.1] *Let* $\mu \geq 1$ *be an integer. Assume that for* (3.1.1), *conditions* (A), (B) *and* $(C_{\mu+2})$ *hold. Then, for each* $1 \leq m \leq \mu$, *there exists a unique solution of* (3.1.1), (3.1.2) *on* (a, b).

Uniqueness implies existence for the third order problem, (3.1.3), was studied by Gray [34]. Gray focused on the m-point nonlocal boundary conditions, (3.1.4), given by

$$y(x_1) = y_1, \quad y(x_2) = y_2, \quad y(x_m) - \sum_{i=3}^{m-1} y(x_i) = y_3,$$

where $a < x_1 < x_2 < \cdots < x_m < b$, and $y_1, y_2, y_3 \in \mathbb{R}$.

Theorem 3.2.4. [[34], Gray, Thm. 3.5] *Let* $m \geq 3$ *be an integer. Assume that for* (3.1.3), *conditions* (A), (B), *and* (Cm) *hold. Then, for each* $3 \leq k \leq m$, $a < x_1 < x_2 < \cdots < x_k < b$, *and* $y_1, y_2, y_3 \in \mathbb{R}$, *there exists a unique solution of* (3.1.3), (3.1.4) *on* (a, b).

Proof. The proof is by induction on m. The boundary conditions, (3.1.4), for $m = 3$ are interpreted as 3-point conjugate conditions and so condition (C3) is precisely condition (C). Thus the assertion is valid in the case, $m = 3$.

So, assume $m > 3$ and assume as the induction hypothesis that the assertion of the theorem is valid for all $3 \leq \mu < m$. Let $a < x_1 < x_2 < \cdots < x_m < b$, and $y_1, y_2, y_3 \in \mathbb{R}$, be given. Let z denote the solution of (3.1.3) on (a, b) satisfying the boundary conditions

$$z(x_2) = y_2, \quad z(x_3) = 0, \quad z(x_m) - \sum_{i=4}^{m-1} z(x_i) = y_3.$$

z exists by the induction hypothesis on m. Moreover, z satisfies the boundary conditions

$$z(x_2) = y_2, \quad z(x_m) - \sum_{i=3}^{m-1} z(x_i) = y_3.$$

Define the set

$$S = \left\{ y(x_1) \mid y(x) \text{ is a solution } (3.1.3) \text{ satisfying} \right.$$

$$\left. y(x_2) = y_2, \quad y(x_m) - \sum_{i=3}^{m-1} y(x_i) = y_3 \right\}.$$

Note that S is nonempty, since $z(x_1) \in S$. In addition, it follows from Continuous Dependence, Theorem 1.1.6, that S is an open subset of $\mathbb{R}$.

It remains to show that S is closed to complete the proof of the theorem. Assume for the sake of contradiction, that there exists $s_0 \in \overline{S} \backslash S$ and assume without loss of generality that there exists a strictly monotone increasing sequence $\{s_q\} \subset S$ such that $\lim_{q \to \infty} s_q = s_0$. By the definition of S, we denote, for each $q \in \mathbb{N}$, by $y_q(x)$ the solution of (3.1.3) satisfying

$$y_q(x_1) = s_q, \ y(x_2) = y_2 \text{ and } y(x_m) - \sum_{i=3}^{m-1} y(x_i) = y_3.$$

By (Cm) and since $s_{q+1} > s_q$, we have

$$y_q(x) < y_{q+1}(x) \text{ on } (a, x_1) \cup (x_1, x_2).$$

Thus, $\{y_q(x)\}$ is not uniformly bounded above on each compact subinterval of each of (a, x_1) or (x_1, x_2).

Let $\tau \in (a, x_1)$ and let $u(x)$ denote the solution of the 3-point conjugate boundary value problem (3.1.3) satisfying the boundary conditions,

$$u(\tau) = 0, \quad u(x_1) = s_0, \quad u(x_2) = y_2.$$

It follows that, for some q sufficiently large, there exist points,

$$a < \tau_1 < x_1 < \tau_2 < x_2,$$

such that

$$y_q(\tau_1) = u(\tau_1), \qquad y_q(\tau_2) = u(\tau_2), \quad y_q(x_2) = u(x_2).$$

Thus, y_q and u are distinct solutions of a 3-point conjugate boundary value problem for (3.1.3) which is a contradiction. Thus, a contradiction is obtained and S is also a closed subset of $\mathbb{R}$ and the theorem is proved.   □

To close this section, to compliment Theorem 3.1.3, we point out Gray's [33, 35] work to obtain uniqueness implies existence results for boundary conditions (3.1.6).

Theorem 3.2.5. [[33], Gray, Thm. 4.5] *Let $m \geq 1$ and $l > 1$, be integers. Assume that for (3.1.3), conditions (A), (B), and (C_{l+m}) hold. Let $a < x_1 < x_2 < \cdots < x_{m+l} < b$, and $y_1, y_2, y_3 \in \mathbb{R}$ be given. Then there exists a solution of (3.1.3), (3.1.6) on (a, b).*

3.3   Nonlocal problems: uniqueness implies uniqueness, II

We now develop the second path (see the introductory remarks to this chapter) for "uniqueness implies uniqueness" results for solutions of nonlocal boundary value problems for (1.1.1). These results will be for nonlocal boundary value problems for (1.1.1), first of orders $n = 3$ and $n = 4$, followed by arbitrary order n.

More specifically, the arguments in this section's development have the flavor of those used in Sections 1.2–1.5. That being the case, the results of this section will involve from Section 1.1 conditions (A), (B) and "compactness condition" (CP), and continuous dependence from Theorem 1.1.6 and Remark 1.1.1, along with other conditions we will impose.   For (1.1.1) of arbitrary order n, many times the nonlocal boundary conditions will be manifested in the form, for $1 \leq k \leq n - 1$,

$$y^{(i-1)}(x_j) = y_{ij},\ 1 \leq i \leq m_j,\ 1 \leq j \leq k,\ y(x_{k+1}) - y(x_{k+2}) = y_n,\ (3.3.1)$$

where $m_1, \ldots, m_k \in \mathbb{N}$, such that $m_1 + \cdots + m_k = n - 1$, $a < x_1 < \cdots < x_{k+2} < b$, and $y_{ij}, y_n \in \mathbb{R}$.

For some of the results of this section, when $k = n - 1$, we will be interested in the question of uniqueness of solutions of (1.1.1) satisfying (3.3.1) implying the uniqueness of solutions of (1.1.1) satisfying (3.3.1), for all $1 \leq k \leq n - 2$. This is in close analogy to the uniqueness question concerning uniqueness of solutions of n-point conjugate boundary value problems implying the uniqueness of solutions of h-point conjugate boundary value problems for (1.1.1), for $2 \leq h \leq n - 1$, which Opial [96] proved to be true when (1.1.1) is a linear equation, and which Jackson [76] showed to be valid when the differential equation is nonlinear; see Theorem 1.2.2.

In the context of nonlocal boundary conditions, Henderson and Kunkel [65] obtained a result analogous to Peano's result for when (1.1.1) is a linear equation. In particular, for $1 \leq k \leq n - 1$, Henderson and Kunkel introduced the property:

(P.k) For all $m_1, \ldots, m_k \in \mathbb{N}$, such that $m_1 + \cdots + m_k = n - 1$, $a < x_1 < \cdots < x_{k+2} < b$, and $y_{ij}, y_n \in \mathbb{R}$, there is at most one solution of (1.1.1)–(3.3.1) on (a, b).

We now state, without proof, when (1.1.1) is a linear equation, the Henderson and Kunkel result for nonlocal problems which is an analogue of the Peano result for conjugate boundary value problems.

Theorem 3.3.1. [[65], Henderson and Kunkel, Thm. 2.2] *Assume that* (1.1.1) *is linear, and assume condition* (P.n−1) *is satisfied. Then, for each* $1 \leq k \leq n - 2$, *condition* (P.k) *is also satisfied.*

And for other results of this section, for fixed $1 \leq k < n - 1$, we will be interested in questions for which condition (P.k) implies condition (P.h) for all $k < h \leq n - 1$. These questions are somewhat in analogy to the converse questions dealt with in Section 1.3 for conjugate boundary value problems and in Section 1.5 for right focal boundary value problems.

We will first focus on relations between certain 4-point nonlocal boundary value problems and 3-point nonlocal boundary value problems for the case when (1.1.1) is of third order; that is for the equation (3.1.3). In particular, the boundary conditions on which we focus include the 4-point nonlocal boundary conditions,

$$y(x_1) = y_1, \quad y(x_2) = y_2, \quad y(x_3) - y(x_4) = y_3, \qquad (3.3.2)$$

$$y(x_1) - y(x_2) = y_1, \quad y(x_3) = y_2, \quad y(x_4) = y_3, \qquad (3.3.3)$$

where $a < x_1 < x_2 < x_3 < x_4 < b$ and $y_1, y_2, y_3 \in \mathbb{R}$, and the 3-point nonlocal boundary conditions,

$$y(x_1) = y_1, \quad y'(x_1) = y_2, \quad y(x_2) - y(x_3) = y_3, \qquad (3.3.4)$$

$$y(x_1) - y(x_2) = y_1, \quad y(x_3) = y_2, \quad y'(x_3) = y_3, \qquad (3.3.5)$$

where $a < x_1 < x_2 < x_3 < b$ and $y_1, y_2, y_3 \in \mathbb{R}$.

Clark and Henderson [13] obtained results concerning uniqueness of solutions of both (3.1.3)-(3.3.2) and (3.1.3)-(3.3.3), when solutions exist, implying uniqueness of solutions of both (3.1.3)-(3.3.4) and (3.1.3)-(3.3.5), when solutions exist, as well as converse results.

Remark 3.3.1. *If solutions of* (3.1.3)-(3.3.2) *or* (3.1.3)-(3.3.3) *are unique, when solutions exist, then solutions of 3-point conjugate boundary value problems* (*see Sections* 1.2 *and* 1.3 *to recall the form of conjugate boundary conditions*), *for* (3.1.3) *are unique, when they exist. And so by Theorem* 1.2.1, *solutions of 2-point conjugate boundary value problems for* (3.1.3) *are unique, when they exist. In addition, by Theorem* 2.2.5 *and Theorem* 2.2.6 *from Section* 2.2, *solutions exist for each 2-point and each 3-point conjugate boundary value problem for* (3.1.3). *Moreover, by Theorem* 1.1.5, *the "compactness condition"* (CP) *also holds.*

Theorem 3.3.2. [[13], Clark and Henderson, Thm. 2.3] *Assume that with respect to* (3.1.3), *conditions* (A) *and* (B) *are satisfied. If solutions for* (3.1.3)-(3.3.2) *are unique, when they exist, then solutions for* (3.1.3)-(3.3.4) *are unique, when they exist.*

Proof. Assume for the purpose of contradiction that, for some $a < x_1 < x_2 < x_3 < b$ and $y_1, y_2, y_3 \in \mathbb{R}$, there exist distinct solutions $y(x)$ and $z(x)$ of (3.1.3)-(3.3.4); that is, $y(x_1) = z(x_1), y'(x_1) = z'(x_1), y(x_2) - y(x_3) = z(x_2) - z(x_3)$. By uniqueness of solutions of initial value problems for (3.1.3), we may assume, without loss of generality, that $y''(x_1) > z''(x_1)$.

Now fix $a < \tau < x_1$. By continuous dependence on 4-point nonlocal boundary conditions, for each $\epsilon > 0$, there is a $\delta > 0$ and there is a solution $z_\delta(x)$ of (3.1.3) such that

$$z_\delta(\tau) = z(\tau), \; z_\delta(x_1) = z(x_1) + \delta, \; z_\delta(x_2) - z_\delta(x_3) = z(x_2) - z(x_3),$$

and $|z_\delta(x) - z(x)| < \epsilon$ on $[\tau, x_3]$. For sufficiently small $\epsilon > 0$, there exist points $\tau < \tau_1 < x_1 < \tau_2 < x_2$ so that

$$z_\delta(\tau_1) = y(\tau_1), \; z_\delta(\tau_2) = y(\tau_2).$$

Also,

$$z_\delta(x_2) - z_\delta(x_3) = z(x_2) - z(x_3) = y(x_2) - y(x_3).$$

So, by hypotheses, $z_\delta(x) \equiv y(x)$, which is a contradiction. We conclude that solutions of (3.1.3)-(3.3.4) are unique, when they exist. $\square$

Of course, there is an analogue for the solutions of (3.1.3)-(3.3.3) and (3.1.3)-(3.3.5).

Theorem 3.3.3. [[13], Clark and Henderson, Thm. 2.4] *Assume that with respect to* (3.1.3), *conditions* (A) *and* (B) *are satisfied. If solutions for* (3.1.3)-(3.3.3) *are unique, when they exist, then solutions for* (3.1.3)-(3.3.5) *are unique, when they exist.*

Our next result is devoted to a question converse to Theorems 3.3.2 and 3.3.3.

Theorem 3.3.4. [[13], Clark and Henderson, Thm. 2.7] *Assume that with respect to* (3.1.3), *conditions* (A) *and* (B) *are satisfied. If solutions for both* (3.1.3)-(3.3.4) *and* (3.1.3)-(3.3.5) *are unique, when they exist, then solutions for both* (3.1.3)-(3.3.2) *and* (3.1.3)-(3.3.3) *are unique, when they exist.*

Proof. We will consider the uniqueness statement for only (3.1.3)-(3.3.2), with the other argument being completely analogous. Assume to the contrary that, for some $a < x_1 < x_2 < x_3 < x_4 < b$ and $y_1, y_2, y_3 \in \mathbb{R}$, there exist distinct solutions, $y(x)$ and $z(x)$, of (3.1.3)-(3.3.2).

By uniqueness of solutions of (3.1.3)-(3.3.4), we may assume, with no loss of generality, that $y'(x_1) < z'(x_1)$ and $y'(x_2) > z'(x_2)$, and $y(x) < z(x)$ on (x_1, x_2). In addition, by Remark 3.3.1 of this section, unique solutions exist for both 3-point conjugate and 2-point conjugate problems for (3.1.3). This implies $y(x) > z(x)$ on $(a, b) \setminus [x_1, x_2]$.

For each real $r \geq 0$, let $y_r(x)$ be the solution of (3.1.3) satisfying the $(2, 1)$ conjugate boundary conditions at x_3 and x_4,

$$y_r(x_3) = y(x_3), \ y_r'(x_3) = y'(x_3) - r, \ y_r(x_4) = y(x_4).$$

It follows from uniqueness of solutions of 3-point conjugate problems for (3.1.3) that, for $s > r \geq 0$,

$$y(x) \leq y_r(x) < y_s(x) \text{ on } (a, x_3).$$

For each $r \geq 0$, let

$$E_r = \{x_1 \leq x \leq x_2 \mid y_r(x) \leq z(x)\}.$$

These sets are compact and nested such that $E_s \subset E_r \subset (x_1, x_2)$ when $s > r > 0$.

For some $r > 0$, suppose that $E_r = \emptyset$, and let η be the least upper bound for those $s > 0$ where $E_s \neq \emptyset$. Then, by continuity with respect to r for solutions of the $(2, 1)$ conjugate boundary value problem , it follows that $E_\eta \neq \emptyset$; because, if $E_\eta = \emptyset$, then $E_{\eta - \epsilon} = \emptyset$ for sufficiently small $\epsilon > 0$. Moreover, $y_\eta(x) = z(x)$ for every $x \in E_\eta$; because, if $y_\eta(x) < z(x)$ for some $x \in E_\eta$ then again by the continuity with respect to r for solutions of $(2, 1)$ conjugate problems, $E_{\eta + \epsilon} \neq \emptyset$ for sufficiently small $\epsilon > 0$.

By the uniqueness for solutions of 3-point conjugate problems for (3.1.3), it follows that E_η consists of at most two distinct points. If τ is one of these points, it necessarily follows that $y_\eta'(\tau) = z'(\tau)$. As a consequence of the hypotheses and the $(2, 1)$ conjugate boundary conditions at x_3 and x_4, we see that

$$y_\eta(\tau) = z(\tau), \ y_\eta'(\tau) = z'(\tau), \ y_\eta(x_3) - y_\eta(x_4) = z(x_3) - z(x_4),$$

and therefore, $y_\eta(x) = z(x)$, for all $a < x < b$. This implies, in turn, by the uniqueness of solutions to 3-point conjugate problems, that $y(x) = z(x)$, for all $a < x < b$, which contradicts the initial assumption that $y(x)$ and $z(x)$ are distinct solutions. Hence, $E_n \neq \emptyset$ for all $n \in \mathbb{N}$, and

$$\bigcap_{n=1}^{\infty} E_n := E \neq \emptyset.$$

Next, we observe that the set E consists of a single point x_0 with $x_1 < x_0 < x_2$. In fact, if $t_1, t_2 \in E$, with $x_1 < t_1 < t_2 < x_2$, then the same type of argument that was used to show the foregoing sets E_n are nonempty leads to the conclusion that the interval $[t_1, t_2]$ must be contained in E. However, $[t_1, t_2] \subset E$ implies that the sequence $\{y_n(x)\}$ is uniformly bounded on $[t_1, t_2]$ which contradicts the part of "compactness condition" (CP) requiring boundedness for some subsequence of $\{y'_{n_j}(x_3)\}$. Thus, we conclude $E = \{x_0\}$ with $x_1 < x_0 < x_2$, and

$$\lim_{n \to \infty} y_n(x_0) := y_0 \leq z(x_0).$$

Now we show this is not possible. There are two cases to be resolved. We first assume $y_0 = z(x_0)$. Then for $\epsilon > 0$, sufficiently small, there is a solution $z(x, \epsilon)$ of the 3-point conjugate boundary value problem for (3.1.3) satisfying

$$z(x_0, \epsilon) = z(x_0) - \epsilon, \; z(x_3, \epsilon) = z(x_3), \; z(x_4, \epsilon) = z(x_4),$$

and $z(x, \epsilon) < z(x)$ on (a, x_3). Let us also note that

$$z(x_3, \epsilon) - z(x_4, \epsilon) = z(x_3) - z(x_4) = y(x_3) - y(x_4).$$

Such a solution $z(x, \epsilon)$, where ϵ is chosen so that $z(x_0, \epsilon) = z(x_0) - \epsilon > y(x_0)$, can be used in place of $z(x)$ in defining the sets $\{E_n\}$ with respect to the given sequence of solutions $\{y_n(x)\}$. Then as previously, it would follow that each of these sets would be nonempty which is impossible.

For the remaining case, let $y(x_0) < y_0 < z(x_0)$. In this case, for $0 \leq \lambda \leq 1$, let $z(x, \lambda)$ denote the solution of (3.1.3) satisfying the 2-point conjugate boundary conditions,

$$z(x_3, \lambda) = \lambda y(x_3) + (1 - \lambda) z(x_3),$$

$$z'(x_3, \lambda) = \lambda y'(x_3) + (1 - \lambda) z'(x_3),$$

$$z(x_4, \lambda) = \lambda y(x_4) + (1 - \lambda) z(x_4).$$

We note that, for each $0 \leq \lambda \leq 1$,

$$z(x_3, \lambda) - z(x_4, \lambda) = y(x_3) - y(x_4).$$

Let

$$L = \{(z(x_3, \lambda), z'(x_3, \lambda), z(x_4, \lambda)) \mid 0 \leq \lambda \leq 1\}.$$

Then L is a line segment in $\mathbb{R}^3$. The function $h : [0,1] \to L$ defined by

$$h(\lambda) = (z(x_3, \lambda), z'(x_3, \lambda), z(x_4, \lambda))$$

is continuous, one-to-one and onto. Next, define $g : L \to \mathbb{R}$ by

$$g((z(x_3, \lambda), z'(x_3, \lambda), z(x_4, \lambda))) = z(x_0, \lambda).$$

By continuous dependence of solutions of (3.1.3) on 2-point conjugate boundary conditions, g is continuous and so $g \circ h : [0,1] \to \mathbb{R}$ is continuous. Now,

$$g \circ h(0) = z(x_0) > y_0 > y(x_0) = g \circ h(1).$$

Hence, there exists $0 < \lambda_0 < 1$ so that

$$g \circ h(\lambda_0) = z(x_0, \lambda_0) = y_0.$$

Now, there is an $\eta > 0$ such that $[x_0 - \eta, x_0 + \eta] \subset (x_1, x_2)$ and such that

$$z(x, \lambda_0) < z(x) \text{ on } [x_0 - \eta, x_0 + \eta].$$

Then with $\{y_n(x)\}$ the same sequence as previously defined, we have

$$\lim_{n \to \infty} y_n(x) > z(x) > z(x, \lambda_0),$$

for all $x \in [x_0 - \eta, x_0 - \eta] \setminus \{x_0\}$, and

$$\lim_{n \to \infty} y_n(x_0) = y_0 = z(x_0, \lambda_0).$$

This is the same contradictory situation as the case $y_0 = z(x_0)$ considered previously.

From this final contradiction, we conclude that $y_0 \leq z(x_0)$ is impossible, and therefore, solutions of (3.1.3)-(3.3.2) are unique. Similarly, solutions of (3.1.3)-(3.3.3) are unique. □

In view of Theorems 3.3.2, 3.3.3 and 3.3.4, we can state the following theorem.

Theorem 3.3.5. *Assume that with respect to (3.1.3), conditions (A) and (B) are satisfied. Then, solutions for both (3.1.3)-(3.3.4) and (3.1.3)-(3.3.5) are unique, when they exist if, and only if, solutions for both (3.1.3)-(3.3.2) and (3.1.3)-(3.3.3) are unique, when they exist.*

Our next focus will involve some extensions by Henderson and Ma [66] of Theorems 3.3.2, 3.3.3 and 3.3.4, and will concern uniqueness of solutions relations among certain 5-point nonlocal boundary value problems, 4-point nonlocal boundary value problems and 3-point nonlocal boundary value problems for the case when (1.1.1) is fourth order; that is for the equation,

$$y^{(4)} = f(x, y, y', y'', y'''), \quad a < x < b, \tag{3.3.6}$$

and the boundary conditions on which we focus include the 5-point nonlocal boundary conditions,

$$y(x_1) = y_1, \ y(x_2) = y_2, \ y(x_3) = y_3, \ y(x_4) - y(x_5) = y_4, \tag{3.3.7}$$

$$y(x_1) - y(x_2) = y_1, \ y(x_3) = y_2, \ y(x_4) = y_3, \ y(x_5) = y_4, \tag{3.3.8}$$

where $a < x_1 < x_2 < x_3 < x_4 < x_5 < b$ and $y_1, y_2, y_3, y_4 \in \mathbb{R}$, and the 4-point nonlocal boundary conditions,

$$y(x_1) = y_1, \ y'(x_1) = y_2, \ y(x_2) = y_3, \ y(x_3) - y(x_4) = y_3, \tag{3.3.9}$$

$$y(x_1) - y(x_2) = y_1, \ y(x_3) = y_2, \ y(x_4) = y_3, \ y'(x_4) = y_4, \tag{3.3.10}$$

$$y(x_1) = y_1, \ y(x_2) = y_2, \ y'(x_2) = y_3, \ y(x_3) - y(x_4) = y_4, \tag{3.3.11}$$

$$y(x_1) - y(x_2) = y_1, \ y(x_3) = y_2, \ y'(x_3) = y_3, \ y(x_4) = y_4, \tag{3.3.12}$$

where $a < x_1 < x_2 < x_3 < x_4 < b$ and $y_1, y_2, y_3, y_4 \in \mathbb{R}$, and the 3-point nonlocal boundary conditions,

$$y(x_1) = y_1, \ y'(x_1) = y_2, \ y''(x_1) = y_3, \ y(x_2) - y(x_3) = y_4, \tag{3.3.13}$$

$$y(x_1) - y(x_2) = y_1, \ y(x_3) = y_2, \ y'(x_3) = y_3, \ y''(x_3) = y_4, \tag{3.3.14}$$

where $a < x_1 < x_2 < x_3 < b$ and $y_1, y_2, y_3, y_4 \in \mathbb{R}$.

Henderson and Ma's [66] first results will be devoted to uniqueness of solutions of (3.3.6)-(3.3.7), when solutions exist, implying the uniqueness of solutions of each of (3.3.6)-(3.3.9), (3.3.6)-(3.3.11), and (3.3.6)-(3.3.13), when solutions exist. (Of course, there is a dual result in terms of uniqueness of solutions of (3.3.6)-(3.3.8) implying the uniqueness of solutions of each of (3.3.6)-(3.3.10), (3.3.6)-(3.3.12), and (3.3.6)-(3.3.14).)

Remark 3.3.2. *If solutions of* (3.3.6)-(3.3.7) *or* (3.3.6)-(3.3.8) *are unique, when solutions exist, then solutions of 4-point conjugate boundary value problems for* (3.3.6) *are unique, when they exist. And so by Theorem 1.2.2, solutions of both 3-point conjugate boundary value problems and 2-point conjugate boundary value problems for* (3.3.6) *are unique, when they exist. In addition, by Theorem 2.3.4 from Section 2.3, solutions exist for each 4-point, each 3-point, and each 2-point conjugate boundary value problem for* (3.3.6). *Moreover, by Theorem 1.1.5, the "compactness condition"* (CP) *also holds.*

Theorem 3.3.6. [[66], Henderson and Ma, Thm. 3] *Assume that with respect to* (3.3.6), *conditions* (A) *and* (B) *are satisfied. If solutions for* (3.3.6)-(3.3.7) *are unique, when they exist, then solutions for* (3.3.6)-(3.3.9) *are unique, when they exist.*

Proof. Our proof is by contradiction. So, we suppose (3.3.6)-(3.3.9) has two solutions, $y(x)$ and $z(x)$, such that

$$z(x_1) = y(x_1),$$
$$z'(x_1) = y'(x_1),$$
$$z(x_2) = y(x_2),$$
$$z(x_3) - z(x_4) = y(x_3) - y(x_4),$$

for some $a < x_1 < x_2 < x_3 < x_4 < b$. By uniqueness of solutions of 2-point conjugate boundary value problems for (3.3.6), $z''(x_1) \neq y''(x_1)$ and $z'(x_2) \neq y'(x_2)$.

We assume, with no loss of generality, that $y(x) > z(x)$ on $(a, x_1) \cup (x_1, x_2)$. Then $y(x) < z(x)$ on (x_2, b). Fix $a < \tau < x_1$. By continuous dependence of solutions on 5-point nonlocal boundary conditions (3.3.7), for $\epsilon > 0$ sufficiently small, there exists a $\delta > 0$ and a solution $z_\delta(x)$ of (3.3.6) satisfying

$$z_\delta(\tau) = z(\tau),$$
$$z_\delta(x_1) = z(x_1) + \delta,$$
$$z_\delta(x_2) = z(x_2) = y(x_2),$$
$$z_\delta(x_3) - z_\delta(x_4) = z(x_3) - z(x_4)$$
$$= y(x_3) - y(x_4)$$

and $|z_\delta^{(i-1)}(x) - z^{(i-1)}(x)| < \epsilon, i = 1, 2, 3, 4$, on $[\tau, x_4]$. For $\epsilon > 0$ small, there exists $\tau < \sigma_1 < x_1 < \sigma_2 < x_2$ so that

$$z_\delta(\sigma_1) = y(\sigma_1),$$
$$z_\delta(\sigma_2) = y(\sigma_2),$$
$$z_\delta(x_2) = y(x_2),$$
$$z_\delta(x_3) - z_\delta(x_4) = y(x_3) - y(x_4).$$

By the uniqueness of solutions of (3.3.6)-(3.3.7), $z_\delta(x) = y(x)$ on (a,b). However, $z_\delta(x_1) = z(x_1) + \delta = y(x_1) + \delta > y(x_1)$, which is a contradiction.
    So solutions of (3.3.6)-(3.3.9) are unique, when they exist.   □

Theorem 3.3.7. [[66], Henderson and Ma, Thm. 3.5] *Assume that with respect to (3.3.6), conditions (A) and (B) are satisfied. If solutions for (3.3.6)-(3.3.7) are unique, when they exist, then solutions for (3.3.6)-(3.3.11) are unique, when they exist.*

Proof. Again for the purpose of contradiction, we suppose (3.3.6)-(3.3.11) has two solutions, $y(x)$ and $z(x)$, such that

$$z(x_1) = y(x_1),$$
$$z(x_2) = y(x_2),$$
$$z'(x_2) = y'(x_2),$$
$$z(x_3) - z(x_4) = y(x_3) - y(x_4),$$

for some $a < x_1 < x_2 < x_3 < x_4 < b$. By uniqueness of solutions of 2-point conjugate boundary value problems for (3.3.6), $z'(x_1) \neq y'(x_1)$ and $z''(x_2) \neq y''(x_2)$.
    Without loss of generality, we assume $y(x) > z(x)$ on $(x_1, b) \backslash \{x_2\}$. Then $y(x) < z(x)$ on (a, x_1). Fix $x_1 < \tau < x_2$. By continuous dependence of solutions of (3.3.6)-(3.3.7) on boundary conditions, for $\epsilon > 0$ sufficiently small, there exists a $\delta > 0$ and a solution $z_\delta(x)$ of (3.3.6) satisfying

$$z_\delta(x_1) = z(x_1) = y(x_1),$$
$$z_\delta(\tau) = z(\tau),$$
$$z_\delta(x_2) = z(x_2) + \delta,$$
$$z_\delta(x_3) - z_\delta(x_4) = z(x_3) - z(x_4)$$
$$= y(x_3) - y(x_4),$$

and $|z_\delta^{(i-1)}(x) - z^{(i-1)}(x)| < \epsilon, i = 1, 2, 3, 4$, on $[\tau, x_4]$. For $\epsilon > 0$ small, there exists $x_1 < \sigma_1 < x_2 < \sigma_2 < x_4$ so that

$$z_\delta(x_1) = y(x_1),$$
$$z_\delta(\sigma_1) = y(\sigma_1),$$
$$z_\delta(\sigma_2) = y(\sigma_2),$$
$$z_\delta(x_3) - z_\delta(x_4) = y(x_3) - y(x_4).$$

But by the uniqueness of solutions of (3.3.6)-(3.3.7), $z_\delta(x) = y(x)$ on (a, b). However, $z_\delta(x_2) = z(x_2) + \delta = y(x_2) + \delta > y(x_2)$, which is a contradiction. So solutions of (3.3.6)-(3.3.11) are unique, when they exist. □

Theorem 3.3.8. [[66], Henderson and Ma, Thm. 3.6] *Assume that with respect to* (3.3.6), *conditions* (A) *and* (B) *are satisfied. If solutions for* (3.3.6)-(3.3.7) *are unique, when they exist, then solutions for* (3.3.6)-(3.3.13) *are unique, when they exist.*

Proof. We suppose (3.3.6)-(3.3.13) has two solutions, $y(x)$ and $z(x)$, satisfying

$$y(x_1) = z(x_1),$$
$$y'(x_1) = z'(x_1),$$
$$y''(x_1) = z''(x_1),$$
$$y(x_2) - y(x_3) = z(x_2) - z(x_3),$$

for some $a < x_1 < x_2 < x_3 < b$. Now $y'''(x_1) \neq z'''(x_1)$, and we may assume $y'''(x_1) > z'''(x_1)$.

By Theorem 3.3.6, solutions of (3.3.6)-(3.3.9) depend continuously on their boundary conditions. Fix $x_1 < \rho < x_2$. For $\epsilon > 0$ small, there is a $\delta > 0$ and a solution $z_\delta(x)$ satisfying

$$z_\delta(x_1) = z(x_1) = y(x_1),$$
$$z_\delta'(x_1) = z'(x_1) + \delta,$$
$$z_\delta(\rho) = z(\rho),$$
$$z_\delta(x_2) - z_\delta(x_3) = z(x_2) - z(x_3)$$
$$= y(x_2) - y(x_3)$$

and $|y^{(i-1)}(x) - z^{(i-1)}(x)| < \epsilon$, $i = 1, 2, 3, 4$, on $[x_1, x_3]$. For $\epsilon > 0$ sufficiently small, there exist points $a < \tau_1 < x_1 < \tau_2 < \rho$, which are in a neighborhood of x_1, such that $y(x)$ and $z_\delta(x)$ satisfy,

$$z_\delta(\tau_1) = y(\tau_1),$$
$$z_\delta(x_1) = y(x_1),$$
$$z_\delta(\tau_2) = y(\tau_2),$$
$$z_\delta(x_2) - z_\delta(x_3) = y(x_2) - y(x_3).$$

So we have $z_\delta(x) = y(x)$ on (a, b) by the uniqueness of solutions of (3.3.6)-(3.3.7). But

$$z'_\delta(x_1) = z'(x_1) + \delta = y'(x_1) + \delta > y'(x_1).$$

This is a contradiction. We conclude (3.3.6)-(3.3.13) has at most one solution.   □

We now provide a type of converse to the combined results obtained in Theorems 3.3.6-3.3.8. We remark that, if solutions of (3.3.6) are unique, when they exist, for each of (3.3.9), (3.3.10), (3.3.11), (3.3.12), (3.3.13) and (3.3.14), then solutions of both 2-point and 3-point conjugate boundary value problems for (3.3.6) are unique. It then follows by the Vidossich continuous dependence result [105] that the "compactness condition" (CP) holds. When, this is coupled with Theorem 1.3.4 and Theorem 2.3.4, it follows that each k-point conjugate boundary value problem for (3.3.6) has a solution, which is unique, for $k = 2, 3, 4$.

We now offer the converse.

Theorem 3.3.9. [[66], Henderson and Ma, Thm. 4.4] *Assume that with respect to (3.3.6), conditions (A) and (B) are satisfied. If solutions of (3.3.6) are unique, when they exist, for each of (3.3.9), (3.3.10), (3.3.11), (3.3.12), (3.3.13) and (3.3.14), then solutions of both (3.3.6)-(3.3.7) and (3.3.6)-(3.3.8) are unique, when they exist.*

Proof. We will provide the proof dealing with (3.3.6)-(3.3.7). As has been the pattern in this development, we assume the conclusion to be false. Then, there exist distinct solutions, $y(x)$ and $z(x)$, and points $a < x_1 < x_2 < x_3 < x_4 < x_5 < b$ such that,

$$y(x_i) = z(x_i), \; i = 1, 2, 3, \text{ and } y(x_4) - y(x_5) = z(x_4) - z(x_5).$$

From the remarks preceding the statement of this theorem, unique solutions exist for 3-point and 4-point conjugate boundary value problems for (3.3.6), and so for each $n \in \mathbb{N}$, let $y_n(x)$ be the solution of the boundary value problem for (3.3.6) satisfying the 3-point conjugate boundary conditions,

$$y_n(x_3) = y(x_3) = z(x_3),$$
$$y_n'(x_3) = y'(x_3) - n,$$
$$y_n(x_4) = y(x_4),$$
$$y_n(x_5) = y(x_5).$$

By uniqueness of solutions of 4-point conjugate problems, for each $n \in \mathbb{N}$,

$$y(x) < y_n(x) < y_{n+1}(x) \text{ on } (a, x_3).$$

Next, for each $n \in \mathbb{N}$, let

$$E_n = \{x : x_1 \le x \le x_2 \mid y_n(x) \le z(x)\}.$$

Because of continuous dependence of solutions of 3-point conjugate boundary value problems and because of the uniqueness of solutions of (3.3.6)-(3.3.9), it follows as in Theorem 3.3.4, that each $E_n \ne \emptyset$. From the compactness of each set and the nesting, $E_{n+1} \subset E_n \subset (x_1, x_2)$, for each $n \in \mathbb{N}$, we have

$$\bigcap_{n=1}^{\infty} E_n := E \ne \emptyset.$$

We claim that the set E consists of a single point x_0 with $x_1 < x_0 < x_2$. For, if $t_1, t_2 \in E$, with $x_1 < t_1 < t_2 < x_2$, then the same type of argument that we used to show the previously defined sets E_n are nonempty leads to the conclusion that the interval $[t_1, t_2]$ must be contained in E. However, $[t_1, t_2] \subset E$ implies that the sequence $\{y_n(x)\}$ is uniformly bounded on $[t_1, t_2]$. But this contradicts the part of "compactness condition" (CP) requiring boundedness for some subsequence of $\{y_{n_j}'(x_3)\}$. We conclude

$$E = \{x_0\},$$

with $x_1 < x_0 < x_2$, and

$$\lim_{n \to \infty} y_n(x_0) \le z(x_0).$$

Next, let $y_0(x)$ denote the solution of the 4-point conjugate boundary value problem for (3.3.6) satisfying

$$y_0(x_0) = \lim_{n \to \infty} y_n(x_0),$$
$$y_0(x_3) = y(x_3) = z(x_3),$$
$$y_0(x_4) = y(x_4),$$
$$y_0(x_5) = y(x_5).$$

By continuous dependence of solutions on 4-point conjugate boundary conditions, there is a subsequence, $\{y_{n_j}(x)\}$, such that $\{y_{n_j}^{(i)}(x)\}$ converges to $y_0^{(i)}(x)$, $i = 0, 1, 2, 3$, on each compact subinterval of (a, b). But this is impossible, since $\{y_{n_j}'(x_3)\}$ diverges.

We conclude that solutions of (3.3.6)-(3.3.7) are unique, when solutions exist. Of course, completely symmetric arguments yield that solutions of (3.3.6)-(3.3.8) are also unique, when solutions exist. $\square$

As a final statement, we present a theorem summarizing the results of Theorems 3.3.6-3.3.9.

Theorem 3.3.10. *Assume that with respect to* (3.3.6), *conditions* (A) *and* (B) *are satisfied. Then, solutions of* (3.3.6) *are unique, when they exist, for each of* (3.3.9), (3.3.10), (3.3.11), (3.3.12), (3.3.13) *and* (3.3.14) *if, and only if, solutions of both* (3.3.6)-(3.3.7) *and* (3.3.6)-(3.3.8) *are unique, when they exist.*

For the remainder of this section, our "uniqueness implies uniqueness" results will be for solutions of nonlocal problems for (1.1.1) of arbitrary order n. In 2007, Eloe and Henderson [26] were concerned with such questions for solutions of (1.1.1) satisfying $(k + 2)$-point nonlocal boundary conditions,

$$y^{(i-1)}(x_j) = y_{ij}, \ 1 \leq i \leq m_j, \ 1 \leq j \leq k,$$
$$a_1 y(x_{k+1}) - a_2 y(x_{k+2}) = y_n, \tag{3.3.15}$$

where $1 \leq k \leq n - 1$, $m_1, \ldots, m_k \in \mathbb{N}$ such that $m_1 + \cdots + m_k = n - 1$, $a < x_1 < \cdots < x_k < x_{k+1} < x_{k+2} < b$, $y_{ij} \in \mathbb{R}, 1 \leq i \leq m_j, 1 \leq j \leq k$, a_1, a_2 are positive real values, and $y_n \in \mathbb{R}$.

We will show that uniqueness of solutions for (1.1.1)-(3.3.15), when $k = n - 1$, implies uniqueness of solutions for (1.1.1)-(3.3.15), when $1 \leq k \leq n - 2$. This will involve first establishing via a theorem and its corollary that uniqueness of solutions of (1.1.1)-(3.3.15), when $k = n - 1$, yields existence of unique solutions of ℓ-point *conjugate boundary value problems*, (1.1.1)-(1.1.2), for $2 \leq \ell \leq n$.

Theorem 3.3.11. [[26], Eloe and Henderson, Thm. 2.1] *Assume that with respect* (1.1.1), *conditions* (A) *and* (B) *are satisfied. If solutions of* (1.1.1)-(3.3.15), *when* $k = n - 1$, *are unique, when they exist, then solutions of* *n-point conjugate boundary value problems* (1.1.1)-(1.1.2) *(that is, when $\ell = n$), are unique.*

Proof. Our approach involves contradiction. That is, we assume there exist distinct solutions $y(x)$ and $z(x)$ of (1.1.1) such that, for some points $a < t_1 < \cdots < t_n < b$,

$$y(t_i) = z(t_i), \quad 1 \le i \le n.$$

Define $w := y - z$. Then $|w(x)|$ has a positive local maximum in (t_{n-1}, t_n).

First, we assume $a_1 = a_2$. There exist points $t_{n-1} < s_1 < s_2 < t_n$ such that

$$w(t_i) = 0, \quad 1 \le i \le n - 1,$$

and

$$w(s_1) = w(s_2).$$

Then, for $a < t_1 < \cdots t_{n-1} < s_1 < s_2 < b$, we have

$$y(t_i) = z(t_i), \quad 1 \le i \le n - 1,$$

and

$$y(s_1) - y(s_2) = z(s_1) - z(s_2).$$

This contradicts the uniqueness of solutions of (1.1.1)-(3.3.15), when $k = n - 1$ and $a_1 = a_2$.

Next, we assume $a_1 > a_2$. Let $s_2 \in (t_{n-1}, t_n)$ be such that $|w|$ attains a local maximum value. Assume, without loss of generality, that $w(x) > 0$ on (t_{n-1}, t_n). Define

$$v(s) = w(s) - \frac{a_2}{a_1} w(s_2).$$

Note that $v(t_{n-1}) < 0$ and $v(s_2) > 0$. Hence, there exists $s_1 \in (t_{n-1}, s_2)$ such that

$$a_1 w(s_1) - a_2 w(s_2) = 0$$

or

$$a_1 y(s_1) - a_2 y(s_2) = a_1 z(s_1) - a_2 z(s_2).$$

And again, we contradict the uniqueness of solutions of (1.1.1)-(3.3.15), when $k = n - 1$.

The case $a_1 < a_2$ is similar. Let s_1 denote a value in (t_{n-1}, t_n) such that $|w|$ attains a local maximum value and define

$$v(s) = \frac{a_1}{a_2} w(s_1) - w(s).$$

Note that $v(s_1) < 0$ and $v(t_n) > 0$. $\qquad \qquad \square$

From Theorem 2.3.4, we have an immediate corollary concerning existence of solutions for ℓ-point conjugate boundary value problems for (1.1.1).

Corollary 3.3.1. [[26], Eloe and Henderson, Cor. 2.2] *Assume that with respect (1.1.1), conditions* (A) *and* (B) *are satisfied. If solutions of* (1.1.1)-(3.3.15), *when* $k = n-1$, *are unique, when they exist, then, given* $2 \leq \ell \leq n$, *each ℓ-point conjugate boundary value problem* (1.1.1)-(1.1.2) *has a unique solution on* (a, b).

We now prove that uniqueness of solutions of (1.1.1)-(3.3.15), when $k = n-1$, implies uniqueness of solutions of (1.1.1)-(3.3.15), when $1 \leq k \leq n-2$.

Theorem 3.3.12. [[26], Eloe and Henderson, Thm. 2.4] *Assume that with respect* (1.1.1), *conditions* (A) *and* (B) *are satisfied, and assume that for* $k = n - 1$, *solutions of* (1.1.1)-(3.3.15) *are unique, when they exist. Then, for each* $1 \leq k \leq n - 2$, *solutions of* (1.1.1)-(3.3.15) *are unique, when they exist.*

Proof. For contradiction purposes, we assume that, for some $1 \leq k \leq n-2$, some boundary value problem (1.1.1)-(3.3.15) has distinct solutions. Let

$$h = \max\{k = 1, \ldots, n - 2 \mid (1.1.1)\text{-}(3.3.15) \text{ has distinct solutions}\}.$$

So, there exist positive integers, $m_1, \ldots, m_h$, such that $m_1 + \cdots + m_h = n-1$, and points $a < x_1 < \cdots < x_h < x_{h+1} < x_{h+2} < b$, for which there are distinct solutions $y(x)$ and $z(x)$ of (1.1.1)-(3.3.15), associated with these $m_1, \ldots, m_h$; that is,

$$y^{(i-1)}(x_j) = z^{(i-1)}(x_j), \ 1 \leq i \leq m_j, \ 1 \leq j \leq h,$$
$$a_1 y(x_{h+1}) - a_2 y(x_{h+2}) = a_1 z(x_{h+1}) - a_2 z(x_{h+2}).$$

Corollary 3.3.1 implies that, for each $1 \leq j \leq h$, x_j is a zero of $y - z$ of exact multiplicity m_j, since y and z are distinct solutions of (1.1.1), (except in the case $h = 1$, for which uniqueness of solutions of initial value problems for (1.1.1) would imply x_1 is a zero of $y - z$ of exact multiplicity $n - 1$).
    Now, let

$$m_{j_0} = \max\{m_j \mid 1 \leq j \leq h\}.$$

Then $m_{j_0} \geq 2$, and we may assume, with no loss of generality, that

$$y^{(m_{j_0})}(x_{j_0}) > z^{(m_{j_0})}(x_{j_0}).$$

Next, we fix $a < \tau < x_1$. By the maximality of h, it follows from continuous dependence on boundary conditions for solutions of (1.1.1)-(3.3.15), when

$k = h + 1$ that, for each $\epsilon > 0$, there is a $\delta > 0$ and there is a solution $z_\delta(x)$ of (1.1.1)-(3.3.15) satisfying at the points $\tau, x_1, \ldots, x_{h+2}$,

$$z_\delta(\tau) = z(\tau),$$
$$z_\delta^{(i-1)}(x_j) = z^{(i-1)}(x_j), \quad 1 \leq i \leq m_j, \quad 1 \leq j \leq h, \quad j \neq j_0,$$
$$z_\delta^{(i-1)}(x_{j_0}) = z^{(i-1)}(x_{j_0}), \quad 1 \leq i \leq m_{j_0} - 2, \quad (\text{if } m_{j_0} > 2),$$
$$z_\delta^{(m_{j_0}-2)}(x_{j_0}) = z^{(m_{j_0}-2)}(x_{j_0}) + \delta,$$
$$a_1 z_\delta(x_{h+1}) - a_2 z_\delta(x_{h+2}) = a_1 z(x_{h+1}) - a_2 z(x_{h+2}),$$

and $|z_\delta(x) - z(x)| < \epsilon$ on $[\tau, x_{h+2}]$.

For sufficiently small $\epsilon > 0$, there exist points $x_{j_0-1} < \rho_1 < x_{j_0} < \rho_2 < x_{j_0+1}$ such that

$$z_\delta^{(i-1)}(x_j) = y^{(i-1)}(x_j), \quad 1 \leq i \leq m_j, \quad 1 \leq j \leq j_0 - 1,$$
$$z_\delta(\rho_1) = y(\rho_1),$$
$$z_\delta^{(i-1)}(x_{j_0}) = y^{(i-1)}(x_{j_0}), \quad 1 \leq i \leq m_{j_0} - 2, \quad (\text{if } m_{j_0} > 2),$$
$$z_\delta(\rho_2) = y(\rho_2),$$
$$z_\delta^{(i-1)}(x_j) = y^{(i-1)}(x_j), \quad 1 \leq i \leq m_j, \quad j_0 + 1 \leq j \leq h,$$

and

$$a_1 z_\delta(x_{h+1}) - a_2 z_\delta(x_{h+2}) = a_1 y(x_{h+1}) - a_2 y(x_{h+2}).$$

By the maximality of h, $z_\delta(x) \equiv y(x)$. This is a contradiction, and the proof is complete. $\qquad\square$

Our next couple of "uniqueness implies uniqueness" results are extensions of Theorem 3.3.12 due to Eloe, Henderson and Khan [29], which involve more general boundary conditions; in 2011, they studied solutions of boundary value problems for (1.1.1) subject to $n - j$ conjugate boundary conditions followed by j nonlocal boundary conditions.

In particular, given $1 \leq j \leq n - 1$, $1 \leq k \leq n - j$, positive integers $m_1, \ldots, m_k$ such that $m_1 + \cdots + m_k = n - j$, points $a < x_1 < x_2 < \cdots < x_k < x_{k+1} < \cdots < x_{k+2j} < b$, real values $y_{il}, 1 \leq i \leq m_l, 1 \leq l \leq k$, and real values $y_n, y_{n-1}, \ldots, y_{n-(j-1)}$, they were concerned first with uniqueness implies uniqueness questions for solutions of (1.1.1) satisfying the conjugate and nonlocal boundary conditions of the type,

$$y^{(i-1)}(x_l) = y_{il}, \ 1 \leq i \leq m_l, \ 1 \leq l \leq k, \text{ conjugate conditions,}$$
$$(a_1 y(x_{k+1}) - a_2 y(x_{k+2}), \ldots, a_{2j-1} y(x_{k+2j-1}) - a_{2j} y(x_{k+2j})) \qquad (3.3.16)$$
$$= (y_n, y_{n-1}, \ldots, y_{n-(j-1)}), \text{ nonlocal conditions,}$$

where $a_1, a_2, \ldots, a_{2j}$ are positive real numbers. We shall refer to the boundary conditions, (3.3.16), as $(k; j)$-*point boundary conditions*. We note that $(k; 0)$-point boundary conditions are k-*point conjugate* boundary conditions.

Ultimately, we will show that for fixed $j_0 \in \{1, \ldots, n-1\}$, uniqueness of solutions for the $(n - j_0; j_0)$-point boundary value problem implies uniqueness of solutions for the $(k; j)$-point boundary value problem, for $0 \le j \le j_0$, $1 \le k \le n - j$.

Theorem 3.3.13. [[29], Eloe, Khan and Henderson, Thm. 2.2] *Let $j \in \{1, \ldots, n-1\}$. Assume that with respect (1.1.1), conditions* (A) *and* (B) *are satisfied, and assume that for $k = n - j$, solutions of the $(n - j; j)$-point boundary value problem* (1.1.1)-(3.3.16) *are unique, when they exist. Then, for each $i = 1, 2, \ldots, j$, solutions of the $(n - j + i; j - i)$-point boundary value problem* (1.1.1)-(3.3.16) *are unique, when they exist.*

Proof. We assume solutions of the $(n - j; j)$-point boundary value problem (1.1.1)-(3.3.16) are unique, when they exist. The proof involves induction on i, and we begin by showing that solutions of the $(n - j + 1; j - 1)$-point boundary value problem are unique. We do this by contradiction. That is, we assume there exist points $a < x_1 < \cdots < x_{n-j+1} < \cdots < x_{n+j-1} < b$ for which there are distinct solutions $y(x)$ and $z(x)$ of the $(n-j+1; j-1)$-point boundary value problem such that

$$y(x_l) = z(x_l), \ 1 \le l \le n - j + 1,$$
$$a_1 y(x_{n-j+1+1}) - a_2 y(x_{n-j+1+2}) = a_1 z(x_{n-j+1+1}) - a_2 z(x_{n-j+1+2}),$$
$$\cdots,$$
$$a_{2j-3} y(x_{n-j+1+2j-3}) - a_{2j-2} y(x_{n-j+1+2j-2})$$
$$= a_{2j-3} z(x_{n-j+1+2j-3}) - a_{2j-2} z(x_{n-j+1+2j-2}).$$

We set $w = y - z$, and then

$$w(x_l) = 0, \ 1 \le l \le n - j + 1,$$
$$a_1 w(x_{n-j+1+1}) - a_2 w(x_{n-j+1+2}) = 0,$$
$$\cdots,$$
$$a_{2j-3} w(x_{n-j+1+2j-3}) - a_{2j-2} y(x_{n-j+1+2j-2}) = 0.$$

If for some $t_1 \in (x_{n-j+1}, x_{n-j+2})$, $w(t_1) = 0$, then we have

$$a w(x_{n-j+1}) - b w(t_1) = 0, \ a, b \in \mathbb{R}.$$

This implies that $y(x)$ and $z(x)$ are distinct solutions of the $(n - j; j)$-point boundary value problem at the points $x_1, \ldots, x_{n-j}, x_{n-j+1}, t_1, x_{n-j+2}, x_{n-j+2}, \ldots, x_{n+j-2}, x_{n+j-1}$, which is a contradiction.

Hence, $w(x) \neq 0$ on (x_{n-j+1}, x_{n-j+2}). Let $w(x) > 0$ on (x_{n-j+1}, x_{n-j+2}). The case $w(x) < 0$ on (x_{n-j+1}, x_{n-j+2}) can be dealt with similarly. Then, there exists $\tau \in [x_{n-j+1}, \frac{x_{n-j+1}+x_{n-j+2}}{2}]$ such that

$$\max\{w(x) : x \in [x_{n-j+1}, \frac{x_{n-j+1} + x_{n-j+2}}{2}]\} = w(\tau) > 0.$$

Define

$$v(x) = \begin{cases} aw(x) - bw(\tau), & \text{if } a > b, \\ bw(x) - aw(\tau), & \text{if } a \leq b. \end{cases}$$

Then, $v(\tau) \geq 0$ and $v(x_{n-j+1}) < 0$. By the Intermediate Value Theorem, there exists $t' \in (x_{n-j+1}, \tau)$ such that $v(t') = 0$ which implies that $aw(t') - bw(\tau) = 0$. Hence, there are distinct solutions of the $(n-j; j)$-point boundary value problem at the points

$$x_1, \ldots, x_{n-j}, t', \tau, x_{n-j+2}, x_{n-j+2}, \ldots, x_{n+j-2}, x_{n+j-1},$$

which is again a contradiction. Hence solutions of the $(n-j+1; j-1)$-point boundary value problem (1.1.1), (3.3.16) are unique.

Now, the remainder of the proof of the theorem follows by induction. $\square$

Corollary 3.3.2. [[29], Eloe, Khan and Henderson, Cor. 2.3] *Assume that with respect (1.1.1), conditions (A) and (B) are satisfied, and assume that for $k = n - j$, solutions of the $(n - j; j)$-point boundary value problem (1.1.1)-(3.3.16) are unique, when they exist. Then, solutions of the $(n; 0)$-point boundary value problem (i.e., the n-point conjugate boundary value problem), are unique, when they exist.*

Then, in view of Theorem 2.3.4 concerning uniqueness of solutions of n-point conjugate boundary value problems for (1.1.1) implying the existence solutions of all ℓ-point conjugate boundary value problems for (1.1.1), $2 \leq \ell \leq n$, we can state a corollary concerning existence of solutions of $(\ell; 0)$-point boundary value problems.

Corollary 3.3.3. [[29], Eloe, Khan and Henderson, Cor. 2.4] *Assume that with respect (1.1.1), conditions (A) and (B) are satisfied, and assume that for $k = n-j$, solutions of the $(n-j; j)$-point boundary value problem (1.1.1)-(3.3.16) are unique, when they exist. Then, for each $2 \leq \ell \leq n$, solutions of the $(\ell; 0)$-point boundary value problem (i.e., the ℓ-point conjugate boundary value problem), exist and are unique.*

Now, we show that uniqueness of solutions of the $(n - j; j)$-point boundary value problem implies uniqueness of solutions of the $(k; j)$-point boundary value problem, for each $1 \leq k \leq n - j - 1$.

Theorem 3.3.14. [[29], Eloe, Khan and Henderson, Thm. 2.5] *Assume that for $k = n - j$, solutions of the $(n - j; j)$-point boundary value problem are unique, when they exist. Then, for each $1 \leq k \leq n - j - 1$, solutions of the $(k; j)$-point boundary value problem are unique, when they exist.*

Proof. Assume that solutions of the $(n - j; j)$-point boundary value problem are unique. And also assume for contradiction that, for some $1 \leq k \leq n - j - 1$, some $(k; j)$-point boundary value problem has distinct solutions.
Let

$$h = \max\{k = 1, \ldots, n - j - 1 \,|\, (k; j)\text{-point problem has distinct solutions}\}.$$

Then, there are positive integers, $m_1, \ldots, m_h$, such that $m_1 + \cdots + m_h = n - j$, points $a < x_1 < \cdots < x_h < \cdots < x_{h+2j} < b$, and positive reals, $a_1, \ldots, a_{2j}$, for which there exist distinct solutions $y(x)$ and $z(x)$ of the associated $(h; j)$-point boundary value problem (1.1.1)-(3.3.16); in particular,

$$y^{(i-1)}(x_l) = z^{(i-1)}(x_l), \ 1 \leq i \leq m_l, \ 1 \leq l \leq h,$$
$$a_1 y(x_{h+1}) - a_2 y(x_{h+2}) = a_1 z(x_{h+1}) - a_2 z(x_{h+2}), \ldots,$$
$$a_{2j-1} y(x_{h+2j-1}) - a_{2j} y(x_{h+2j}) = a_{2j-1} z(x_{h+2j-1}) - a_{2j} z(x_{h+2j}).$$

Since $h \leq n - j - 1$, then some $m_l \geq 2$. Let

$$m_{l_0} = \max\{m_l \,|\, 1 \leq l \leq h\} \geq 2.$$

(At this point, we need to argue that each x_l is a zero of $y - z$ of exact multiplicity $m_l, 1 \leq l \leq h$. This argument is done by induction on i, and in fact, the proof of this theorem is actually completed by induction on j.
For $j = 1$, if any of the next higher order derivatives vanish at x_l, then y and z are distinct solutions of an $(h; 0)$-point boundary value problem (that is, an h-point conjugate boundary value problem). Complete the proof of this theorem below for $j = 1$. Now for $j > 1$, if any of the next higher order derivatives vanish at x_l, then y and z are distinct solutions of an $(h; j - 1)$-point boundary value problem.)
So, we complete this proof by assuming that each x_l, is a zero of $y - z$ of exact multiplicity $m_l, 1 \leq l \leq h$.
Thus, we assume, with no loss of generality, that

$$y^{(m_{l_0})}(x_{l_0}) > z^{(m_{l_0})}(x_{l_0}).$$

Now fix $a < \tau < x_1$. By the maximality of h, solutions of the $(h+1;j)$-point boundary value problems (1.1.1)-(3.3.16) at the points $\tau, x_1, \ldots, x_h, \ldots x_{h+2j}$ are unique. Hence, it follows from continuous dependence on boundary conditions that, for each $\epsilon > 0$, there is a $\delta > 0$ and there is a solution $z_\delta(x)$ of the $(h+1;j)$-point boundary value problem (1.1.1)-(3.3.16), satisfying the conditions,

$$z_\delta(\tau) = z(\tau),$$
$$z_\delta^{(i-1)}(x_l) = z^{(i-1)}(x_l) = y^{(i-1)}(x_l), \quad 1 \le i \le m_l, \quad 1 \le l \le h, \quad l \ne l_0,$$
$$z_\delta^{(i-1)}(x_{l_0}) = z^{(i-1)}(x_{l_0}) = y^{(i-1)}(x_{l_0}), \quad 1 \le i \le m_{l_0} - 2, \quad (\text{if } m_{l_0} > 2),$$
$$z_\delta^{(m_{l_0}-2)}(x_{l_0}) = z^{(m_{l_0}-2)}(x_{l_0}) + \delta = y^{(m_{l_0}-2)}(x_{l_0}) + \delta,$$
$$a_1 z_\delta(x_{h+1}) - a_2 z_\delta(x_{h+2}) = a_1 z(x_{h+1}) - a_2 z(x_{h+2})$$
$$= a_1 y(x_{h+1}) - a_2 y(x_{h+2}),$$
$$\cdots,$$
$$a_{2j-1} z_\delta(x_{h+2j-1}) - a_{2j} z_\delta(x_{h+2j}) = a_{2j-1} z(x_{h+2j-1}) - a_{2j} z(x_{h+2j})$$
$$= a_{2j-1} y(x_{h+2j-1}) - a_{2j} y(x_{h+2j}),$$

and $|z_\delta(x) - z(x)| < \epsilon$ on $[\tau, x_{h+2j}]$. For sufficiently small $\epsilon > 0$, there exist points $x_{l_0-1} < \rho_1 < x_{l_0} < \rho_2 < x_{l_0+1}$ such that

$$z_\delta^{(i-1)}(x_l) = y^{(i-1)}(x_l), \quad 1 \le i \le m_l, \quad 1 \le l \le l_0 - 1,$$
$$z_\delta(\rho_1) = y(\rho_1),$$
$$z_\delta^{(i-1)}(x_{l_0}) = y^{(i-1)}(x_{l_0}), \quad 1 \le i \le m_{l_0} - 2, \quad (\text{if } m_{l_0} > 2),$$
$$z_\delta(\rho_2) = y(\rho_2),$$
$$z_\delta^{(i-1)}(x_l) = y^{(i-1)}(x_l), \quad 1 \le i \le m_l, \quad l_0 + 1 \le l \le h,$$
$$a_1 z_\delta(x_{h+1}) - a_2 z_\delta(x_{h+2}) = a_1 y(x_{h+1}) - a_2 y(x_{h+2}),$$
$$\cdots,$$
$$a_{2j-1} z_\delta(x_{h+2j-1}) - a_{2j} z_\delta(x_{h+2j}) = a_{2j-1} y(x_{h+2j-1}) - a_{2j} y(x_{h+2j}).$$

If $m_{l_0} > 2$, $z_\delta(x)$ and $y(x)$ are distinct solutions of the $(h+2;j)$-point boundary value problem at the points $x_1, \ldots, x_{l_0-1}, \rho_1, x_{l_0}, \rho_2, x_{l_0+1}, \ldots, x_h, \ldots, x_{h+2j}$, which is a contradiction because of the maximality of h. If $m_{l_0} = 2$, $z_\delta(x)$ and $y(x)$ are distinct solutions of the $(h+1;j)$-point boundary value problem at the points

$$x_1, \ldots, x_{l_0-1}, \rho_1, \rho_2, x_{l_0+1}, \ldots, x_h, \ldots, x_{h+2j},$$

which is again a contradiction. The proof is complete. $\square$

From Theorem 3.3.13 and Theorem 3.3.14, we have as an immediate corollary to the main uniqueness result for $(k;j)$-point boundary value problems.

Corollary 3.3.4. [[29], Eloe, Khan and Henderson, Cor. 2.6] *Let* $j_0 \in \{1, \ldots, n-1\}$. *Assume that with respect* (1.1.1), *conditions* (A) *and* (B) *are satisfied, and assume that solutions of* (1.1.1)-(3.3.16), *for* $k = n - j_0$, $j = j_0$, *are unique, when they exist. Then, for each* $1 \leq j \leq j_0$, $1 \leq k \leq n - j$, *solutions of the* $(k; j)$-*point boundary value problem are unique, when they exist.*

The next several "uniqueness implies uniqueness" results, due to Eloe, Henderson and Khan [30], extend the preceding result, Corollary 3.3.4, to solutions of (1.1.1) subject to j nonlocal boundary conditions, followed by $n - 2j$ conjugate boundary conditions, followed by an additional j nonlocal boundary conditions.

More precisely, for $n \geq 3$, and $j \geq 1$, given $1 \leq k \leq n - 2j$, positive integers $m_1, \ldots, m_k$ such that $m_1 + \cdots + m_k = n - 2j$, points $a < t_1 < \cdots < t_{2j} < x_1 < x_2 < \cdots < x_k < s_1 < \cdots < s_{2j} < b$, real values $y_i, 1 \leq i \leq j$, $y_{il}, 1 \leq i \leq m_l, 1 \leq l \leq k$, and real values $y_{n-(i-1)}, 1 \leq i \leq j$, they were concerned with uniqueness implies uniqueness questions for solutions of (1.1.1) satisfying the nonlocal and conjugate and nonlocal boundary conditions of the type,

$$a_i y(t_{2i-1}) - b_i y(t_{2i}) = y_i, \ 1 \leq i \leq j, \ j \text{ nonlocal conditions,}$$

$$y^{(i-1)}(x_l) = y_{il}, 1 \leq i \leq m_l, 1 \leq l \leq k,$$

$$\text{(3.3.17)}$$

$$k\text{-point}, n - 2 \text{ conjugate conditions,}$$

$$c_i y(s_{2i-l}) - d_i y(s_{2i}) = y_{n-(i-1)}, \ 1 \leq i \leq j, \ j \text{ nonlocal conditions,}$$

where a_i, b_i, c_i, d_i, $1 \leq i \leq j$, are positive real numbers. We shall refer to the boundary conditions, (3.3.17), as $(j; k; j)$-*point boundary conditions*. We note that $(0; k; 0)$-point boundary conditions are k-*point conjugate* boundary conditions.

We remark that not only do the boundary conditions (3.3.17) extend those given in (3.3.16), but they are related also to those studied by Gray [35], which we denoted by (3.1.6).

A goal here is to show that for fixed $j \geq 1$ and $k = n - 2j$, uniqueness of solutions for the $(j; n-2j; j)$-point boundary value problem implies uniqueness of solutions for the $(j; k; j)$-point boundary value problem, for $1 \leq k \leq n - 2j - 1$. We now establish that for $k = n - 2j$, uniqueness of solutions of the $(j; n-2j; j)$-point boundary value problem (1.1.1)-(3.3.17), implies uniqueness of solutions of the $(j - i; n - 2j + i; j)$-point boundary value problem (1.1.1)-(3.3.17), for $i = 1, 2, \ldots, j$.

Theorem 3.3.15. [[30], Eloe, Khan and Henderson, Thm. 2.2] *Assume that with respect to (1.1.1), conditions (A) and (B) are satisfied. Let $j \geq 1$. Assume that for $k = n - 2j$, solutions of the $(j; n - 2j; j)$-point boundary value problem (1.1.1)-(3.3.17) are unique, when they exist. Then, for each $i = 1, 2, \ldots, j$, solutions of the $(j - i; n - 2j + i; j)$-point boundary value problem (1.1.1)-(3.3.17) are unique, when they exist.*

Proof. Assume uniqueness of solutions of the $(j; n - 2j; j)$-point problem (1.1.1)-(3.3.17). We first show that solutions of the $(j-1; n-2j+1, j)$-point boundary value problem are unique. Assume this not to be true. Then, there exist points $a < t_1 < \cdots < t_{2j-2} < x_1 < \cdots < x_{n-2j+1} < s_1 < \cdots < s_{2j} < b$ for which there exist distinct solutions, $y(x)$ and $z(x)$, of the $(j-1; n-2j+1, j)$-point boundary value problem such that

$$a_i y(t_{2i-1}) - b_i y(t_{2i}) = a_i z(t_{2i-1}) - b_i z(t_{2i}), \ i = 1, 2, \ldots, j - 1,$$
$$y(x_1) = z(x_1),$$
$$y(x_l) = z(x_l), \ 2 \leq l \leq n - 2j + 1,$$
$$c_i y(s_{2i-1}) - d_i y(s_{2i}) = c_i z(s_{2i-1}) - d_i z(s_{2i}), \ i = 1, 2, \ldots, j.$$

We define $w := y - z$. Then, we have

$$a_i w(t_{2i-1}) - b_i w(t_{2i}) = 0, \ i = 1, 2, \ldots, j - 1,$$
$$w(x_1) = 0,$$
$$w(x_l) = 0, \ 2 \leq l \leq n - 2j + 1,$$
$$c_i w(s_{2i-1}) - d_i w(s_{2i}) = 0, \ i = 1, 2, \ldots, j.$$

If there exists some point $p_1 \in (t_{2j-2}, x_1)$ such that $w(p_1) = 0$, then we have

$$a_j w(p_1) - b_j w(x_1) = 0, \ a_j, b_j \in \mathbb{R}.$$

This implies that $y(x)$ and $z(x)$ are distinct solutions of the $(j; n - 2j; j)$-point boundary value problem at the points $t_1, \ldots, t_{2j-2}, p_1, x_1, \ldots, x_{n-2j}, s_1, \ldots, s_{2j}$, which is a contradiction.

Hence, $w(x) \neq 0$ on (t_{2j-2}, x_1). Let $w(x) > 0$ on (t_{2j-2}, x_1). The case $w(x) < 0$ on (t_{2j-2}, x_1) is dealt with similarly. Then,

$$\max\{w(x) : x \in [t_{2j-2}, x_1]\} = w(\tau_1) > 0.$$

Define

$$v(x) = \begin{cases} a_j w(x) - b_j w(\tau_1), & \text{if } a_j \geq b_j, \\ b_j w(x) - a_j w(\tau_1), & \text{if } a_j \leq b_j. \end{cases}$$

Then, $v(\tau_1) > 0$ and $v(x_1) < 0$. Applying the Mean Value Theorem, there exists $p' \in (\tau_1, x_1)$ such that $v(p') = 0$ which implies that $a_j w(p') -$

$b_j w(\tau_1) = 0$. Hence, there are distinct solutions of the $(j; n - 2j; j)$-point boundary value problem at the points

$$t_1, \ldots, t_{2j-2}, \tau_1, p_1', x_2, \ldots, x_{n-2j}, s_1, \ldots, s_{2j},$$

which is again a contradiction. Hence, solutions of the $(j - 1; n - 2j + 1; j)$−point boundary value problem (1.1.1)-(3.3.17) are unique.

Repeating the argument above, but using the uniqueness of solutions of the $(j-1; n-2j+1; j)$-point problems, we obtain uniqueness of solutions of the $(j - 2; n - 2j + 2; j)$-point problems (1.1.1)-(3.3.17). Continuing in this manner, we obtain uniqueness of solutions of the $(j - i; n - 2j + i; j)$-point boundary value problem, for each $i = 1, 2, \ldots, j$. □

Corollary 3.3.5. [[30], Eloe, Khan and Henderson, Cor. 2.1] *Assume the hypotheses of Theorem* 3.3.15 *are satisfied. Then, solutions of the* $(0; n - j; j)$-*point boundary value problem* (1.1.1)-(3.3.17) *are unique, when they exist.*

Via an analogous argument to the proof given for Theorem 3.3.15, we can state a dual result.

Theorem 3.3.16. [[30], Eloe, Khan and Henderson, Thm. 2.3] *Assume the hypotheses of Theorem* 3.3.15 *are satisfied. Then, for each* $i = 1, 2, \ldots, j$, *solutions of the* $(j; n - 2j + 1; j - i)$-*point boundary value problem* (1.1.1)-(3.3.17) *are unique, when they exist.*

Corollary 3.3.6. [[30], Eloe, Khan and Henderson, Cor. 2.2] *Assume the hypotheses of Theorem* 3.3.15 *are satisfied. Then, solutions of the* $(j; n - j; 0)$-*point boundary value problem* (1.1.1)-(3.3.17) *are unique, when they exist.*

Corollary 3.3.7. [[30], Eloe, Khan and Henderson, Cor. 2.3] *Assume the hypotheses of Theorem* 3.3.15 *are satisfied. Then, solutions of the* $(0; n; 0)$-*point boundary value problem* (1.1.1)-(3.3.17) *(that is, of the n-point conjugate boundary value problem for* (1.1.1)), *are unique, when they exist.*

In view of Theorem 2.3.4, we have an important corollary concerning existence of solutions for ℓ-point conjugate boundary value problems for (1.1.1).

Corollary 3.3.8. [[30], Eloe, Khan and Henderson, Cor. 2.4] *Assume the hypotheses of Theorem* 3.3.15 *are satisfied. Then, each ℓ-point conjugate* *boundary value problem for* (1.1.1), *(that is, each* $(0; \ell; 0)$*-point problem* (1.1.1)-(3.3.17)), *for* $2 \leq \ell \leq n$, **has** *a unique solution on* (a, b).

We now present our main result for this sequence concerning uniqueness of solutions of (1.1.1)-(3.3.17).

Theorem 3.3.17. [[30], Eloe, Khan and Henderson, Thm. 2.4] *Assume the hypotheses of Theorem 3.3.15 are satisfied. Then, for each* $1 \leq k \leq n - 2j - 1$, *solutions of the* $(j; k; j)$*-point boundary value problem* (1.1.1)-(3.3.17) *are unique, when they exist.*

Proof. Assume that solutions of the $(j; n - 2j; j)$-point problem (1.1.1)-(3.3.17) are unique. But for contradiction purposes, we assume that, for some $1 \leq k \leq n - 2j - 1$, some $(j; k; j)$-point boundary value problem (1.1.1)-(3.3.17) has distinct solutions. Let

$$h = \max\{k = 1, \ldots, n - 2j - 1 \,|\, \text{ some } (j; k; j)\text{-point problem has}$$

distinct solutions\}.

Then, there are positive integers, $m_1, \ldots, m_h$, such that $m_1 + \cdots + m_h = n - 2j$, and points $a < t_1 < \cdots < t_{2j} < x_1 < \cdots < x_h < s_1 < \cdots < s_{2j} < b$, for which there exist distinct solutions $y(x)$ and $z(x)$ of the $(j; h; j)$-point boundary value problem (1.1.1)-(3.3.17), for these $m_1, \ldots, m_h$; that is,

$$a_i y(t_{2i-1}) - b_i y(t_{2i}) = a_i z(t_{2i-1}) - b_i z(t_{2i}), \ i = 1, 2, \ldots, j,$$
$$y^{(i-1)}(x_l) = z^{(i-1)}(x_l), \ 1 \leq i \leq m_l, \ 1 \leq l \leq h,$$
$$c_i y(s_{2i-1}) - d_i y(s_{2i}) = c_i z(s_{2i-1}) - d_i z(s_{2i}), \ i = 1, 2, \ldots, j.$$

Now, $h \leq n - 2j - 1$, and so some $m_l \geq 2$. Let

$$m_{l_0} = \max\{m_l \,|\, 1 \leq l \leq h\}.$$

Then, $m_{l_0} \geq 2$. Also, since x_l is a zero of $y - z$ of exact multiplicity $m_l, 1 \leq l \leq h$, and y and z are distinct solutions of (1.1.1), we may assume, without loss of generality, that

$$y^{(m_{l_0})}(x_{l_0}) > z^{(m_{l_0})}(x_{l_0}).$$

Next, we fix a point $a < \tau < x_1$. By the maximality of h, solutions of the $(j; h + 1; j)$-point problems (1.1.1)-(3.3.17) at the points $t_1, \ldots, t_{2j}, \tau, x_1, \ldots, x_h, s_1, \ldots, s_{2j}$ are unique, and hence, solutions depend continuously on such boundary conditions; that is, for each $\epsilon > 0$, there is a $\delta > 0$ and there is a solution $z_\delta(x)$ of the $(j; h + 1; j)$-point boundary value problem (1.1.1)-(3.3.17), (corresponding to $k = h + 1$), satisfying at the points $t_1, \ldots, t_{2j}, \tau, x_1, \ldots, x_h, s_1, \ldots, s_{2j}$,

$$a_i z_\delta(t_{2i-1}) - b_i z_\delta(t_{2i}) = a_i z(t_{2i-1}) - b_i z(t_{2i}) = a_i y(t_{2i-1}) - b_i y(t_{2i}),$$
$$i = 1, 2, \ldots, j,$$

$$z_\delta(\tau) = z(\tau),$$
$$z_\delta^{(i-1)}(x_l) = z^{(i-1)}(x_l) = y^{(i-1)}(x_l), \quad 1 \le i \le m_l, \quad 1 \le l \le h, \quad l \ne l_0,$$
$$z_\delta^{(i-1)}(x_{l_0}) = z^{(i-1)}(x_{l_0}) = y^{(i-1)}(x_{l_0}), \quad 1 \le i \le m_{l_0} - 2, \quad \text{(if } m_{l_0} > 2\text{)},$$
$$z_\delta^{(m_{l_0}-2)}(x_{l_0}) = z^{(m_{l_0}-2)}(x_{l_0}) + \delta = y^{(m_{l_0}-2)}(x_{l_0}) + \delta,$$
$$c_i z_\delta(s_{2i-1}) - d_i z_\delta(s_{2i}) = c_i z(s_{2i-1}) - d_i z(s_{2i}) = c_i y(s_{2i-1}) - d_i y(s_{2i}),$$
$$i = 1, 2, \ldots, j,$$

and $|z_\delta(x) - z(x)| < \epsilon$ on $[t_1, s_{2j}]$. For sufficiently small $\epsilon > 0$, there exist points $x_{l_0-1} < \rho_1 < x_{l_0} < \rho_2 < x_{l_0+1}$ such that

$$a_i z_\delta(t_{2i-1}) - b_i z_\delta(t_{2i}) = a_j y(t_{2i-1}) - b_i y(t_{2i}), \quad i = 1, 2, \ldots, j,$$
$$z_\delta^{(i-1)}(x_l) = y^{(i-1)}(x_l), \quad 1 \le i \le m_l, \quad 1 \le l \le l_0 - 1,$$
$$z_\delta(\rho_1) = y(\rho_1),$$
$$z_\delta^{(i-1)}(x_{l_0}) = y^{(i-1)}(x_{l_0}), \quad 1 \le i \le m_{l_0} - 2, \quad \text{(if } m_{l_0} > 2\text{)},$$
$$z_\delta(\rho_2) = y(\rho_2),$$
$$z_\delta^{(i-1)}(x_l) = y^{(i-1)}(x_l), \quad 1 \le i \le m_l, \quad l_0 + 1 \le l \le h,$$
$$c_i z_\delta(s_{2i-1}) - d_i z_\delta(s_{2i}) = c_i y(s_{2i-1}) - d_i y(s_{2i}), \quad i = 1, 2, \ldots, j.$$

Thus, $z_\delta(x)$ and $y(x)$ are distinct solutions of the $(j; h+1; j)$−point boundary value problem at the points $t_1, \ldots, t_{2j}, x_1, \ldots, x_{l_0-1}, \rho_1, \rho_2, x_{l_0+1}, \ldots, x_h, s_1, \ldots, s_{2j}$, which is a contradiction because of the maximality of h. The proof is complete. $\qquad \square$

From Theorems 3.3.15 and 3.3.17, we have two immediate corollaries.

Corollary 3.3.9. [[30], Eloe, Khan and Henderson, Cor. 2.5] *Assume the hypotheses of Theorem 3.3.15 are satisfied. Then, for $1 \le k \le n - 2j$ and $1 \le i \le j$, solutions of the $(j; k + i; j - i)$-point boundary value problem are unique, when they exist.*

Corollary 3.3.10. [[30], Eloe, Khan and Henderson, Cor. 2.6] *Assume the hypotheses of Theorem 3.3.15 are satisfied. Then, for $1 \le k \le n - 2j$ and $1 \le i \le j$, solutions of the $(j - i; k + i; j)$-point boundary value problem are unique, when they exist.*

The last two results for nonlocal boundary value problems of this section are due to Henderson [60], with the motivation in part being the "uniqueness implies uniqueness" results by Henderson and Jackson [62] presented in Section 1.2 in Theorems 1.2.3 and 1.2.4.

For $n \geq 3$, we will be concerned with uniqueness of solutions of $(k + 2)$-point nonlocal boundary value problems for (1.1.1). Specifically, given $1 \leq k \leq n - 1$, positive integers $m_1, \ldots, m_k$ such that $m_1 + \cdots + m_k = n - 1$, points $a < x_1 < x_2 < \cdots < x_k < x_{k+1} < x_{k+2} < b$, real values $y_{ij}, 1 \leq i \leq m_j, 1 \leq j \leq k$, and $y_n \in \mathbb{R}$, we are concerned with uniqueness of solutions of (1.1.1) satisfying the $(k + 2)$-point nonlocal boundary conditions,

$$
\begin{aligned}
y^{(i-1)}(x_j) &= y_{ij}, \ 1 \leq i \leq m_j, \ 1 \leq j \leq k, \\
y(x_{k+1}) - y(x_{k+2}) &= y_n.
\end{aligned}
\tag{3.3.18}
$$

Remark. Sometimes the positive integers, $m_1, \ldots, m_k$, will be referred to as *indices* associated with (1.1.1)-(3.3.18).

As a reminder, our results will involve from Section 1.1 conditions (A) and (B) and "compactness condition" (CP), and continuous dependence from Theorem 1.1.6 and Remark 1.1.1, along with other conditions we will impose.

Theorem 3.3.18. [[60], Henderson, Thm. 3] *Assume that with respect to* (1.1.1), *conditions* (A) *and* (B) *are satisfied. Let* $1 \leq h - 1 < h < n - 1$, *and assume that, for both* $k = h-1$ *and* $k = h$, *there is at most one solution of* (1.1.1)-(3.3.18) *on* (a, b). *Then, for each* $1 \leq k \leq h$, *there is at most one solution of* (1.1.1)-(3.3.18) *on* (a, b).

Proof. We assume the conclusion of the Theorem is false. Then h must satisfy $1 < h - 1 < h < n - 1$, and there is a largest integer κ with $1 \leq \kappa < h - 1$ for which (1.1.1)-(3.3.18), for $k = \kappa$, has two distinct solutions on (a, b). We note that $\kappa \leq n - 4$. From the maximality of κ and $\kappa < h - 1 < h$, it follows that, for both $k = \kappa + 1$ and $k = \kappa + 2$, solutions of (1.1.1)-(3.3.18) on (a, b) are unique, when they exist.

So, let $y(x)$ and $z(x)$ be distinct solutions of (1.1.1) such that, for some points $a < x_1 < \cdots < x_\kappa < x_{\kappa+1} < x_{\kappa+2} < b$, $y(x) - z(x)$ has a zero at x_j of exact order $p_j \geq 1, 1 \leq j \leq \kappa$, and in addition $\sum_{j=1}^{\kappa} p_j \geq n - 1$, and $y(x_{\kappa+1}) - y(x_{\kappa+2}) = z(x_{\kappa+1}) - z(x_{\kappa+2})$. By condition (B), each $p_j < n$, and since $\kappa \leq n - 4$, at least one $p_j > 1$. Let $1 \leq r \leq \kappa$ be such that

$$
p_r = \max\{p_j \mid 1 \leq j \leq \kappa\}.
$$

Assume $y^{(p_r)}(x_r) > z^{(p_r)}(x_r)$, and fix x_0 and $x_{\kappa+3}$ such that $a < x_0 < x_1$ and $x_{\kappa+2} < x_{\kappa+3} < b$. There are two cases to consider:

(a) $p_j = 1$, for all $1 \leq j \leq \kappa$, $j \neq r$. Now $\kappa \leq n - 4$, and so it follows that $p_r \geq 4$. Let $s \geq 1$ be an integer such that

$$2 \leq q_r := p_r - 2s \text{ and } (\kappa - 1) + q_r = n - 3.$$

We fix $x_\kappa < \xi < x_{\kappa+1}$. Solutions of (1.1.1)-(3.3.18), for $k = \kappa + 1$, are unique, and hence depend continuously on boundary conditions. So, there exists an $\epsilon_0 > 0$ such that for $0 < \epsilon < \epsilon_0$, there exists a solution $y_\epsilon(x)$ of (1.1.1)-(3.3.18), for $k = \kappa + 1$, satisfying

$$\begin{aligned}
y_\epsilon(x_j) &= y(x_j), \ 1 \leq j \leq \kappa, \ j \neq r, \\
y_\epsilon^{(i)}(x_r) &= y^{(i)}(x_r), \ 0 \leq i \leq q_r - 1, \\
y_\epsilon^{(q_r)}(x_r) &= y^{(q_r)}(x_r) - \epsilon, \\
y_\epsilon(\xi) &= y(\xi), \\
y_\epsilon(x_{\kappa+1}) - y_\epsilon(x_{\kappa+2}) &= y(x_{\kappa+1}) - y(x_{\kappa+2}),
\end{aligned}$$

and $|y_\epsilon^{(i)}(x) - y^{(i)}(x)|$ is uniformly small on the compact interval $[x_0, x_{\kappa+3}]$, for each $0 \leq i \leq n - 1$. So, for $0 < \epsilon < \epsilon_0$ sufficiently small, $y_\epsilon(x) - z(x)$ has a zero at x_j, $1 \leq j \leq \kappa$, $j \neq r$, has a zero in each of the open intervals (x_{r-1}, x_r) and (x_r, x_{r+1}), and has a zero of order q_r at x_r. Moreover, $y_\epsilon(x_{\kappa+1}) - y_\epsilon(x_{\kappa+2}) = z(x_{\kappa+1}) - z(x_{\kappa+2})$. Hence, for $0 < \epsilon < \epsilon_0$ sufficiently small, $y_\epsilon(x)$ and $z(x)$ are distinct solutions of (1.1.1) satisfying the same boundary conditions (3.3.18), for $k = \kappa + 2$. This is a contradiction to the uniqueness of solutions for such problems.

(b) $p_{j_0} > 1$, for some $1 \leq j_0 \leq \kappa$ and $j_0 \neq r$. There are similarities in the arguments for this case with those of case (a). In this case, there are integers $q_j, 1 \leq j \leq \kappa$, and an integer $s \geq 1$ such that,

$$\begin{aligned}
&1 \leq q_j \leq p_j, \ 1 \leq j \leq \kappa, \ j \neq r, \\
&0 \leq q_r \leq p_r - 2s, \\
&\sum_{j=1}^{\kappa} q_j = n - 3.
\end{aligned}$$

Again, we fix $x_\kappa < \xi < x_{\kappa+1}$. By uniqueness of solutions of (1.1.1)-(3.3.18), for $k = \kappa + 1$, solutions depend continuously on boundary conditions. In particular, there exists an $\epsilon_0 > 0$ such that, for $0 < \epsilon < \epsilon_0$, there exists a solution $y_\epsilon(x)$ of (1.1.1)-(3.3.18), for $k = \kappa + 1$, satisfying

$$y_\epsilon^{(i)}(x_j) = y^{(i)}(x_j),\ 0 \le i \le q_j - 1,\ 1 \le j \le \kappa,\ j \ne r,$$
$$y_\epsilon^{(i)}(x_r) = y^{(i)}(x_r),\ 0 \le i \le q_r - 1,\ \text{(omitted if } q_r = 0),$$
$$y_\epsilon^{(q_r)}(x_r) = y^{(q_r)}(x_r) - \epsilon,$$
$$y_\epsilon(\xi) = y(\xi),$$
$$y_\epsilon(x_{\kappa+1}) - y_\epsilon(x_{\kappa+2}) = y(x_{\kappa+1}) - y(x_{\kappa+2}),$$

and $|y_\epsilon^{(i)}(x) - y^{(i)}(x)|$ is uniformly small on the compact interval $[x_0, x_{\kappa+3}]$, for each $0 \le i \le n - 1$. Hence, for $0 < \epsilon < \epsilon_0$ sufficiently small, $y_\epsilon(x) - z(x)$ has a zero of order q_j at x_j, $1 \le j \le \kappa, j \ne r$, has a zero in each of the open intervals (x_{r-1}, x_r) and (x_r, x_{r+1}), and, if $q_r > 0$, has a zero of order q_r at x_r. Furthermore, $y_\epsilon(x_{\kappa+1}) - y_\epsilon(x_{\kappa+2}) = z(x_{\kappa+1}) - z(x_{\kappa+2})$. That is, for $0 < \epsilon < \epsilon_0$ sufficiently small, $y_\epsilon(x)$ and $z(x)$ are distinct solutions of (1.1.1) satisfying the same boundary conditions (3.3.18), for $k = \kappa+1$, when $q_r = 0$, and are distinct solutions of (1.1.1) satisfying the same boundary conditions (3.3.18), for $k = \kappa + 2$, when $q_r > 0$. In either case we have a contradiction, since in each case, such boundary value problems have at most one solution. The proof is complete. $\square$

The next result deals with uniqueness of solutions of (1.1.1)-(3.3.18), for $k = n - 2$, implying the uniqueness of solutions of (1.1.1)-(3.3.18), for all $1 \le k \le n - 2$.

Theorem 3.3.19. [[60], Henderson, Thm. 4] *Assume that with respect to (1.1.1), conditions* (A) *and* (B) *are satisfied. In addition, assume that, for $k = n - 2$, there is at most one solution of (1.1.1)-(3.3.18) on (a, b). Then, for each $1 \le k \le n - 2$, there is at most one solution of (1.1.1)-(3.3.18) on (a, b).*

Proof. In view of Theorem 3.3.18, it suffices to show that solutions of (1.1.1)-(3.3.18) for $k - n - 3$, are unique, when they exist. As is customary, we assume to the contrary that there are distinct solutions $y(x)$ and $z(x)$ of (1.1.1) such that $y(x) - z(x)$ has a zero of exact order $p_j \ge 1$ at $x_j, 1 \le j \le n - 3$, and $y(x_{n-2}) - y(x_{n-1}) = z(x_{n-2}) - z(x_{n-1})$, where $a < x_1 < \cdots < x_{n-3} < x_{n-2} < x_{n-1} < b$ and $\sum_{j=1}^{n-3} p_j \ge n - 1$.

There are two cases to consider:
(i) Each p_j is an odd integer, or one p_j is an even integer and all other $p_j = 1$. For either of these situations, we let $1 \le r \le n - 3$ be such that

$$p_r = \max\{p_j \mid 1 \le j \le n - 3\}.$$

Then $p_r \geq 3$. We may assume $y^{(p_r)}(x_r) > z^{(p_r)}(x_r)$. This time, choose $a < x_0 < x_1$ and $x_{n-1} < x_n < b$, and let $\delta > 0$ be such that $x_r + \delta < x_{r+1}$, $x_{n-3} < x_{n-2} - \delta$, and $x_{n-1} + \delta < x_n$.

Given $\epsilon > 0$, let $y_\epsilon(x)$ be the solution of the initial value problem for (1.1.1) satisfying the initial conditions,

$$y_\epsilon^{(i)}(x_r) = y^{(i)}(x_r),\ 0 \leq i \leq n-1,\ i \neq p_r - 1,$$
$$y_\epsilon^{(p_r-1)}(x_r) = y^{(p_r-1)}(x_r) - \epsilon.$$

Then by continuous dependence on initial conditions, it follows that

$$\lim_{\epsilon \to 0^+} y_\epsilon^{(i)}(x) = y^{(i)}(x)$$

uniformly on $[x_0, x_n]$, for each $0 \leq i \leq n-1$. Then, for sufficiently small $\epsilon > 0$, $y_\epsilon(x) - z(x)$ has a zero in a neighborhood of x_j, for each $1 \leq i \leq n-3$ with $j \neq r$, $y_\epsilon(x) - z(x)$ has a zero of order $p_r - 1$ at x_r, $y_\epsilon(x) - z(x)$ has a zero in $(x_r, x_r + \delta)$, and $y_\epsilon(\tau_1) - y_\epsilon(\tau_2) = z(\tau_1) - z(\tau_2)$, for some $x_{n-2} - \delta < \tau_1 < \tau_2 < x_{n-1} + \delta$. Thus $y_\epsilon(x)$ and $z(x)$ are distinct solutions of (1.1.1) satisfying the same boundary conditions (3.3.18), for $k = n-2$. This contradicts the uniqueness of solutions for such problems.

(ii) This case is the complement of case (i); that is, there exists $1 \leq r \leq n-3$, such that p_r is even and such that $\sum_{j=1, j \neq r}^{n-3} p_j \geq n-3$.

We fix $a < x_0 < x_1$, $x_{n-1} < x_n < b$, and $x_{n-3} < \xi < x_{n-2}$, and we assume $y^{(p_r)}(x_r) > z^{(p_r)}(x_r)$. As in the proof of case (b) of Theorem 3.3.18, due to the continuous dependence of solutions of (1.1.1)-(3.3.18), for $k = n-2$, on boundary conditions, it follows that, for sufficiently small $\epsilon > 0$, there is a solution $y_\epsilon(x)$ of (1.1.1) satisfying the boundary conditions,

$$y_\epsilon^{(i)}(x_j) = y^{(i)}(x_j),\ 0 \leq i \leq q_j - 1,\ 1 \leq j \leq n-3,\ j \neq r,$$
$$y_\epsilon(x_r) = y(x_r) - \epsilon,$$
$$y_\epsilon(\xi) = y(\xi),$$
$$y_\epsilon(x_{n-2}) - y_\epsilon(x_{n-1}) = y(x_{n-2}) - y(x_{n-1}),$$

and $\lim_{\epsilon \to 0^+} y_\epsilon^{(i)}(x) = y^{(i)}(x)$ uniformly on $[x_0, x_n]$, for each $0 \leq i \leq n-1$. So, for sufficiently small $\epsilon > 0$, $y_\epsilon(x) - z(x)$ has a zero in each of the open intervals (x_{r-1}, x_r) and (x_r, x_{r+1}), $y_\epsilon(x) - z(x)$ has a zero of order q_j at x_j, for $1 \leq j \leq n-3$, $j \neq r$, and $y_\epsilon(x_{n-2}) - y_\epsilon(x_{n-1}) = z(x_{n-2}) - z(x_{n-1})$. This again contradicts the uniqueness of solutions of (1.1.1)-(3.3.18), for $k = n-2$. The proof of the theorem is complete. $\square$

3.4 Nonlocal problems: uniqueness implies existence, II

We shall close the chapter by considering the uniqueness implies existence methods and results for the types of nonlocal problems that have been considered in Section 3.3.

We begin with the problems studied by Clark and Henderson [13].

Theorem 3.4.1. [[13], Clark and Henderson, Thm. 3.2] *Assume that with respect to* (3.1.3), *conditions* (A) *and* (B) *are satisfied. Assume solutions for* (3.1.3), (3.3.2) *are unique. Let* $a < x_1 < x_2 < x_3 < b, y_1, y_2, y_3 \in \mathbb{R}$, *be given. Then there exists a solutions for* (3.1.3), (3.3.4) *on* (a, b).

Proof. To outline the highlights of the proof, let $\tau \in (a, x_1)$ and let z denote the solution of (3.1.3) on (a, b) satisfying the boundary conditions

$$z(\tau) = 0, \quad z(x_1) = y_1, \quad z(x_2) - z(x_3) = y_3.$$

Theorem 3.2.2 implies the existence of z.

Define the set

$$S = \{y'(x_1) \mid y \text{ is a solution of } (3.1.3) \text{ satisfying } y(x_1) = z(x_1),$$
$$\text{and } y(x_2) - y(x_3) = z(x_2) - z(x_3)\}.$$

Again S is nonempty and open. To show S is closed, argue by contradiction and assume $s_0 \in \overline{S} \setminus S$ and assume the existence of a strictly monotone increasing sequence $\{s_q\} \subset S$ such that $\lim_{q \to \infty} s_q = s_0$.

By the definition of S, we denote, for each $q \in \mathbb{N}$, by $y_q(x)$ the solution of (3.1.3) satisfying

$$y_q(x_1) = y_1, \ y_q'(x_1) = s_q, \text{ and } y_q(x_2) - y_q(x_3) = z(x_2) - z(x_3) = y_3.$$

By uniqueness of solutions of (3.1.3), (3.3.2), we have

$$y_q(x) > y_{q+1}(x) \text{ on } (a, x_1), \text{ and } y_q(x) < y_{q+1}(x) \text{ on } (x_1, x_2).$$

The "compactness condition" (CP), and the fact that $s_0 \notin S$ implies that $\{y_q(x)\}$ is not uniformly bounded below on each compact subinterval of (a, x_1), and $\{y_q(x)\}$ is not uniformly bounded above on each compact subinterval of (x_1, x_2).

Recalling that both 2-point and 3-point conjugate problems for (3.1.3) have unique solutions, let $u(x)$ be the solution of the 2-point conjugate boundary value problem for (3.1.3) satisfying,

$$u(x_1) = z(x_1), \ u'(x_1) = s_0, \ u(x_2) = 0.$$

It follows that there exists q sufficiently large and $a < \tau_1 < x_1 < \tau_2 < x_2$ such that

$$y_q(\tau_1) = u(\tau_1) \text{ and } y_q(\tau_2) = u(\tau_2).$$

Moreover,

$$y_q(x_1) = z(x_1) = u(x_1),$$

and the uniqueness of solutions of the 3-point conjugate boundary value problem for (3.1.3) is violated. Hence, S is closed and the proof is complete.
□

In complete analogy, one can show the following.

Theorem 3.4.2. [[13], Clark and Henderson, Thm. 3.2] *Assume that with respect to (3.1.3), conditions* (A) *and* (B) *are satisfied. Assume solutions for (3.1.3), (3.3.3) are unique. Let $a < x_1 < x_2 < x_3 < b, y_1, y_2, y_3 \in \mathbb{R}$, be given. Then there exists a solutions for (3.1.3), (3.3.5) on (a, b).*

Henderson and Ma [67] considered the fourth order equation, (3.3.6) with 5-point nonlocal boundary conditions, (3.3.7), (3.3.8), 4-point non-local conditions, (3.3.9), (3.3.10), (3.3.11), (3.3.12), and 3-point nonlocal conditions, (3.3.13), (3.3.14). In particular, they showed that the unique-ness of solutions of the 5-point nonlocal boundary value problems, (3.3.6), (3.3.7) implies the existence of solutions for the 5-point problems, (3.3.6), (3.3.7), the 4-point problems, (3.3.6), (3.3.9), (3.3.6), (3.3.11), and the 3-point problems, (3.3.6), (3.3.13). In complete analogy, the uniqueness of solutions of the 5-point nonlocal boundary value problems, (3.3.6), (3.3.8) implies the existence of solutions for the 5-point problems, (3.3.6), (3.3.8), the 4-point problems, (3.3.6), (3.3.10), (3.3.6), (3.3.12), and the 3-point problems, (3.3.6), (3.3.14).

Remark 3.4.1. *Note, under the assumption that solutions for (3.3.6)-(3.3.7) are unique, then solutions for each of (3.3.6)-(3.3.9), (3.3.6)-(3.3.11) and (3.3.6)-(3.3.13) are unique by Theorems 3.3.6, 3.3.7 and 3.3.8, respec-tively. Similarly, under the assumption that solutions for (3.3.6)-(3.3.8) are unique, then solutions for each of (3.3.6)-(3.3.10), (3.3.6)-(3.3.12) and (3.3.6)-(3.3.14) are unique. It is also the case that under either assumption, solutions for (3.3.6)-(3.3.7) are unique or solutions for (3.3.6)-(3.3.8) are unique, then solutions of 4-point conjugate boundary value problems for (3.3.6) are unique, when they exist. That is, if $y(x)$ and $z(x)$ are both so-lutions of (3.3.6) such that, for some points $a < x_1 < x_2 < x_3 < x_4 < b$,*

$y(x_i) = z(x_i)$, $i = 1, 2, 3, 4$, *then by the Intermediate Value Theorem, there exist* $x_1 < \tau_1 < \tau_2 < x_2 < x_3 < \sigma_1 < \sigma_2 < x_4$ *such that, both* $y(\tau_1) - y(\tau_2) = z(\tau_1) - z(\tau_2)$, $y(x_i) = z(x_i)$, $i = 2, 3, 4$, *and* $y(x_i) = z(x_i)$, $i = 1, 2, 3$, $y(\sigma_1) - y(\sigma_2) = z(\sigma_1) - z(\sigma_2)$. *So, if either solutions for (3.3.6)-(3.3.7) are unique or solutions for (3.3.6)-(3.3.7) are unique holds, then* $y(x) = z(x)$. *Thus, if either* (A), (B) *and solutions for (3.3.6)-(3.3.7) are unique, or* (A), (B) *and solutions for (3.3.6)-(3.3.7) are unique hold, then each k-point conjugate boundary value problem for (3.3.6), $k = 2, 3, 4$, has a unique solution.*

Theorem 3.4.3. [[67], Henderson and Ma, Thm. 3.3] *Assume that with respect to (3.3.6), conditions* (A) *and* (B) *are satisfied. If solutions for (3.3.6)-(3.3.7) are unique, when they exist, then solutions for (3.3.6)-(3.3.7) exist.*

Proof. Let $a < x_1 < x_2 < x_3 < x_4 < x_5 < b$ and $y_1, y_2, y_3, y_4 \in \mathbb{R}$ be given. It follows from Remark 3.4.1 that 4-point, 3-point, and 2-point, conjugate boundary value problems for (3.3.6) have unique solutions.

Let $z(x)$ be the solution of (3.3.6) satisfying the 4-point conjugate boundary conditions at x_2, x_3, x_4 and x_5,

$$z(x_2) = y_2, \ z(x_3) = y_3, \ z(x_4) = y_4, \ z(x_5) = 0.$$

Observe that $z(x_4) - z(x_5) = y_4$. Next, define the set

$S = \{y(x_1) \mid y(x)$ is a solution of equation (3.3.6) satisfying

$$y(x_2) = y(x_2), \ y(x_3) = z(x_3), \ y(x_4) - y(x_5) = z(x_4) - z(x_5)\}.$$

S is nonempty, since $z(x_1) \in S$ and S is open by Theorem 1.1.6. Thus, it remains to be shown that S is closed.

Assume for the sake of contradiction that $s_0 \in \overline{S} \backslash S$ and let $\{s_q\} \subset S$ denote a strictly monotone increasing sequence such that $\lim_{q \to \infty} s_q = s_0$.

By the definition of S, for each $q \in \mathbb{N}$, denote by $y_q(x)$ the solution of equation (3.3.6) satisfying

$$y_q(x_1) = s_q, \ y_q(x_2) = z(x_2), \ y_q(x_3) = z(x_3),$$
$$y_q(x_4) - y_q(x_5) = z(x_4) - z(x_5).$$

By the uniqueness of solutions of (3.3.6)-(3.3.7), it follows that

$$y_q(x) < y_{q+1}(x) \text{ on } (a, x_2).$$

Since $s_0 \notin S$, it follows from the "compactness condition" (CP) that $\{y_q(x)\}$ is not uniformly bounded above on each compact subinterval of (a, x_1) or (x_1, x_2).

Let $u(x)$ denote the solution of the 3-point conjugate boundary problem for equation (3.3.6) satisfying,

$$u(x_1) = s_0, \ u'(x_1) = 0, \ u(x_2) = y_2, \ u(x_3) = y_3.$$

Due to the monotonicity and unboundedness properties of the sequence $\{y_q\}$, it follows that there exists q sufficiently large and points $a < \tau_1 < x_1 < \tau_2 < x_2$ such that

$$y_q(\tau_1) = u(\tau_1), \ y_q(\tau_2) = u(\tau_2).$$

Moreover,

$$y_q(x_2) = z(x_2) = u(x_2), \ y_q(x_3) = z(x_3) = u(x_3).$$

Thus, y_q and u are distinct solutions of a 4-point conjugate boundary value problems for (3.3.6). Thus, S is closed and the proof is complete.

$\square$

Theorem 3.4.4. [[67], Henderson and Ma, Thms. 3.4, 3.5 and 3.6] *Assume that with respect to (3.3.6), conditions (A) and (B) are satisfied. If solutions for (3.3.6)-(3.3.7) are unique, when they exist, then solutions for each of (3.3.6)-(3.3.9), (3.3.6)-(3.3.11), and (3.3.6)-(3.3.13) exist.*

Proof. To prove that solutions of (3.3.6)-(3.3.9) exist, let $a < x_1 < x_2 < x_3 < x_4 < b$, and $y_1, y_2, y_3, y_4 \in \mathbb{R}$ be given. For the proof of this existence theorem, z is taken to be the solution the nonlocal 5-point boundary value problem, (3.3.6)-(3.3.7), with conditions

$$z(\tau) = 0, \ z(x_1) = y_1, \ z(x_2) = y_3, \ z(x_3) - z(x_4) = y_4,$$

where $a < \tau < x_1$. The existence of z is given by Theorem 3.4.3. S is dehned as the set

$$S = \{y'(x_1)| \ u(x) \text{ is a solution of } (3.3.6) \text{ satisfying}$$
$$y(x_1) = z(x_1), y(x_2) = z(x_2), y(x_3) - y(x_4) = z(x_3) - z(x_4)\}.$$

S is nonempty and open; thus, it remains to show S is closed. Assume for the sake of contradiction that S is not closed and let $s_0 \in \overline{S} \setminus S$. Assume $\{s_q\} \subset S$ is a strictly monotone increasing sequence such that $\lim_{q \to \infty} s_q = s_0$, then $y_q(x)$ denotes the solution of equation (3.3.6) satisfying

$$y_q(x_1) = z(x_1), \ y_q'(x_1) = s_q, \ y_q(x_2) = z(x_2),$$
$$y_q(x_3) - y_q(x_4) = z(x_3) - z(x_4).$$

Then $\{y_q\}$ is unbounded and monotone decreasing on (a, x_1) and unbounded and monotone increasing on (x_1, x_2). To complete the contradiction, let u denote the solution of (3.3.6) satisfying the 3-point conjugate boundary conditions

$$u(x_1) = z(x_1), \ u'(x_1) = s_0, \ u(x_2) = z(x_2), \ u(x_3) = z(x_3).$$

Then there exists q sufficiently large and points $a < \tau_1 < x_1 < \tau_2 < x_2$ such that

$$y_q(\tau_1) = u(\tau_1), \ y_q(\tau_2) = u(\tau_2).$$

Since,

$$y_q(x_1) = z(x_1) = u(x_1), \ y_q(x_2) = z(x_2) = u(x_2),$$

the uniqueness of solutions of 4-point conjugate boundary value problems for (3.3.6) is violated.

To prove that solutions of (3.3.6)-(3.3.11) exist, let z denote the solution the nonlocal 5-point boundary value problem, (3.3.6)-(3.3.7), with conditions

$$z(x_1) = y_1, \ z(\tau) = 0, \ z(x_2) = y_3, \ z(x_3) - z(x_4) = y_4,$$

where $x_1 < \tau < x_2$. S is defined as the set

$$S = \{y'(x_2)|\ y(x) \text{ is a solution of (3.3.6) satisfying}$$
$$y(x_1) = z(x_1), y(x_2) = z(x_2), y(x_3) - y(x_4) = z(x_3) - z(x_4)\}$$

and u is the solution of (3.3.6) satisfying the 3-point conjugate boundary conditions

$$u(x_1) = z(x_1), \ u(x_2) = z(x_2), \ u'(x_2) = s_0, \ u(x_3) = z(x_3).$$

To prove that solutions of (3.3.6)-(3.3.13) exist, let $a < x_1 < x_2 < x_3 < b$, and $y_1, y_2, y_3, y_4 \in \mathbb{R}$ be given and let z denote the solution the nonlocal 4-point boundary value problem, (3.3.6)-(3.3.11), with conditions

$$z(\tau) = 0, \ z(x_1) = y_1, \ z'(x_1) = 0, \ z(x_2) - z(x_3) - y_4,$$

where $a < \tau < x_1$. S is defined as the set

$$S = \{y''(x_1)|\ y(x) \text{ is a solution of (3.3.6) satisfying}$$
$$y(x_1) = z(x_1), y'(x_1) = z'(x_1), y(x_2) - y(x_3) = z(x_2) - z(x_3)\}$$

and u is the solution of (3.3.6) satisfying the 2-point conjugate boundary conditions

$$u(x_1) = z(x_1), \ u'(x_1) = z'(x_1), \ u''(x_1) = s_0, \ u(x_3) = z(x_3). \qquad \square$$

Similar results are valid for the boundary value problems with the non-local terms at the left.

Theorem 3.4.5. [[67], Henderson and Ma, Thm. 3.7] *Assume that with respect to (3.3.6), conditions (A) and (B) are satisfied. If solutions for (3.3.6)-(3.3.8) are unique, when they exist, then solutions for each of (3.3.6)-(3.3.8), (3.3.6)-(3.3.10), (3.3.6)-(3.3.12), and (3.3.6)-(3.3.14) exist.*

We now consider nonlocal boundary value problems for the general nth order equation (1.1.1) and begin with the problems considered by Eloe and Henderson [26] in 2007. Recall the boundary conditions (3.3.15), which we restate here for convenience:

$$y^{(i-1)}(x_j) = y_{ij}, \ 1 \leq i \leq m_j, \ 1 \leq j \leq k,$$
$$a_1 y(x_{k+1}) - a_2 y(x_{k+2}) = y_n$$

where $1 \leq k \leq n - 1$, $m_1, \ldots, m_k \in \mathbb{N}$ such that $m_1 + \cdots + m_k = n - 1$, $a < x_1 < \cdots < x_k < x_{k+1} < x_{k+2} < b$, $y_{ij} \in \mathbb{R}, 1 \leq i \leq m_j, 1 \leq j \leq k$, a_1, a_2 are positive real values, and $y_n \in \mathbb{R}$.

The goal here is to show that uniqueness of solutions for (1.1.1)-(3.3.15), when $k = n - 1$, implies existence of solutions for (1.1.1)-(3.3.15), when $1 \leq k \leq n-1$. Recall, it is already shown in Theorem 3.3.12 that uniqueness of solutions for (1.1.1)-(3.3.15), when $k = n - 1$, implies uniqueness of solutions for (1.1.1)-(3.3.15), when $1 \leq k \leq n - 2$.

Theorem 3.4.6. [[26], Eloe and Henderson, Thm. 3.3] *Assume that with respect (1.1.1), conditions (A) and (B) are satisfied, and assume that for $k = n - 1$, solutions of (1.1.1)-(3.3.15) are unique, when they exist. Then, for each $1 \leq k \leq n - 1$, solutions of (1.1.1)-(3.3.15) exist.*

Proof. Let $1 < k < n-1$, positive integers $m_1, \ldots, m_k$ such that $m_1 + \cdots + m_k = n - 1$, points $a < x_1 < \cdots < x_k < x_{k+1} < x_{k+2} < b$, real values $y_{ij}, 1 \leq i \leq m_j, 1 \leq j \leq k$, and $y_n \in \mathbb{R}$ be given.

In view of Corollary 3.3.1, there exists a unique solution $z(x)$ of (1.1.1) satisfying the $(k + 2)$-point conjugate boundary conditions (1.1.2) at the points $x_1, \ldots, x_{k+2}$ (or alternatively, if $m_1 = 1$, $z(x)$ satisfies the $(k + 1)$-point conjugate conditions at the points $x_2, \ldots, x_{k+2}$),

$$z^{(i-1)}(x_1) = y_{i1}, \quad 1 \leq i \leq m_1 - 1,$$
$$z^{(i-1)}(x_j) = y_{ij}, \quad 1 \leq i \leq m_j, \quad 2 \leq j \leq k,$$
$$z(x_{k+1}) = \frac{y_n}{a_1},$$
$$z(x_{k+2}) = 0.$$

(We note that $k + 2 \leq n$, for all cases, except when $k = n - 1$, in which case $m_1 = \cdots = m_k = 1$.) Observe that

$$a_1 z(x_{k+1}) - a_2 z(x_{k+2}) = y_n.$$

Next, define the set

$$S = \{y^{(m_1-1)}(x_1) \mid y \text{ is a solution of } (1.1.1) \text{ satisfying}$$
$$y^{(q-1)}(x_1) = y_{i1}, \ 1 \leq q \leq m_1 - 1,$$
$$y^{(i-1)}(x_j) = y_{ij}, 1 \leq i \leq m_j, 2 \leq j \leq k, \text{and}$$
$$a_1 y(x_{k+1}) - a_2 y(x_{k+2}) = y_n\}.$$

From the observation immediately above, $z^{(m_1-1)}(x_1) \in S$, and so $S \neq \emptyset$.

Next, choose $s_0 \in S$. Then, there is a solution $y_0(x)$ of (1.1.1) satisfying

$$y_0^{(i-1)}(x_1) = y_{i1}, \quad 1 \leq i \leq m_1 - 1,$$
$$y_0^{(m_1-1)}(x_1) = s_0,$$
$$y_0^{(i-1)}(x_j) = y_{ij}, \quad 1 \leq i \leq m_j, \quad 2 \leq j \leq k,$$
$$a_1 y_0(x_{k+1}) - a_2 y_0(x_{k+2}) = y_n.$$

By Theorem 1.1.6, there exists a $\delta > 0$ such that, for each $0 \leq |s - s_0| < \delta$, there is a solution $y_s(x)$ of (1.1.1) satisfying

$$y_s^{(i-1)}(x_1) = y_{i1}, \quad 1 \leq i \leq m_1 - 1,$$
$$y_s^{(m_1-1)}(x_1) = s,$$
$$y_s^{(i-1)}(x_j) = y_{ij}, \quad 1 \leq i \leq m_j, \quad 2 \leq j \leq k,$$
$$a_1 y_s(x_{k+1}) - a_2 y_s(x_{k+2}) = y_n;$$

in other words, $s \in S$; that is $(s_0 - \delta, s_0 + \delta) \subset S$, and S is an open subset of $\mathbb{R}$.

So, it remains to show that S is also a closed subset of $\mathbb{R}$. Assume by contradiction that there exists $s_0 \in \overline{S} \setminus S$ and let $\{s_q\} \subset S$ denote a strictly monotone increasing sequence such that $\lim_{q \to \infty} s_q = s_0$. By the definition of S, denote, for each $q \in \mathbb{N}$, by $y_q(x)$ the solution of (1.1.1) satisfying

$$y_q^{(i-1)}(x_1) = y_{i1}, \quad 1 \leq i \leq m_1 - 1,$$
$$y_q^{(m_1-1)}(x_1) = s_q,$$
$$y_q^{(i-1)}(x_j) = y_{ij}, \quad 1 \leq i \leq m_j, \quad 2 \leq j \leq k,$$
$$a_1 y_q(x_{k+1}) - a_2 y_q(x_{k+2}) = y_n.$$

By uniqueness of solutions of (1.1.1)-(3.3.15), for each $1 \leq k \leq n - 1$, due to Theorem 3.3.12, then either

(a) $y_q(x) < y_{q+1}(x)$ on $(a, x_2) \setminus \{x_1\}$, if m_1 is odd, or

(b) $y_q(x) > y_{q+1}(x)$ on (a, x_1) and $y_q(x) < y_{q+1}(x)$ on (x_1, x_2), if m_1 is even.

Case (a) is addressed here; case (b) is treated analogously. So, for case (a), it follows from the "compactness condition" (CP) and the fact that $s_0 \notin S$, that $\{y_q(x)\}$ is not uniformly bounded above on each compact subinterval of each of (a, x_1) and (x_1, x_2).

Let $u(x)$ denote the solution of (1.1.1) satisfying $(k+1)$-point conjugate boundary conditions (1.1.2) at the points $x_1, \ldots, x_{k+1}$,

$$u^{(i-1)}(x_1) = y_{i1}, \quad 1 \le i \le m_1 - 1, \ (\text{if } m_1 > 1),$$
$$u^{(m_1-1)}(x_1) = s_0,$$
$$u^{(i-1)}(x_j) = y_{ij}, \quad 1 \le i \le m_j, \quad 2 \le j \le k,$$
$$u(x_{k+1}) = 0.$$

It now follows from the unboundedness condition on $\{y_q(x)\}$ that there exists q sufficiently large and points $a < \tau_1 < x_1 < \tau_2 < x_2$ such that

$$y_q(\tau_1) = u(\tau_1),$$
$$y_q^{(i-1)}(x_1) = y_{i1} = u^{(i-1)}(x_1), \quad 1 \le i \le m_1 - 1,$$
$$y_q(\tau_2) = u(\tau_2),$$
$$y_q^{(i-1)}(x_j) = y_{ij} = u^{(i-1)}(x_j), \quad 1 \le i \le m_j, \quad 2 \le j \le k.$$

In particular, $y_q(x)$ and $u(x)$ are distinct solutions of the same $(k+2)$-point (or if $m_1 = 1$, the same $(k+1)$-point) conjugate boundary value problem (1.1.1), (1.1.2). This contradicts Corollary 3.3.1; thus, S is closed and the proof is complete. □

Henderson [60] considered a related uniqueness implies existence type question for a special case of (3.3.15). Recall the $(k+2)$-point nonlocal boundary conditions, (3.3.18), restated here, for convenience.

$$y^{(i-1)}(x_j) = y_{ij}, \ 1 \le i \le m_j, \ 1 \le j \le k,$$
$$y(x_{k+1}) - y(x_{k+2}) = y_n.$$

Note that Theorem 3.4.6 says, in rough terms, that uniqueness of solutions of the boundary value problem, (1.1.1)-(3.3.15), for $k = n - 1$ implies the existence of solutions of the boundary value problem, (1.1.1)-(3.3.15), for $1 \le k \le n - 1$. In [60], the implication of uniqueness of solutions of the boundary value problem, (1.1.1)-(3.3.18), for $k = n - 2$ and existence of solutions of the boundary value problem, (1.1.1)-(3.3.18), for $k = 2$ are studied. This line of questioning was motivated by the original work of Henderson and Jackson [62].

Theorem 3.4.7. [[60], Henderson, Thm. 5] *Assume that with respect to* (1.1.1), *conditions* (A) *and* (B) *are satisfied. Assume that, for* $k = n - 2$, *there is at most one solution of* (1.1.1)-(3.3.18) *on* (a, b). *Assume that for* $k = 2$, *there exists a solution of* (1.1.1)-(3.3.18) *on* (a, b). *Then, for each* $1 \leq k \leq n - 1$, *there exists a solution of* (1.1.1)-(3.3.18) *on* (a, b).

Proof. Note that as a consequence of Theorem 3.3.19, solutions of (1.1.1)-(3.3.18), for each $1 \leq k \leq n - 2$, are unique when they exist. Moreover, due to Theorem 3.4.6, it suffices to verify that solutions of (1.1.1)-(3.3.18), for $k = n - 1$, are unique when they exist.

This proof will consist of three steps. First, by an induction on k, it is shown that for each $3 \leq k \leq n - 4$, there exist solutions of (1.1.1)-(3.3.18) on (a, b). Then a similar induction is to obtain existence of solutions of (1.1.1)-(3.3.18) for $k = n - 3$ in the particular case that the index $m_1 = 1$. Then using this fact and methods similar to those employed by Jackson [75] and Clark and Henderson [13], it is shown that solutions of (1.1.1)-(3.3.18), for $k = n - 1$, are unique when they exist.

For the first step of the proof, let $2 < \kappa \leq n-4$ be given and assume that, for each ℓ with $2 \leq \ell < \kappa$, there exist solutions of (1.1.1)-(3.3.18), for $k = \ell$, on (a, b). Let $m_1, \ldots, m_\kappa$ be positive integers such that $m_1 + \cdots + m_\kappa = n - 1$, let $a < x_1 < \cdots < x_\kappa < x_{\kappa+1} < x_{\kappa+2} < b$ be given, and let $y_{ij} \in \mathbb{R}, 1 \leq i \leq m_j, 1 \leq j \leq \kappa$, and $y_n \in \mathbb{R}$ be given. By induction on m_1 and m_2, we prove that there exist solutions of (1.1.1)-(3.3.18), for $k = \kappa$, for the corresponding boundary points and boundary values above.

Let $m_1 = m_2 = 1$ and let $m_3, \ldots, m_\kappa$ be any positive integers such that $1 + 1 + m_3 + \cdots + m_\kappa = n - 1$. Let $z(x)$ be the solution, given by the induction hypothesis, of (1.1.1)-(3.3.18), for $k = \kappa - 1$ and for indices $i_1 + \cdots + i_{\kappa-1} = n - 1$, where $i_1 = 2$ and $i_j = m_{j+1}$, for $2 \leq j \leq \kappa - 1$, satisfying

$$
\begin{aligned}
z(x_2) &= y_{12}, \quad z'(x_2) = 0, \\
z^{(i-1)}(x_j) &= y_{ij}, \quad 1 \leq i \leq m_j, \ 3 \leq j \leq \kappa, \\
z(x_{\kappa+1}) &- z(x_{\kappa+2}) = y_n.
\end{aligned}
$$

Define

$$
S = \{y(x_1) \mid y \text{ is a solution of } (1.1.1) \text{ satisfying}
$$
$$
y(x_2) = z(x_2), y^{(i-1)}(x_j) = z^{(i-1)}(x_j), 1 \leq i \leq m_j, 3 \leq j \leq \kappa,
$$
$$
y(x_{\kappa+1}) - y(x_{\kappa+2}) = z(x_{\kappa+1}) - z(x_{\kappa+2})\}.
$$

As usual, $S \neq \emptyset$ since $z(x_1) \in S$ and Theorem 1.1.6 implies S is open. And so, to complete the first step of the proof, it remains to show that S is closed.

Assume S is not closed, assume $s_0 \in \overline{S} \setminus S$ and assume there is a strictly monotone increasing sequence $\{s_q\} \subset S$ which converges to s_0. Let y_q denote a solution of (1.1.1) satisfying

$$y_q(x_1) = s_q, \; y_q(x_2) = y_{12},$$
$$y_q^{(i-1)}(x_j) = y_{ij}, \; 1 \leq i \leq m_j, \; 3 \leq j \leq \kappa,$$
$$y_q(x_{\kappa+1}) - y(x_{\kappa+2}) = y_n.$$

From the uniqueness results of Theorem 3.3.19, it follows that $y_q(x) < y_{q+1}(x)$ on (a, x_2), for each $q \geq 1$. Since $s_0 \notin S$, it follows from the "compactness condition" (CP) that $\{y_q(x)\}$ is not bounded above on any nondegenerate subinterval of (a, x_2).

Now, let $u(x)$ be the solution, given by the induction hypotheses, of (1.1.1)-(3.3.18), for $k = \kappa - 1$ and for the indices, $i_1 + \cdots + i_{\kappa-1} = n - 1$, where $i_1 = 1, i_2 = m_3 + 1$, and $i_j = m_{j+1}$, for $3 \leq j \leq \kappa - 1$, satisfying

$$u(x_1) = s_0,$$
$$u^{(i-1)}(x_3) = z^{(i-1)}(x_3), \; 1 \leq i \leq m_3,$$
$$u^{(m_3)}(x_3) = 0,$$
$$u^{(i-1)}(x_j) = z^{(i-1)}(x_j), \; 1 \leq i \leq m_j, \; 4 \leq j \leq \kappa,$$
$$u(x_{\kappa+1}) - u(x_{\kappa+2}) = z(x_{\kappa+1}) - z(x_{\kappa+2}).$$

Since $\{y_q(x)\}$ is not bounded above on either (a, x_1) or (x_1, x_2), and $y_q(x_1) < u(x_1)$, for all $q \geq 1$, it follows that there exists q sufficiently large and $a < \tau_1 < x_1 < \tau_2 < x_2$ such that

$$y_q(\tau_1) = u(\tau_1), \quad y_q(\tau_1) = u(\tau_1).$$

Moreover,

$$y_q^{(i-1)}(x_j) = u^{(i-1)}(x_j), \; 1 \leq i \leq m_j, \; 3 \leq j \leq \kappa,$$
$$y_q(x_{\kappa+1}) - y_q(x_{\kappa+2}) = u(x_{\kappa+1}) - u(x_{\kappa+2}),$$

which contradicts the uniqueness of solutions of (1.1.1)-(3.3.18), for $k = \kappa$ and for indices $i_1 + \cdots + i_\kappa = n - 1$, where $i_1 = i_2 = 1$, and $i_j = m_j, 3 \leq j \leq \kappa$. Therefore, S is a closed there exist a solution of (1.1.1)-(3.3.18), for $k = \kappa$ and for indices $m_1 = m_2 = 1$ and any positive integers $m_3, \ldots, m_\kappa$ such that $1 + 1 + m_3 + \cdots + m_\kappa = n - 1$.

To continue the first step, perform an induction on m_2. Let $m_1 = 1$, let $m_2 > 1$ be fixed, and assume for all $1 \leq \ell < m_2$ and all positive integers

$m_{j\ell}, 3 \leq j \leq \kappa$, for which $1 + \ell + m_{3\ell} + \cdots + m_{\kappa\ell} = n - 1$, that there exist solutions of (1.1.1)-(3.3.18), for $k = \kappa$ and for these indices. Now, let $m_3, \ldots, m_\kappa$ be any fixed positive integers such that $1 + m_2 + m_3 + \cdots + m_\kappa = n - 1$, and also let $z(x)$ be the solution (given by the majorant induction hypotheses) of (1.1.1)-(3.3.18), for $k = \kappa - 1$ and for indices $i_1 + \cdots + i_{\kappa-1} = n - 1$, where $i_1 = m_2 + 1$ and $i_j = m_{j+1}$, for $2 \leq j \leq \kappa - 1$, satisfying

$$z^{(i-1)}(x_2) = y_{i2}, \ 1 \leq i \leq m_2,$$
$$z^{(m_2)}(x_2) = 0,$$
$$z^{(i-1)}(x_j) = y_{ij}, \ 1 \leq i \leq m_j, \ 3 \leq j \leq \kappa,$$
$$z(x_{\kappa+1}) - z(x_{\kappa+2}) = y_n;$$

$$S = \{y(x_1) \mid y \text{ is a solution of (1.1.1) satisfying}$$
$$y^{(i-1)}(x_j) = z^{(i-1)}(x_j), \ 1 \leq i \leq m_j, \ 2 \leq j \leq \kappa,$$
$$y(x_{\kappa+1}) - y(x_{\kappa+2}) = z(x_{\kappa+1}) - z(x_{\kappa+2})\}.$$

S is a nonempty open subset of the reals. To show S is closed assume there exists $s_0 \in \overline{S} \setminus S$, and a strictly monotone increasing sequence $\{s_q\} \subset S$ converging to s_0. Let $y_q(x)$ denote the corresponding solution of (1.1.1) satisfying the boundary conditions

$$y(x_1) = s_q,$$
$$y^{(i-1)}(x_j) = z^{(i-1)}(x_j), \ 1 \leq i \leq m_j, \ 2 \leq j \leq \kappa,$$
$$y(x_{\kappa+1}) - y(x_{\kappa+2}) = z(x_{\kappa+1}) - z(x_{\kappa+2}).$$

Then, by Theorem 3.3.19, $\{y_q(x)\}$ is strictly increasing pointwise on (a, x_2), and since $s_0 \notin S$, $\{y_q(x)\}$ is not bounded above on any nondegenerate compact subinterval of (a, x_2).

Let $u(x)$ be the solution, given by the induction hypothesis on m_2, of (1.1.1)-(3.3.18), for $k = \kappa$ and for indices $i_1 + i_2 + \cdots + i_\kappa = n - 1$, where $i_1 = 1, i_2 = m_2 - 1, i_3 = m_3 + 1$ and $i_j = m_j$, for $4 \leq j \leq \kappa$, satisfying

$$u(x_1) = s_0,$$
$$u^{(i-1)}(x_2) = z^{(i-1)}(x_2), \ 1 \leq i \leq m_2 - 1,$$
$$u^{(i-1)}(x_3) = z^{(i-1)}(x_3), \ 1 \leq i \leq m_3,$$
$$u^{(m_3)}(x_3) = 0,$$
$$u^{(i-1)}(x_j) = z^{(i-1)}(x_j), \ 1 \leq i \leq m_j, \ 4 \leq j \leq \kappa, \ (\text{omitted if } \kappa = 3),$$
$$u(x_{\kappa+1}) - u(x_{\kappa+2}) = z(x_{\kappa+1}) - z(x_{\kappa+2}).$$

An argument similar to the previous one implies that there exist q sufficiently large such that $y_q(x)$ and $u(x)$ are distinct solutions of the same problem (1.1.1)-(3.3.18), for $k = \kappa + 1$ and for indices $i_1 + i_2 + \cdots + i_{\kappa+1} = n - 1$, where $i_1 = 1, i_2 = 1, i_3 = m_2 - 1$ and $i_j = m_{j-1}, 4 \leq j \leq \kappa + 1$. Since $\kappa + 1 \leq n - 3$, this contradicts the conclusion of Theorem 3.3.19. Thus S is also closed, and there exist a solution of (1.1.1)-(3.3.18), for $k = \kappa$ and for the special indices $m_1 = 1$ and any positive integers $m_2, \ldots, m_\kappa$ such that $1 + m_2 + \cdots + m_\kappa = n - 1$.

Continuing the first step of the proof, perform an induction on m_1. Let $m_1 > 1$ be a fixed positive integer, and assume, for all $1 \leq \ell < m_1$ and all positive integers $m_{2\ell}, \ldots, m_{\kappa\ell}$ for which $\ell + m_{2\ell} + \cdots + m_{\kappa\ell} = n - 1$, there exist solutions of (1.1.1)-(3.3.18), for $k = \kappa$ and for these indices. This time, let $m_2 = 1$ and $m_3, \ldots, m_\kappa$ be any fixed positive integers such that $m_1 + 1 + m_3 + \cdots + m_\kappa = n - 1$, and let $z(x)$ be the solution, given by the latter induction hypotheses on m_1, of (1.1.1)-(3.3.18), for $k = \kappa$ and for indices $i_1 + \cdots + i_\kappa = n - 1$, where $i_1 = m_1 - 1, i_2 = 2$ and $i_j = m_j$, for $3 \leq j \leq \kappa$, satisfying

$$z^{(i-1)}(x_1) = y_{i1}, \ 1 \leq i \leq m_1 - 1,$$
$$z(x_2) = y_{1,2}, \ z'(x_2) = 0,$$
$$z^{(i-1)}(x_j) = y_{ij}, \ 1 \leq i \leq m_j, \ 3 \leq j \leq \kappa,$$
$$z(x_{\kappa+1}) - z(x_{\kappa+2}) = y_n.$$

And this time, define

$$S = \{y^{(m_1-1)}(x_1) \mid y(x) \text{ is a solution of (1.1.1) satisfying}$$
$$y^{(i-1)}(x_1) = z^{(i-1)}(x_1), 1 \leq i \leq m_1 - 1, \ y(x_2) = z(x_2),$$
$$y^{(i-1)}(x_j) = z^{(i-1)}(x_j), \ 1 \leq i \leq m_j, \ 3 \leq j \leq \kappa,$$
$$y(x_{\kappa+1}) - y(x_{\kappa+2}) = z(x_{\kappa+1}) - z(x_{\kappa+2})\},$$

S is a nonempty open subset of the reals and to show S is closed assume that s_0 is a limit point of S with $s_0 \notin S$, that $\{s_q\} \subset S$ is a strictly increasing sequence converging to s_0, and for $q \in \mathbb{N}$, y_q is the corresponding solution of (1.1.1) such that

$$y_q^{(i-1)}(x_1) = z^{(i-1)}(x_1), 1 \leq i \leq m_1 - 1, \ y_q(x_2) = z(x_2),$$
$$y_q^{(i-1)}(x_j) = z^{(i-1)}(x_j), \ 1 \leq i \leq m_j, \ 3 \leq j \leq \kappa,$$
$$y_q(x_{\kappa+1}) - y_q(x_{\kappa+2}) = z(x_{\kappa+1}) - z(x_{\kappa+2}).$$

By Theorem 3.3.19, it follows that either

(i) $y_q(x) > y_{q+1}(x)$ on (a, x_1) and $y_q(x) < y_{q+1}(x)$ on (x_1, x_2) for each $q \geq 1$, when m_1 is even, or

(ii) $y_q(x) < y_{q+1}(x)$ on $(a, x_2) \setminus \{x_1\}$ for each $\nu \geq 1$, when m_1 is odd.

In turn, it follows from the "compactness condition" (CP) and the assumption that $s_0 \notin S$, that either

(i) $\{y_q(x)\}$ is not bounded below on any nondegenerate compact subinterval of (a, x_1) and is not bounded above on any nondegenerate compact subinterval of (x_1, x_2), when m_1 is even, or

(ii) $\{y_q(x)\}$ is not bounded above on any nondegenerate compact subinterval of $(a, x_2) \setminus \{x_1\}$, when m_1 is odd.

Let $u(x)$ be the solution of $(1.1.1)$-$(3.3.18)$, for $k = \kappa - 1$ and for indices $i_1 + i_2 + \cdots + i_{\kappa-1} = n - 1$, where $i_1 = m_1, i_2 = m_3 + 1$ and $i_j = m_{j+1}$, for $3 \leq j \leq \kappa - 1$, satisfying

$$u^{(i-1)}(x_1) = z^{(i-1)}(x_1), \ 1 \leq i \leq m_1 - 1,$$
$$u^{(m_1-1)}(x_1) = r_0,$$
$$u^{(i-1)}(x_3) = z^{(i-1)}(x_3), \ 1 \leq i \leq m_3,$$
$$u^{(m_3)}(x_3) = 0,$$
$$u^{(i-1)}(x_j) = z^{(i-1)}(x_j), \ 1 \leq i \leq m_j, \ 4 \leq j \leq \kappa,$$
$$u(x_{\kappa+1}) - u(x_{\kappa+2}) = z(x_{\kappa+1}) - z(x_{\kappa+2}).$$

The unboundedness conditions on $\{y_q(x)\}$ observed in (i) or (ii), and the fact that, for $q \geq 1, u^{(i-1)}(x_1) = y_q^{(i-1)}(x_1), \ 1 \leq i \leq m_1 - 1$, while $u^{(m_1-1)}(x_1) > y_q^{(m_1-1)}(x_1)$, implies there exists q sufficiently large and $a < \tau_1 < x_1 < \tau_2 < x_2$ such that

$$y_q(\tau_1) = u(\tau_1), \quad y_q(\tau_1) = u(\tau_1).$$

Thus, there exists q sufficiently large such that $y_q(x)$ and $u(x)$ are distinct solutions of the same problem $(1.1.1)$-$(3.3.18)$, for $k = \kappa + 1$ and for indices $i_1 + i_2 + \cdots + i_{\kappa+1} = n - 1$, where $i_1 = 1, i_2 = m_1 - 1, i_3 = 1$ and $i_{j+1} = m_j$, for $3 \leq j \leq \kappa$. But, since $\kappa + 1 \leq n - 3$, Theorem 3.3.19 is contradicted. We conclude that S is closed, $S = \mathbb{R}$, and there exist solutions of $(1.1.1)$-$(3.3.18)$, for $k = \kappa$ and for the special indices $m_2 = 1$ and any positive integers $m_1, m_3, \ldots, m_\kappa$ such that $m_1 + 1 + m_3 + \cdots + m_\kappa = n - 1$.

For the final stage of the induction, we recall that $m_1 > 1$, and we now assume that $m_2 > 1$ and for all $1 \leq \ell < m_2$ and for all positive integers $m_{3\ell}, \ldots, m_{\kappa\ell}$ for which $m_1 + \ell + m_{3\ell} + \cdots + m_{\kappa\ell} = n - 1$, there exist solutions of $(1.1.1)$-$(3.3.18)$, for $k = \kappa$ and for these indices. With m_1 and m_2 as above, let $m_3, \ldots, m_\kappa$ be any fixed positive integers such that $m_1 + m_2 + m_3 + \cdots + m_\kappa = n - 1$, and let $z(x)$ be the solution, given by the induction hypotheses on m_1, of $(1.1.1)$-$(3.3.18)$, for $k = \kappa$ and for indices

$i_1 + \cdots + i_\kappa = n - 1$, where $i_1 = m_1 - 1, i_2 = m_2 + 1$ and $i_j = m_j$, for $3 \leq j \leq \kappa$, satisfying

$$z^{(i-1)}(x_1) = y_{i1}, \ 1 \leq i \leq m_1 - 1,$$
$$z^{(i-1)}(x_2) = y_{i2}, \ 1 \leq i \leq m_2,$$
$$z^{(m_2)}(x_2) = 0,$$
$$z^{(i-1)}(x_j) = y_{ij}, \ 1 \leq i \leq m_j, \ 3 \leq j \leq \kappa,$$
$$z(x_{\kappa+1}) - z(x_{\kappa+2}) = y_n;$$

Finally, define

$$S = \{y^{(m_1-1)}(x_1)| \ y(x) \text{ is a solution of (1.1.1) with}$$
$$y^{(i-1)}(x_1) = z^{(i-1)}(x_1), 1 \leq i \leq m_1 - 1,$$
$$y^{(i-1)}(x_j) = z^{(i-1)}(x_j), 1 \leq i \leq m_j, 2 \leq j \leq \kappa,$$
$$y(x_{\kappa+1}) - y(x_{\kappa+2}) = z(x_{\kappa+1}) - z(x_{\kappa+2})\},$$

and we again conclude that S is a nonempty open subset of the reals.

To show that S is also closed, assume not, let $s_0 \in \overline{S} \backslash S$ and $\{s_q\} \subset S$ be strictly increasing and converging to s_0. Let $\{y_q(x)\}$ be the corresponding sequence of solutions of (1.1.1) satisfying

$$y^{(i-1)}(x_1) = z^{(i-1)}(x_1), 1 \leq i \leq m_1 - 1,$$
$$y^{(m_1-1)}(x_1) = s_q,$$
$$y^{(i-1)}(x_j) = z^{(i-1)}(x_j), 1 \leq i \leq m_j, 2 \leq j \leq \kappa,$$
$$y(x_{\kappa+1}) - y(x_{\kappa+2}) = z(x_{\kappa+1}) - z(x_{\kappa+2}).$$

As in the last previous step above, $\{y_q(x)\}$ satisfies property (i), when m_1 is even, and satisfies property (ii), when m_1 is odd. If $u(x)$, this time, is the solution of (1.1.1)-(3.3.18), for $k = \kappa$, given by the last inductive assumption on m_2, for indices $i_1 | i_2 | \quad | i_\kappa = n \quad 1$, where $i_1 - m_1, i_2 - m_2 - 1, i_3 = m_3 + 1$ and $i_j = m_j$, for $4 \leq j \leq \kappa$, satisfying

$$u^{(i-1)}(x_1) = z^{(i-1)}(x_1), \ 1 \leq i \leq m_1 - 1,$$
$$u^{(m_1-1)}(x_1) = s_0,$$
$$u^{(i-1)}(x_2) = z^{(i-1)}(x_2), \ 1 \leq i \leq m_2 - 1,$$
$$u^{(i-1)}(x_3) = z^{(i-1)}(x_3), \ 1 \leq i \leq m_3,$$
$$u^{(m_3)}(x_3) = 0,$$
$$u^{(i-1)}(x_j) = z^{(i-1)}(x_j), \ 1 \leq i \leq m_j, \ 4 \leq j \leq \kappa,$$
$$u(x_{\kappa+1}) - w(x_{\kappa+2}) = z(x_{\kappa+1}) - z(x_{\kappa+2}),$$

then it follows that some q sufficiently large $y_q(x)$ and $u(x)$ are distinct solutions of the same problem (1.1.1)-(3.3.18), for $k = \kappa + 2$ and for indices

$i_1 + i_2 + \cdots + i_{\kappa+2} = n - 1$, where $i_1 = 1, i_2 = m_1 - 1, i_3 = 1, i_4 = m_2 - 1$ and $i_{j+2} = m_j$, for $3 \leq j \leq \kappa$. Since $\kappa + 2 \leq n - 2$, this contradicts either one of the hypotheses of this Theorem or the conclusion of Theorem 3.3.19. Therefore, S is closed, $S = \mathbb{R}$; hence, by induction, this completes the part of the proof that, for each $3 \leq k \leq n - 4$, there exist solutions of (1.1.1)-(3.3.18) on (a, b).

The second step of the proof is to obtain existence of solutions of (1.1.1)-(3.3.18) for $k = n - 3$ in the case that $m_1 = 1$. This proof duplicates the induction argument used above in proving the existence of solutions of (1.1.1)-(3.3.18), for $3 \leq k \leq n - 4$ and for all indices $m_1 + \cdots + m_k = n - 1$, with $m_1 = 1$. So, it is omitted here. We point out that in this second step, when $k = n - 3$, the proof that the corresponding nonempty open set S is also closed depends on contradicting the uniqueness of solutions of (1.1.1)-(3.3.18), for $k = n - 3$ when the index $m_2 = 1$ (still with $m_1 = 1$), and on contradicting the uniqueness of solutions of (1.1.1)-(3.3.18), for $k = n - 2$ when the index $m_2 > 1$ (still with $m_1 = 1$).

For the third and final step of the proof, it is shown that solutions of (1.1.1)-(3.3.18), for $k = n - 1$, are unique, when they exist. Once that has been established, then by way of Theorem 3.4.6, the proof of this theorem will be complete. Assume for purposes of contradiction that $y(x)$ and $z(x)$ are distinct solutions of the same problem (1.1.1)-(3.3.18), for $k = n - 1$, and that

$$y(x_j) = z(x_j), 1 \leq j \leq n - 1, \quad y(x_n) - y(x_{n+1}) = z(x_n) - z(x_{n+1}),$$

where $a < x_1 < \cdots < x_{n-1} < x_n < x_{n+1} < b$. Since solutions of (1.1.1)-(3.3.18), for $k = n - 2$, are unique, when they exist, $y'(x_j) \neq z'(x_j), 1 \leq j \leq n - 1$. Assume that the x_j's are successive zeros of $y(x) - z(x)$, $1 \leq j \leq n - 1$, that $y(x) < z(x)$ on (x_{n-4}, x_{n-3}) and on (x_{n-2}, x_{n-1}), and that $y(x) > z(x)$ on (x_{n-3}, x_{n-2}). For each $p \geq 1$, let $y_p(x)$ denote the solution of (1.1.1)-(3.3.18), for $k = n - 3$ and for indices $m_1 + \cdots + m_{n-3} = n - 1$, where $m_1 = 1, m_j = 1$, for $2 \leq j \leq n - 4$, and $m_{n-3} = 3$, satisfying

$$y_p(x_j) = y(x_j), \ 1 \leq j \leq n - 3,$$
$$y_p'(x_{n-3}) = y'(x_{n-3}),$$
$$y_p''(x_{n-3}) = y''(x_{n-3}) + p,$$
$$y_p(x_n) - y_p(x_{n+1}) = y(x_n) - y(x_{n+1}).$$

Since solutions of (1.1.1)-(3.3.18), for $k = n - 2$, are unique, it follows that, for each $p \geq 1$, $y_{p+1}(x) > y_p(x) > y(x)$ on $(x_{n-4}, x_n) \setminus \{x_{n-3}\}$. For each $p \geq 1$, define

$$E_p := \{x \mid x_{n-2} \leq x \leq x_{n-1} \text{ and } y_p(x) \leq z(x)\}.$$

It follows from Continuous Dependence, Theorem 1.1.6 that solutions of (1.1.1)-(3.3.18), for $k = n - 3$, depend continuously upon the type of boundary conditions used in defining each $y_p(x)$. From this continuity and the fact that solutions of (1.1.1)-(3.3.18), for $k = n - 2$, are unique, when they exist, we conclude that each E_p is nonempty. Therefore, $E_{p+1} \subset E_p \subset (x_{n-2}, x_{n-1})$, for each $p \geq 1$, and each E_p is nonempty and compact. Hence

$$\bigcap_{p=1}^{\infty} E_p := E \subset (x_{n-2}, x_{n-1})$$

is nonempty and compact.

Next, if t_1, $t_2 \in E$ with $x_{n-2} < t_1 < t_2 < x_{n-1}$, then an argument similar to the one used in showing that each E_p is nonempty leads to the conclusion that $[t_1, t_2] \subset E$. In this case, the sequence $\{y_p(x)\}$ is uniformly bounded on $[t_1, t_2]$. Due to the "compactness condition" (CP), this is impossible, since $y_p''(x_{n-3}) \to +\infty$ as $p \to +\infty$. Hence, $E = \{x_0\}$, where $x_{n-2} < x_0 < x_{n-1}$, and $\lim_{p\to\infty} y_p(x_0) := y_0 \leq z(x_0)$. We will proceed to show that each of the cases, $y_0 = z(x_0)$ and $y_0 < z(x_0)$, is impossible. This will show the existence of two distinct solutions of the same problem (1.1.1)-(3.3.18), for $k = n - 1$, is not possible.

Assume first that $y_0 = z(x_0)$. Let $0 < \epsilon < z(x_0) - y(x_0)$. Then there is an $\eta > 0$ such that the solution $z(x; \eta)$ of (1.1.1)-(3.3.18), for $k = n - 3$, satisfying the boundary conditions of the form,

$$z(x_j; \eta) = z(x_j),\ 1 \leq j \leq n - 3,$$
$$z'(x_{n-3}; \eta) = z'(x_{n-3}),$$
$$z''(x_{n-3}; \eta) = z''(x_{n-3}) - \eta,$$
$$z(x_n; \eta) - z(x_{n+1}; \eta) = z(x_n) - z(x_{n+1}),$$

also satisfies

$$y(x_0) < z(x_0) - \epsilon < z(x_0; \eta) < z(x_0) = y_0.$$

The solution $z(x; \eta)$ can be used in place of $z(x)$ in defining the sequence of sets $\{E_p\}$ with respect to the original sequence of solutions $\{y_p(x)\}$. The same arguments would again yield that each E_p is nonempty, but this is impossible. Thus, the case, $y_0 = z(x_0)$, is not possible.

Assume second that $y(x_0) < y_0 < z(x_0)$. This time, for $0 \leq \lambda \leq 1$, let $z(x; \lambda)$ be the solution of (1.1.1)-(3.3.18), for $k = n - 3$, satisfying the boundary conditions of the form,

$$z(x_j; \lambda) = \lambda y(x_j) + (1 - \lambda)z(x_j),\ 1 \leq j \leq n - 3,$$
$$z'(x_{n-3}; \lambda) = \lambda y'(x_{n-3}) + (1 - \lambda)z'(x_{n-3}),$$
$$z''(x_{n-3}; \lambda) = \lambda y''(x_{n-3}) + (1 - \lambda)z''(x_{n-3}),$$
$$z(x_n; \lambda) - z(x_{n+1}; \lambda) = \lambda(y(x_n) - y(x_{n+1})) + (1 - \lambda)(z(x_n) - z(x_{n+1})).$$

Again, by Continuous Dependence, Theorem 1.1.6, it follows that there exists λ_0 such that $0 < \lambda_0 < 1$ and $z(x_0; \lambda_0) = y_0$. In turn, there exists $\delta > 0$ such that $[x_0 - \delta, x_0 + \delta] \subset (x_{n-2}, x_{n-1})$ and such that $z(x; \lambda_0) < z(x)$ on $[x_0 - \delta, x_0 + \delta]$. Then, with $\{y_p(x)\}$ again the same sequence defined earlier, we have

$$\lim_{p \to \infty} y_p(x) > z(x; \lambda_0)$$

on $[x_0 - \delta, x_0 + \delta] \setminus \{x_0\}$, and

$$\lim_{p \to \infty} y_p(x_0) = y_0 = z(x_0; \lambda_0).$$

An argument similar to the one used in the first case shows this is impossible, and so the case, $y_0 < z(x_0)$, is also impossible. The proof is complete.

□

The boundary conditions (3.3.15) have the conjugate boundary conditions stacked at the left and one nonlocal condition at the right. Eloe, Henderson and Khan [29] produced an inductive argument and considered the boundary conditions, (3.3.16), in which conjugate boundary conditions are stacked at the left and j nonlocal condition are specified at the right. Again, we restate the boundary conditions, (3.3.16), for convenience:

$$y^{(i-1)}(x_l) = y_{il}, \ 1 \le i \le m_l, \ 1 \le l \le k, \ \text{conjugate conditions},$$
$$(a_1 y(x_{k+1}) - a_2 y(x_{k+2}), \ldots, a_{2j-1} y(x_{k+2j-1}) - a_{2j} y(x_{k+2j}))$$
$$= (y_n, y_{n-1}, \ldots, y_{n-(j-1)}), \ \text{nonlocal conditions},$$

where, $1 \le j \le n - 1$, $1 \le k \le n - j$, are positive integers $m_1, \ldots, m_k$ such that $m_1 + \cdots + m_k = n - j$, $a < x_1 < x_2 < \cdots < x_k < x_{k+1} < \cdots < x_{k+2j} < b$, $y_{il} \in \mathbb{R}, 1 \le i \le m_l, 1 \le l \le k$, and real values $y_n, y_{n-1}, \ldots, y_{n-(j-1)}$, and $a_1, a_2, \ldots, a_{2j}$ are positive real numbers. Recall as well the terminology, $(k; j)$-*point boundary conditions*.

We now present our uniqueness implies existence result for the $(k; j)$-point boundary value problems.

Theorem 3.4.8. [[29], Eloe, Khan and Henderson, Thm. 3.3] *Let $j_0 \in \{0, \ldots, n-1\}$. Assume that solutions of (1.1.1)-(3.3.16), when $k = n - j_0$, $j = j_0$, are unique. Then, for each $1 \le j \le j_0$, $1 \le k \le n - j$, positive integers $m_1, \ldots, m_k$ such that $m_1 + \cdots + m_k = n - j$, points $a < x_1 < \cdots < x_{k+2j} < b$, real values $y_{il}, 1 \le i \le m_l, 1 \le l \le k$, $y_n, y_{n-1}, \ldots, y_{n-(j-1)} \in \mathbb{R}$, and $a_1, a_2, \ldots, a_{2j}$, positive real numbers, there exists a unique solution of the $(k; j)$-point BVP, (1.1.1)-(3.3.16).*

Proof. Let $1 \leq j \leq j_0$, $1 \leq k \leq n - j$, positive integers $m_1, \ldots, m_k$ such that $m_1 + \cdots + m_k = n - j$, points $a < x_1 < \cdots < x_{k+2j} < b$, real values $y_{il}, 1 \leq i \leq m_l, 1 \leq l \leq k$, $y_n, y_{n-1}, \ldots, y_{n-(j-1)} \in \mathbb{R}$, and $a_1, a_2, \ldots, a_{2j}$, positive real numbers, be given.

Since solutions of the $(n - j_0; j_0)$-point BVP, (1.1.1)-(3.3.16), are unique, it follows from Corollary 3.3.3 that solutions of the $(l; 0)$-point conjugate BVP, for $2 \leq l \leq n$ are unique; thus, solutions of the $(l; 0)$-point conjugate BVP, for $2 \leq l \leq n$ exist by Theorem 2.3.4.

Let $1 \leq j \leq j_0$ and $1 \leq k \leq n - j$. Let $z(x)$ be the unique solution of (1.1.1) satisfying $(k + j + 1; 0)$-point conjugate boundary conditions

$$z^{(i-1)}(x_1) = y_{i1}, \quad 1 \leq i \leq m_1 - 1,$$
$$z^{(i-1)}(x_l) = y_{il}, \quad 1 \leq i \leq m_l, \quad 2 \leq l \leq k,$$
$$z(x_{k+1}) = \frac{y_n}{a_1},$$
$$z(x_{k+3}) = \frac{y_{n-1}}{a_3},$$
$$\cdots,$$
$$z(x_{k+2j-1}) = \frac{y_{n-(j-1)}}{a_{2j-1}},$$
$$z(x_{k+2j}) = 0.$$

Note that in the case, $m_1 = 1$, z satisfies a $(k + j; 0)$-point problem with boundary conditions beginning at x_2. From the last two conditions

$$z(x_{k+2j}) = 0, \quad z(x_{k+2j-1}) = \frac{y_{n-(j-1)}}{a_{2j-1}},$$

we obtain

$$a_{2j-1}z(x_{k+j}) - a_{2j}z(t_1) = y_{n-(j-1)}.$$

Define the set

$$S = \left\{ y^{(m_1-1)}(x_1) \mid y \text{ is a solution of (1.1.1) satisfying} \right.$$

$$y^{(i-1)}(x_1) = y_{i1}, \, 1 \leq i \leq m_1 - 1,$$
$$y^{(i-1)}(x_l) = y_{il}, \, 1 \leq i \leq m_l, \, 2 \leq l \leq k,$$
$$y(x_{k+1}) = \frac{y_n}{a_1}, y(x_{k+3}) = \frac{y_{n-1}}{a_3}, \ldots, \, y(x_{k+2j-3}) = \frac{y_{n-(j-2)}}{a_{2j-3}},$$
$$\left. a_{2j-1}y(x_{k+2j-1}) - a_{2j}y(x_{k+2j}) = y_{n-(j-1)} \right\}.$$

Clearly, $z^{(m_1-1)}(x_1) \in S$, and so S is a nonempty subset of $\mathbb{R}$.

The argument that S is open is a standard application of Theorem 1.1.6.

Now, we show that S is also a closed subset of $\mathbb{R}$. To do this, assume that S is not closed and there exists an $s_0 \in \bar{S} \setminus S$ and a strictly monotone increasing sequence $\{s_q\} \subset S$ such that $\lim_{q \to \infty} s_q = s_0$. By the definition of S, for each $q \in \mathbb{N}$, there exists a unique solution $y_q(x)$ of (1.1.1) satisfying

$$y_q^{(i-1)}(x_1) = y_{i1}, \quad 1 \le i \le m_1 - 1,$$
$$y_q^{(m_1-1)}(x_1) = s_q,$$
$$y_q^{(i-1)}(x_l) = y_{il}, \quad 1 \le i \le m_l, \quad 2 \le l \le k,$$
$$y_q(x_{k+1}) = \frac{y_n}{a_1}, y_q(x_{k+3}) = \frac{y_{n-1}}{a_3}, \dots, y_q(x_{k+2j-3}) = \frac{y_{n-(j-2)}}{a_{2j-3}},$$
$$a_{2j-1} y_q(x_{k+2j-1}) - a_{2j} y_q(x_{k+2j}) = y_{n-(j-1)}.$$

To determine the monotone behavior of $\{y_q\}$ set $v = y_q - y_{q+1}$; then,

$$v^{(i-1)}(x_1) = 0, \quad 1 \le i \le m_1 - 1,$$
$$v^{(m_1-1)}(x_1) = s_q - s_{q+1} < 0,$$
$$v^{(i-1)}(x_l) = 0, \quad 1 \le i \le m_l, \quad 2 \le l \le k,$$
$$v(x_{k+1}) = 0, v(x_{k+3}) = 0, \dots, v(x_{k+2j-3}) = 0,$$
$$a_{2j-1} v(x_{k+2j-1}) - a_{2j} v(x_{k+2j}) = y_{n-(j-1)} = 0.$$

By the uniqueness of solution of the $(k + j - 1; 1)$-point boundary value problem, it follows that

(i) $y_q(x) < y_{q+1}(x)$ on $(a, x_2) \setminus \{x_1\}$, if m_1 is odd,

or

(ii) $y_q(x) > y_{q+1}(x)$ on (a, x_1) and $y_q(x) < y_{q+1}(x)$ on (x_1, x_2), if m_1 is even.

Details are given for case (a); details for case (b) are completely analogous. So, for case (a), by the "compactness condition" (CP) and the fact that $s_0 \notin S$, then $\{y_q(x)\}$ is not uniformly bounded above on each compact subinterval of each of (a, x_1) and (x_1, x_2).

Now, let $u(x)$ be the solution of (1.1.1) satisfying $(k + j; 0)$-point conjugate boundary conditions (1.1.2) at the points

$$x_1, \dots, x_k, x_{k+1}, x_{k+3}, \dots, x_{k+2j-1};$$

in particular, assume u satisfies the conditions,

$$u^{(i-1)}(x_1) = y_{i1}, \quad 1 \le i \le m_1 - 1, \text{ (if } m_1 > 1),$$
$$u^{(m_1-1)}(x_1) = s_0,$$
$$u^{(i-1)}(x_l) = y_{il}, \quad 1 \le i \le m_l, \quad 2 \le l \le k,$$
$$u(x_{k+1}) = \frac{y_n}{a_1}, \quad u(x_{k+3}) = \frac{y_{n-1}}{a_3},$$
$$\dots,$$
$$u(x_{k+2j-1}) = \frac{y_{n-(j-1)}}{a_{2j-1}}.$$

From the monotonicity and unboundedness property of the sequence $\{y_q(x)\}$, it follows that there exists q sufficiently large and points $a < \tau_1 < x_1 < \tau_2 < x_2$ such that $y_q(\tau_1) = u(\tau_1)$, $y_q(\tau_2) = u(\tau_2)$. Hence,

$$y_q(\tau_1) = u(\tau_1),$$
$$y_q^{(i-1)}(x_1) = y_{i1} = u^{(i-1)}(x_1), \quad 1 \le i \le m_1 - 1,$$
$$y_q(\tau_2) = u(\tau_2),$$
$$y_q^{(i-1)}(x_l) = y_{il} = u^{(i-1)}(x_l), \quad 1 \le i \le m_l, \quad 2 \le l \le k$$
$$y_q(x_{k+1}) = \frac{y_n}{a_1} = u(x_{k+1}), \quad y_q(x_{k+3}) = \frac{y_{n-1}}{a_3} = u(x_{k+3}),$$

$$\cdots,$$

$$y_q(x_{k+2j-3}) = \frac{y_{n-(j-2)}}{a_{2j-3}} = u(x_{k+2j-3}).$$

Thus, $y_q(x)$ and $u(x)$ are distinct solutions of the same $(k+j+1;0)$-point (or if $m_1 = 1$, the same $(k+j;0)$-point) conjugate boundary value problem (1.1.1), (1.1.2) which contradicts Corollary 3.3.3. Thus, S is also a closed subset of $\mathbb{R}$.

As a consequence of S being a nonempty subset of $\mathbb{R}$ which is both open and closed, we have $S \equiv \mathbb{R}$. By choosing $y_{m_11} \in S$, there is a corresponding solution $y(x)$ of (1.1.1) such that

$$y^{(i-1)}(x_1) = y_{i1}, \quad 1 \le i \le m_1 - 1,$$
$$y^{(m_1-1)}(x_1) = y_{m_11},$$
$$y^{(i-1)}(x_l) = y_{il}, \quad 1 \le i \le m_l, \quad 2 \le l \le k,$$
$$y(x_{k+1}) = \frac{y_n}{a_1}, y(x_{k+3}) = \frac{y_{n-1}}{a_3},$$

$$\cdots,$$

$$y(x_{k+2j-3}) = \frac{y_{n-(j-2)}}{a_{2j-3}}, \quad a_{2j-1}y(x_{k+2j-1}) - a_{2j}y(x_{k+2j}) = y_{n-(j-1)}$$

which is the desired solution of the $(k+j-1;1)$-point boundary value problem. Since $1 \le j \le j_0$ and $1 \le k \le n - j_0$ implies $1 \le k+j-1 \le n-1$, we have shown existence for each of the $(k;1)$-point boundary value problems, $1 \le k \le n-1$.

Note, if $j_0 = 1$, the proof is complete. If $j_0 > 1$, the proof is completed by induction on j. Details for $j_0 > 1$, $j = 2$, are provided here.

Let $j_0 > 1$, let $2 \le j \le j_0$ and let $k+j \le n-1$. Let $z_1(x)$ be the unique solution of the $(k+j;1)$-point boundary value problem,

$$z_1^{(i-1)}(x_1) = y_{i1}, \quad 1 \le i \le m_1 - 1,$$
$$z_1^{(i-1)}(x_l) = y_{il}, \quad 1 \le i \le m_l, \quad 2 \le l \le k,$$
$$z_1(x_{k+1}) = \frac{y_n}{a_1},$$

$$\cdots,$$

$$z_1(x_{k+2j-3}) = \frac{y_{n-(j-2)}}{a_{2j-3}},$$
$$z_1(x_{k+2j-2}) = 0,$$
$$a_{2j-1}z_1(x_{k+2j-1}) - a_{2j}z_1(x_{k+2j}) = y_{n-(j-1)}.$$

From the two conditions

$$z(x_{k+2j-2}) = 0, \quad z(x_{k+2j-3}) = \frac{y_{n-(j-2)}}{a_{2j-3}},$$

we obtain

$$a_{2j-3}z(x_{k+2j-3}) - a_{2j-2}z(x_{k+2j-2}) = y_{n-(j-1)}.$$

Define the set

$$S_1 = \Big\{ y^{(m_1-1)}(x_1) \mid u \text{ is a solution of (1.1.1) satisfying}$$

$$y^{(i-1)}(x_1) = y_{i1}, \ 1 \le i \le m_1 - 1,$$

$$y^{(i-1)}(x_l) = y_{il}, \ 1 \le i \le m_l, \ 2 \le l \le k,$$

$$y(x_{k+1}) = \frac{y_n}{a_1}, y(x_{k+3}) = \frac{y_{n-1}}{a_3}, \ldots, y(x_{k+2j-5}) = \frac{y_{n-(j-3)}}{a_{2j-5}},$$

$$a_{2j-3}y(x_{k+2j-3}) - a_{2j-2}y(x_{k+2j-2}) = y_{n-(j-2)},$$

$$a_{2j-1}y(x_{k+2j-1}) - a_{2j}y(x_{k+2j}) = y_{n-(j-1)} \Big\}.$$

Clearly, $z_1^{(m_1-1)}(x_1) \in S_1$, and so S_1 is a nonempty subset of $\mathbb{R}$. A construction, completely analogous to the above argument implies $S_1 = \mathbb{R}$. Hence, $y_{m_11} \in S_1$, and there is a corresponding solution, $y(x)$ of (1.1.1) such that

$$y^{(i-1)}(x_1) = y_{i1}, \quad 1 \le i \le m_1 - 1,$$

$$y^{(m_1-1)}(x_1) = y_{m_11},$$

$$y^{(i-1)}(x_l) = y_{il}, \quad 1 \le i \le m_l, \quad 2 \le l \le k,$$

$$y(x_{k+1}) = \frac{y_n}{a_1},$$

$$\ldots,$$

$$y(x_{k+2j-5}) = \frac{y_{n-(j-3)}}{a_{2j-5}},$$

$$a_{2j-3}y(x_{k+2j-3}) - a_{2j-2}y(x_{k+2j-2}) = y_{n-(j-2)},$$

$$a_{2j-1}y(x_{k+2j-1}) - a_{2j}y(x_{k+2j}) = y_{n-(j-1)},$$

which is the desired solution of the $(k + j - 2; 2)$-point boundary value problem. Again, since $k + j - 2 \le n - 2$ then the above argument gives existence of solutions for each $(k; 2)$-point boundary value problem, $1 \le k \le n - 2$. The proof of the theorem is then completed by induction on j. $\qquad\square$

We close the chapter by briefly considering the nonlocal boundary conditions considered by Eloe, Henderson and Khan [30] in which nonlocal conditions are given at the right and at the left with conjugate boundary conditions stacked between. In particular, recall the $(j; k; j)$-point boundary conditions given by (3.3.17), restated here for convenience. For $n \ge 3$, and $j \geq 1$, given $1 \leq k \leq n - 2j$, positive integers $m_1, \ldots, m_k$ such that $m_1 + \cdots + m_k = n - 2j$, points $a < t_1 < \cdots < t_{2j} < x_1 < x_2 < \cdots < x_k < s_1 < \cdots < s_{2j} < b$, real values $y_i, 1 \leq i \leq j, y_{il}, 1 \leq i \leq m_l, 1 \leq l \leq k$, and real values $y_{n-(i-1)}, 1 \leq i \leq j$, consider the $(j; k; j)$-point boundary conditions, (3.3.17), given by

$a_i y(t_{2i-1}) - b_i y(t_{2i}) = y_i, \ 1 \leq i \leq j, \ j$ nonlocal conditions,

$y^{(i-1)}(x_l) = y_{il}, 1 \leq i \leq m_l, 1 \leq l \leq k$, k-point, $n - 2$ conjugate conditions,

$c_i y(s_{2i-l}) - d_i y(s_{2i}) = y_{n-(i-1)}, \ 1 \leq i \leq j, \ j$ nonlocal conditions,

where a_i, b_i, c_i, d_i, $1 \leq i \leq j$, are positive real numbers.

Theorem 3.4.9. [[30], Eloe, Khan and Henderson, Thm. 3.2] *Assume that with respect to (1.1.1), conditions (A) and (B) are satisfied. Let $j \geq 1$. Assume that for $k = n - 2j$, solutions of the $(j; n - 2j; j)$-point boundary value problem (1.1.1)-(3.3.17) are unique, when they exist. Then, for each $1 \leq k \leq n - 2j - 1$, solutions of the $(j; k; j)$-point boundary value problem (1.1.1)-(3.3.17) are unique, when they exist.*

Proof. Let $1 \leq k \leq n - 2j$, positive integers $m_1, \ldots, m_k$ such that $m_1 + \cdots + m_k = n - 2j$, points $a < t_1 < \cdots < t_{2j} < x_1 < \cdots < x_k < s_1 < \cdots < s_{2j} < b$, real values $y_i, 1 \leq i \leq j, y_{il}, 1 \leq i \leq m_l, 1 \leq l \leq k$ and $y_{n-i}, 0 \leq i \leq j - 1$, be given.

Since solutions of the $(j; n - 2j; j)$-point boundary value problem, (1.1.1)-(3.3.17), are unique, it follows by Corollary 3.3.8, solutions of the $(0; l; 0)$-point boundary value (l-point conjugate boundary value problem) for $2 \leq l \leq n$, are also unique. Let $z(x)$ be the unique solution of (1.1.1) satisfying the $(k + 2j + 2)$-point conjugate boundary conditions (1.1.2) at the points $t_1, p_1, t_2, \ldots, t_j, x_1, \ldots, x_k, s_1, \ldots, s_{j+1}$ if $m_1 > 1$, $m_k > 1$ (or alternatively, if $m_1 = 1, m_k = 1$, $z(x)$ satisfies the $(k + 2j)$-point conjugate boundary conditions and if one of $m_1 = 1, m_k = 1$ hold, then $z(x)$ satisfies the $(k + 2j + 1)$-point conjugate boundary conditions); that is,

$$z(t_1) = \frac{y_1}{a_1}, \ z(p_1) = 0,$$

$$z(t_i) = \frac{y_i}{a_i}, \ 2 \leq i \leq j,$$

$$z^{(i-1)}(x_1) = y_{i1}, \quad 1 \leq i \leq m_1 - 1,$$

$$z^{(i-1)}(x_l) = y_{il}, \quad 1 \leq i \leq m_l, \quad 2 \leq l \leq k - 1,$$

$$z^{(i-1)}(x_k) = y_{ik}, \quad 1 \leq i \leq m_k - 1,$$

$$z(s_i) = \frac{y_{n-(i-1)}}{c_i}, 1 \leq i \leq j - 1,$$

$$z(s_j) = \frac{y_{n-(j-1)}}{c_j}, \ z(s_{j+1}) = 0.$$

From the first and the last lines, we obtain
$$a_1 z(t_1) - b_1 z(p_1) = y_1,\ c_j z(s_j) - d_j z(s_{j+1}) = y_{n-(j-1)}.$$

Now, define the set
$$S = \Big\{ (y^{(m_1-1)}(x_1), y^{(m_k-1)}(x_k)) \mid y \text{ is a solution of (1.1.1) satisfying}$$

$$a_1 y(t_1) - b_1 y(p_1) = y_1,\ y(t_i) = \frac{y_i}{a_i},\ 2 \le i \le j,$$

$$y^{(i-1)}(x_1) = y_{i1},\ 1 \le i \le m_1 - 1,$$

$$y^{(i-1)}(x_l) = y_{il},\ 1 \le i \le m_l,\ 2 \le l \le k-1,$$

$$y^{(i-1)}(x_k) = y_{ik},\ 1 \le i \le m_k - 1,$$

$$y(s_i) = \frac{y_{n-(i-1)}}{c_i},\ 1 \le i \le j-1,\ c_j y(s_j) - d_j y(s_{j+1}) = y_{n-(j-1)} \Big\}.$$

Clearly, $(z^{(m_1-1)}(x_1), z^{(m_k-1)}(x_k)) \in S$, and so, $S \ne \emptyset$.

Next, choose $(\rho_0, \sigma_0) \in S$. Then, there is a solution $u(x)$ of (1.1.1) satisfying

$$a_1 u(t_1) - b_1 u(p_1) = y_1,\ u(t_i) = \frac{y_i}{a_i},\ 2 \le i \le j,$$

$$u^{(i-1)}(x_1) = y_{i1},\ 1 \le i \le m_1 - 1,$$

$$u^{(m_1-1)}(x_1) = \rho_0,$$

$$u^{(i-1)}(x_l) = y_{il},\ 1 \le i \le m_l,\ 2 \le l \le k-1,$$

$$u^{(i-1)}(x_k) = y_{ik},\ 1 \le i \le m_k - 1,$$

$$u^{(m_k-1)}(x_k) = \sigma_0,$$

$$u(s_i) = \frac{y_{n-(i-1)}}{c_i},\ 1 \le i \le j-1,\ c_j u(s_j) - d_j u(s_{j+1}) = y_{n-(j-1)}.$$

By the uniqueness of solutions of the $(1; k+2j-2; 1)$-point boundary value problem by Corollary 3.3.10, and in view of Continuous Dependence, Theorem 1.1.6, there exists a $\delta > 0$ such that, for each $|\rho - \rho_0| < \delta$, $|\sigma - \sigma_0| < \delta$, there is a solution $u_{\rho\sigma}(x)$ of (1.1.1) satisfying

$$a_1 u_{\rho\sigma}(t_1) - b_1 u_{\rho\sigma}(p_1) = y_1,\ u_{\rho\sigma}(t_i) = \frac{y_i}{a_i},\ 2 \le i \le j,$$

$$u_{\rho\sigma}^{(i-1)}(x_1) = y_{i1},\ 1 \le i \le m_1 - 1,$$

$$u_{\rho\sigma}^{(m_1-1)}(x_1) = \rho,$$

$$u_{\rho\sigma}^{(i-1)}(x_l) = y_{il},\ 1 \le i \le m_l,\ 2 \le l \le k-1,$$

$$u_{\rho\sigma}^{(i-1)}(x_k) = y_{ik},\ 1 \le i \le m_k - 1,$$

$$u_{\rho\sigma}^{(m_k-1)}(x_k) = \sigma,$$

$$u_{\rho\sigma}(s_i) = \frac{y_{n-(i-1)}}{c_i},\ 1 \le i \le j-1,\ c_j u_{\rho\sigma}(s_j) - d_j u_{\rho\sigma}(s_{j+1}) = y_{n-(j-1)}$$

and
$$|u_{\rho\sigma}(x) - u_0(x)| < \delta,\quad t_1 \le x \le s_{j+1}],$$

which implies that $(u_{\rho\sigma}^{(m_1-1)}(x_1), u_{\rho\sigma}^{(m_k-1)}(x_k)) \in S$; that is, $(\rho, \sigma) \in S$. Hence,

$$\{(\rho, \sigma)| : |\rho - \rho_0| < \delta, \ |\sigma - \sigma_0| < \delta\} \subset S$$

and S is an open, nonempty subset of $\mathbb{R}^2$.

To show that S is also a closed subset of $\mathbb{R}^2$, assume that S is not closed and assume there exists $r_0 = (p_0, q_0) \in \overline{S} \setminus S$. Let $\{r_\nu\} = \{(p_\nu, q_\nu)\} \subset S$ such that

$$\lim_{\nu \to \infty} r_\nu = \lim_{\nu \to \infty} (p_\nu, q_\nu) = (p_0, q_0) = r_0.$$

One can assume that each sequence $\{p_\nu\}, \{q_\nu\}$ is monotone, and for the sake of this argument, assume that each of $\{p_\nu\}$ and $\{q_\nu\}$ is monotone nondecreasing; the arguments for the other three cases, $\{p_\nu\}$ nondecreasing and $\{q_\nu\}$ nonincreasing, $\{p_\nu\}$ nonincreasing and $\{q_\nu\}$ nondecreasing, and each of $\{p_\nu\}, \{q_\nu\}$ nonincreasing are analogous.

So assume $p_\nu < p_{\nu+1} \leq p_0$, $q_\nu < q_{\nu+1} \leq q_0$ and assume one of the inequalities, $p_{\nu+1} \leq p_0$, $q_{\nu+1} \leq q_0$, is strict. By the definition of S, for each term $r_\nu, \nu \in \mathbb{N}$, there exists a unique solution $y_\nu(x)$ of (1.1.1) satisfying

$$a_1 y_\nu(t_1) - b_1 y_\nu(p_1) = y_1, \ y_\nu(t_i) = \frac{y_i}{a_i}, \ 2 \leq i \leq j,$$

$$y_\nu^{(i-1)}(x_1) = y_{i1}, \quad 1 \leq i \leq m_1 - 1,$$

$$y_\nu^{(m_1-1)}(x_1) = p_\nu,$$

$$y_\nu^{(i-1)}(x_l) = y_{il}, \quad 1 \leq i \leq m_l, \quad 2 \leq l \leq k - 1,$$

$$y_\nu^{(i-1)}(x_k) = y_{ik}, \ 1 \leq i \leq m_k - 1,$$

$$y_\nu^{(m_k-1)}(x_k) = q_\nu,$$

$$y_\nu(s_i) = \frac{y_{n-(i-1)}}{c_i}, \ 1 \leq i \leq j - 1, \ c_j y_\nu(s_j) - d_j y_\nu(s_{j+1}) = y_{n-(j-1)}.$$

Set $w_\nu = y_\nu - y_{\nu+1}$. Then

$$a_1 w_\nu(t_1) - b_1 w_\nu(p_1) = 0, \ w_\nu(t_i) = 0, \ 2 \leq i \leq j,$$

$$w_\nu^{(i-1)}(x_1) = 0, \quad 1 \leq i \leq m_1 - 1,$$

$$w_\nu^{(m_1-1)}(x_1) = p_\nu - p_{\nu+1} \leq 0,$$

$$w_\nu^{(i-1)}(x_l) = 0, \quad 1 \leq i \leq m_l, \quad 2 \leq l \leq k - 1,$$

$$w_\nu^{(i-1)}(x_k) = 0, \ 1 \leq i \leq m_k - 1,$$

$$w_\nu^{(m_k-1)}(x_k) = q_\nu - q_{\nu+1} \leq 0,$$

$$w_\nu(s_i) = 0, \ 1 \leq i \leq j - 1, \ c_j w_\nu(s_j) - d_j w_\nu(s_{j+1}) = 0.$$

First assume $p_{\nu+1} < p_0$ and $q_{\nu+1} < q_0$. By the uniqueness of solutions of the $(1; k + 2j - 2; 1)$-point boundary value problem, there exists $\epsilon_\nu > 0$ such that

(i) $u_\nu(x) < u_{\nu+1}(x)$ on $(x_1 - \epsilon_\nu, x_1) \cup (x_1, x_2)$, if m_1 is odd,

(ii) $u_\nu(x) > u_{\nu+1}(x)$ on $(x_1 - \epsilon_\nu, x_1)$ and $u_\nu(x) < u_{\nu+1}(x)$ on (x_1, x_2), if m_1 is even,

(iii) $u_\nu(x) < u_{\nu+1}(x)$ on $(x_{k-1}, x_k) \cup (x_k, x_k + \epsilon_\nu)$, if m_k is odd,

(iv) $u_\nu(x) > u_{\nu+1}(x)$ on (x_{k-1}, x_k) and $u_\nu(x) < u_{\nu+1}(x)$ on $(x_k, x_k + \epsilon_\nu)$, if m_k is even.

For the sake of this argument, we shall assume that m_1 and m_k are odd; the other cases are argued analogously. We also note that either $u_\nu(x) < u_{\nu+1}(x)$ on (t_j, x_1) or $u_\nu(x) < u_{\nu+1}(x)$ on (x_k, s_1). If neither of these inequalities hold, then there exist $t_j < \hat{t} < x_1$ and $x_k < \hat{s} < s_1$ such that $u_\nu(\hat{t}) - u_{\nu+1}(\hat{t}) = 0 = u_\nu(\hat{s}) - u_{\nu+1}(\hat{s})$ violating the uniqueness of solutions of $(1; k + 2j; 1)$-point boundary value problems. For the sake of this argument, let us assume that $u_\nu(x) < u_{\nu+1}(x)$ on (t_j, x_1). The sequence $\{r_\nu\}$ converges to r_0 and $r_0 \notin S$. From the "compactness condition" (CP), the sequence $\{u_\nu(x)\}$ is not uniformly bounded on any compact subset of each of $(t_j, x_1), (x_1, x_2)$, and (x_{k-1}, x_k).

Now, let $u(x)$ be the unique solution of the $(0; k + 2j; 0)$-point conjugate boundary value problem, (1.1.1), (1.1.2) satisfying at the points $t_1, p_1, t_2, \ldots, t_j, x_1, \ldots, x_k, s_1, \ldots, s_j$,

$$u(t_1) = \frac{y_1}{a_1}, \quad u(p_1) = 0,$$
$$u(t_i) = \frac{y_i}{a_i}, \quad 2 \le i \le j,$$
$$u^{(i-1)}(x_1) = y_{i1}, \quad 1 \le i \le m_1 - 1, \text{ (if } m_1 > 1),$$
$$u^{(m_1-1)}(x_1) = p_0,$$
$$u^{(i-1)}(x_l) = y_{il}, \quad 1 \le i \le m_l, \quad 2 \le l \le k - 1,$$
$$u^{(i-1)}(x_k) = y_{ik}, \quad 1 \le i \le m_k - 1, \text{ (if } m_k > 1),$$
$$u^{(m_k-1)}(x_k) = q_0,$$
$$w(s_i) = \frac{y_{n-(i-1)}}{c_i}, \quad 1 \le i \le j - 1.$$

From the monotonicity and unboundedness property of the sequence $\{y_\nu(x)\}$, it follows that there exists for ν sufficiently large and points

$$t_j < \tau_1 < x_1 < \tau_2 < x_2, \quad x_{k-1} < \rho_1 < x_k,$$

such that

$$y_\nu(\tau_1) = u(\tau_1), \quad y_\nu(\tau_2) = u(\tau_2), \quad y_\nu(\rho_1) = u(\rho_1).$$

In particular,

$$a_1 y_\nu(t_1) - b_1 y_\nu(p_1) = y_1 = au(t_1) - b_1 u(p_1),$$
$$y_\nu(t_i) = \frac{y_i}{a_i} = u(t_i),\ 2 \le i \le j,$$
$$y_\nu(\tau_1) = w(\tau_1),$$
$$y_\nu^{(i-1)}(x_1) = y_{i1} = u^{(i-1)}(x_1),\quad 1 \le i \le m_1 - 1,$$
$$y_\nu(\tau_2) = u(\tau_2),$$
$$y_\nu^{(i-1)}(x_l) = y_{il} = u^{(i-1)}(x_l),\quad 1 \le i \le m_l,\quad 2 \le l \le k - 1,$$
$$y_\nu(\rho_1) = u(\rho_1),$$
$$y_\nu^{(i-1)}(x_k) = y_{ik} = u^{(i-1)}(x_k),\quad 1 \le i \le m_k - 1,$$
$$y_\nu(s_i) = \frac{y_{n-(i-1)}}{c_i} = u(s_i),\ 1 \le i \le j - 1.$$

Thus, $y_\nu(x)$ and $u(x)$ are distinct solutions of the same $(1; k + 2j + 1; 0)$-point (or if $m_1 = 1$ and $m_k = 1$, the same $(1; k + 2j + 2; 0)$-point) boundary value problem which contradicts Corollary 3.3.9.

If $q_{\nu+1} = q_0$ (and keeping with the assumptions that m_1, m_k odd), then

$$y_\nu(x) < y_{\nu+1}(x),\quad t_j < x < x_2.$$

Now u is already constructed and as before, there exists ν sufficiently large and $t_j < \tau_1 < x_1 < \tau_2 < x_2$, such that

$$y_\nu(\tau_1) = u(\tau_1),\quad y_\nu(\tau_2) = u(\tau_2).$$

Then,

$$a_1 y_\nu(t_1) - b_1 y_\nu(p_1) = y_1 = au(t_1) - b_1 u(p_1),$$
$$y_\nu(t_i) = \frac{y_i}{a_i} = u(t_i),\ 2 \le i \le j,$$
$$y_\nu(\tau_1) = u(\tau_1),$$
$$y_\nu^{(i-1)}(x_1) = y_{i1} = u^{(i-1)}(x_1),\quad 1 \le i \le m_1 - 1,$$
$$y_\nu(\tau_2) = u(\tau_2),$$
$$y_\nu^{(i-1)}(x_l) = y_{il} = u^{(i-1)}(x_l),\quad 1 \le i \le m_l,\quad 2 \le l \le k,$$
$$y_\nu(s_i) = \frac{y_{n-(i-1)}}{c_i} = u(s_i),\ 1 \le i \le j - 1,$$

and again, Corollary 3.3.9 is contradicted.

Thus, S is closed and so, $S \equiv \mathbb{R}^2$. This argument shows that solutions of the boundary value problem, (1.1.1)-(3.3.17), exist in the case, $j = 1$. The proof is completed by induction on j and the details are analogous to those produced for the case $j = 1$. $\qquad\square$

Chapter 4

Boundary Value Problems for Finite Difference Equations

Known results for ordinary differential equations have long been one of the sources of questions in analogy for finite difference equations. In the context of boundary value problems, a great deal of activity was launched by Hartman [44] in 1978 with his paper devoted to disconjugacy criteria for linear difference equations. Eventually, many researchers turned their attention to questions dealing with boundary value problems for nonlinear difference equations, and papers during the early stages of interest in nonlinear difference equations include those by Agarwal [1, 2], Eloe [22, 23], Hankerson [36, 37] and Peterson [100]. And research devoted to boundary value problems for nonlinear difference equations remains quite high with well over 100 papers published in that area during the years 2012–2015.

In this chapter, we are concerned with uniqueness of solutions implying existence of solutions of boundary value problems for nonlinear difference equations.

For $a \in \mathbb{R}$, let the interval $[a, \infty)$ be the discrete set,

$$[a, \infty) := \{a, a + 1, \ldots\},$$

and if $b = a + n$, for some $n \in \mathbb{N}$, let the closed interval $[a, b]$ be the discrete set,

$$[a, b] := \{a, a + 1, \ldots, b\},$$

and let the intervals $[a, b), (a, b]$ and (a, b) denote the analogous discrete sets. More specifically, for $n \geq 2$, we will be concerned with uniqueness of solutions implying the existence of solutions of certain boundary value problems for the nth order nonlinear difference equation,

$$u(m + n) = f(m, u(m), u(m + 1), \ldots, u(m + n - 1)), \qquad (4.0.1)$$

where throughout,

(DA) $f(m, r_1, \ldots, r_n) : [a, \infty) \times \mathbb{R}^n \to \mathbb{R}$ is continuous, and the equation $r_n = f(m, r_0, \ldots, r_{n-1})$ can be solved for r_0 as a continuous function of $r_1, \ldots, r_n$, for each $m \in [a, \infty)$.

Given $m_0 \in [a, \infty)$, an *initial value problem (discrete)* consists of the equation (4.0.1) along with the conditions,

$$u(m_0 + i - 1) = s_i, \quad 1 \leq i \leq n, \tag{4.0.2}$$

where $s_1, \ldots, s_n \in \mathbb{R}$.

Remark 4.0.1. *We observe here that condition* (DA) *implies that* (4.0.1) *is an nth order difference equation on any subinterval of* $[a, \infty)$, *that solutions of initial value problems* (4.0.1)-(4.0.2) *are unique and exist on* $[a, \infty)$, *and that solutions of* (4.0.1)-(4.0.2) *depend continuously on initial conditions.*

In our forthcoming uniqueness hypothesis, we will use some of the terminology introduced by Hartman [44] and still very much employed by researchers in the area of difference equations. For a function $u : [a, \infty) \to \mathbb{R}$, Hartman defined $m_0 \in [a, \infty)$, in the case that $m_0 = a$, to be a *node* of u if $u(a) = 0$, and $m_0 > a$ to be a *node* of u if $u(m_0) = 0$ or $u(m_0 - 1)u(m_0) < 0$. Furthermore, Hartman defined $m_0 = a$ to be a *generalized zero* of u if $u(a) = 0$, and $m_0 > a$ to be a *generalized zero* of u if $u(m_0) = 0$ or there is an integer $j \geq 1$ such that $(-1)^j u(m_0 - j)u(m_0) > 0$ and if $j > 1$, $u(m_0 - j + 1) = \cdots = u(m_0 - 1) = 0$. We note that, if m_0 is a node of u, then m_0 is a generalized zero of u.

For notation, if m_0 is a generalized zero of u, we will write,

"m_0 is a *gz* of u."

4.1 Conjugate boundary value problems: uniqueness implies existence

The results of this section were obtained by Henderson [53] in 1989 for solutions of (4.0.1) satisfying conjugate boundary conditions (discrete). These types of boundary value problems for (4.0.1), for which we address the question of uniqueness of solutions implying their existence, are analogous in some sense to the conjugate boundary value problems (1.1.1)-(1.1.2) of Chapter 1.

Definition 4.1.1. *Given $m_1 \in [a, \infty)$ and $m_2, \ldots, m_n \in \mathbb{N}$, let $s_1, \ldots, s_n \in [a, \infty)$ be defined by $s_1 = m_1$ and $s_i = s_{i-1} + m_i, 2 \leq i \leq n$. A boundary value problem for (4.0.1) satisfying the conditions*

$$u(s_i) = y_i, \quad 1 \leq i \leq n, \tag{4.1.1}$$

where $y_i \in \mathbb{R}, 1 \leq i \leq n$, is called an $(m_1, \ldots, m_n)$ conjugate boundary value problem for (4.0.1).

Of course, this section's main result for solutions of (4.0.1)-(4.1.1) is motivated by the Hartman [41, 43] and Klaasen [85] results that were summarized and presented in Theorem 2.3.4 in Section 2.3.

Our uniqueness assumption on solutions of (4.0.1)-(4.1.1) takes the form:

(DC) Given $m_1 \in [a, \infty)$ and $m_1, \ldots, m_n \in \mathbb{N}$, with $s_1 = m_1$ and $s_i = s_{i-1} + m_i, 2 \leq i \leq n$, if $u(m)$ and $v(m)$ are solutions of (4.0.1) such that $u(s_1) = v(s_1)$ and $u(m) - v(m)$ has a gz at $s_i, 2 \leq i \leq n$, then it follows that $u(m) = v(m)$ on $[s_1, s_n]$ (hence on $[a, \infty)$).

Our first result is immediate from condition (DA) and Remark 4.0.1, and the fact that the hypotheses concern a finite number of bounded sequences of values at the successive points $m_0, \ldots, m_0 + n - 1$.

Theorem 4.1.1. [[53], Henderson, Thm. 1] *Assume that with respect to (4.0.1), condition (DA) is satisfied. If there exist a sequence $\{y_k(m)\}$ of solutions of (4.0.1), an interval $[m_0, m_0 + n - 1]$, and an $M > 0$ such that $|y_k(m)| \leq M$, for all $m \in [m_0, m_0 + n - 1]$ and for all $k \geq 1$, then there is a subsequence $\{y_{k_j}(m)\}$ that converges pointwise on $[a, \infty)$ to a solution of (4.0.1).*

If, in addition, we also assume condition (DC), then continuous dependence of solutions on initial conditions and an application of the Brouwer Theorem on Invariance of Domain imply that solutions of conjugate problems (4.0.1)-(4.1.1) depend continuously on boundary conditions (in particular, on boundary values). This next result is an analogue of Theorem 1.1.6 in Section 1.1 of Chapter 1. Its proof is easier than the proof of Theorem 1.1.6 because of the discrete setting here.

Theorem 4.1.2. [[53], Henderson, Thm. 2] *Assume that with respect to (4.0.1), conditions (DA) and (DC) are satisfied. Given a solution $u(m)$ of (4.0.1), points $s_1 < s_2 < \cdots < s_n$ in $[a, \infty)$, an interval $[s_1, b]$, where $b \geq s_n$, and an $\epsilon > 0$, there is a $\delta = \delta(\epsilon, [s_1, b]) > 0$ such that, if $|u(s_i) - y_i| < \delta$,*

$1 \leq i \leq n$, then there exists a solution $v(m)$ of (4.0.1) satisfying $v(s_i) = y_i$, $1 \leq i \leq n$, and $|u(m) - v(m)| < \epsilon$, for all $m \in [s_1, b]$.

From Theorem 4.1.2, and in the presence of solutions of **some** $(m_1, \ldots, m_n)$ conjugate boundary value problems for (4.0.1), we can state a result which is somewhat analogous to Theorem 4.1.1, as well as somewhat analogous to the "compactness condition" (CP).

Theorem 4.1.3. [[53], Henderson, Thm. 3] *Assume that with respect to (4.0.1), conditions* (DA) *and* (DC) *are satisfied, and suppose that, given $m_1 \in [a, \infty)$ and* **some** *$m_2, \ldots, m_n \in \mathbb{N}$, there exist unique solutions of (4.0.1)-(4.1.1) at the corresponding points $s_1, \ldots, s_n$. If there exist a sequence $\{y_k(m)\}$ of solutions of (4.0.1) and an $M > 0$ such that $|y_k(s_i)| \leq M$, $1 \leq i \leq n$, for all $k \in \mathbb{N}$, then there exists a subsequence $\{y_{k_j}(m)\}$ that converges pointwise on $[a, \infty)$. In particular, for this subsequence, if $\lim_{j \to \infty} y_{kj}(s_i) := y_i$, $1 \leq i \leq n$, then $\{y_{k_j}(m)\}$ converges pointwise on $[a, \infty)$ to the solution $y(m)$ of the $(m_1, \ldots, m_n)$ conjugate boundary value problem (4.0.1)-(4.1.1) satisfying*

$$y(s_i) = y_i, \quad 1 \leq i \leq n.$$

We now prove that hypotheses (DA) and (DC) imply the existence of solutions of (4.0.1)-(4.1.1). The method involves shooting in conjunction with inductions on $m_2, \ldots, m_n$. To illustrate the inductive pattern of shooting in the proof, we will give specific details in the inductive steps on the indices m_n and m_{n-1}. Then, a general inductive step is outlined in the latter part of the proof.

Theorem 4.1.4. [[53], Henderson, Thm. 4] *Assume that with respect to (4.0.1), conditions* (DA) *and* (DC) *are satisfied. Then, given $s_1, \ldots, s_n \in [a, \infty)$ and $y_1, \ldots, y_n \in \mathbb{R}$, there exists a unique solution of (4.0.1)-(4.1.1) on $[a, \infty)$.*

Proof. In the proof, we will verify that, for each $m_1 \in [a, \infty)$ and $m_2, \ldots, m_n \in \mathbb{N}$, each $(m_1, \ldots, m_n)$ conjugate boundary value problem for (4.0.1) has a unique solution on $[a, \infty)$. We remark that the uniqueness of all such solutions follows from condition (DC). The proof is by induction on $m_2, \ldots, m_n$.

Throughout the proof, let $y_i \in \mathbb{R}, 1 \leq i \leq n$, be given.

Now, let $m_1 \in [a, \infty)$ be given, let $m_2 = \cdots = m_n = 1$, let $s_1 = m_1$, and let $s_i = s_{i-1} + m_i, 2 \leq i \leq n$. Also, let $u(m)$ be the solution of the initial value problem for (4.0.1) on $[a, \infty)$ satisfying

$$u(s_i) = y_i, \ \ 1 \le i \le n.$$

Next, let $m_n > 1$. Assume, given $m_1 \in [a, \infty)$ and $m_2 = \cdots = m_{n-1} = 1$, that, for each $1 \le h < m_n$, there exists a unique solution of each $(m_1, 1, \ldots, 1, h)$ conjugate boundary value problem for (4.0.1) on $[a, \infty)$. With $m_1 \in [a, \infty)$ given, let $m_2 = \cdots = m_{n-1} = 1$, and $s_1 = m_1$, $s_i = s_{i-1} + m_i$, $2 \le i \le n$, and let $v_1(m)$ be the solution of the $(m_1, 1, \ldots, 1, m_n - 1)$ conjugate boundary value problem for (4.0.1) satisfying

$$v_1(s_i) = y_i, \ \ 1 \le i \le n - 1,$$
$$v_1(s_n - 1) = 0.$$

Define $S_1 := \{r \in \mathbb{R} \mid \text{there is a solution } y(m) \text{ of } (4.0.1) \text{ satisfying } y(s_i) = v_1(s_i), 1 \le i \le n - 1, \text{ and } y(s_n) = r\}$. $S_1 \ne \emptyset$, since $v_1(s_n) \in S_1$. Furthermore, by Theorem 4.1.2, S_1 is an open subset of $\mathbb{R}$.

We claim that S_1 is also a closed subset of $\mathbb{R}$. We assume the claim to be false. Then, there exist a point $r_0 \in \overline{S}_1 \setminus S_1$ and a strictly monotone sequence $\{r_k\} \subset S_1$ such that $\lim_{k \to \infty} r_k = r_0$. With no loss of generality, we may assume $r_k \uparrow r_0$. From the definition of S_1, for each $k \ge 1$, there is a solution $y_k(m)$ of (4.0.1) satisfying

$$y_k(s_i) = v_1(s_i), \ \ 1 \le i \le n - 1,$$
$$y_k(s_n) = r_k.$$

Condition (DC) implies, for each $k \ge 1$, $y_k(m) < y_{k+1}(m)$ on (s_{n-1}, ∞). Now, by the induction hypothesis, there are unique solutions of $(m_1, 1, \ldots, 1, m_n - 1)$ conjugate problems. So, by Theorem 4.1.3, in conjunction with $r_0 \notin S_1$, it follows that $y_k(s_n - 1) \uparrow \infty$, as $k \to \infty$. In addition, by Theorem 4.1.1, there exists $m_0 \in (s_n, s_n + n - 1]$ such that $y_k(m_0) \uparrow \infty$, as $k \to \infty$.

Now, we let $u(m)$ be the solution of the $(m_1 + 1, 1, \ldots, 1, m_n - 1)$ conjugate problem for (4.0.1), with boundary conditions at the points $s_2, \ldots, s_{n-1}, s_{n-1} + 1, s_n$, satisfying

$$u(s_i) = v_1(s_i), \ \ 2 \le i \le n - 1,$$
$$u(s_{n-1} + 1) = 0,$$
$$u(s_n) = r_0.$$

From $y_k(m) \uparrow \infty$, for $m = s_n - 1, m_0$, while $y_k(s_n) = r_k < r_0 = u(s_n)$, for all $k \ge 1$, it follows that, for some $K \ge 1$, $u(m) - y_K(m)$ has a gz at s_n and also a gz (or zero) at some $n_0 \in (s_n, m_0]$. We also have $u(s_i) - y_K(s_i) = 0$, $2 \leq i \leq n - 1$. By condition (DC), $u(m) = y_K(m)$ on $[a, \infty)$. This is a contradiction.

Therefore, S_1 is also closed and $S_1 = \mathbb{R}$. We choose $y_n \in S_1$, and hence there is a solution $y(m)$ of (4.0.1) satisfying

$$y(s_i) = y_i, \quad 1 \leq i \leq n.$$

Specifically, given $m_1 \in [a, \infty)$, $m_2 = \cdots = m_{n-1} = 1$ and $m_n \geq 1$, each $(m_1, 1, \ldots, 1, m_n)$ conjugate boundary value problem (4.0.1)-(4.1.1) has a unique solution on $[a, \infty)$.

The next part of the proof involves induction on m_{n-1}. Our assumptions are that $m_{n-1} > 1$ and that, given $m_1 \in [a, \infty), m_2 = \cdots = m_{n-2} = 1$, and $m_n \geq 1$, there exists a unique solution of each $(m_1, 1, \ldots, 1, l, m_n)$ conjugate boundary value problem, for $1 \leq l < m_{n-1}$, for (4.0.1) on $[a, \infty)$. With $m_1 \in [a, \infty)$ given, let $m_2 = \cdots = m_{n-2} = m_n = 1$, and $s_1 = m_1, s_i = s_{i-1} + m_i$, $2 \leq i \leq n$, and then let $v_2(m)$ be the solution of the $(m_1, 1, \ldots, 1, m_{n-1} - 1, 1)$ conjugate problem for (4.0.1), with conditions at the points $s_1, \ldots, s_{n-2}, s_{n-1} - 1, s_{n-1}$, satisfying

$$v_2(s_i) = y_i, \ 1 \leq i \leq n - 2,$$
$$v_2(s_{n-1} - 1) = 0,$$
$$v_2(s_{n-1}) = y_{n-1}.$$

Define $S_2 := \{r \in \mathbb{R} \mid \text{there is a solution } y(m) \text{ of (4.0.1) satisfying } y(s_i) = v_2(s_i), 1 \leq i \leq n - 1, \text{ and } y(s_n) = r\}$. $S_2 \neq \emptyset$, since $v_2(s_n) \in S_2$. Again, Theorem 4.1.2 implies S_2 is an open subset of $\mathbb{R}$.

We now claim that S_2 is also closed. We again assume S_2 is not closed. Then, there exist $r_0 \subset \overline{S_2} \setminus S_2$ and a strictly monotone sequence $\{r_k\} \subset S_2$ such that $\lim_{k \to \infty} r_k = r_0$, and we may assume again that $r_k \uparrow r_0$. As in the previous case, let $y_k(m)$ denote the associated solution of (4.0.1) satisfying

$$y_k(s_i) = v_2(s_i), \ 1 \leq i \leq n - 1,$$
$$y_k(s_n) = r_k.$$

From condition (DC), for each $k \geq 1$, $y_k(m) > y_{k+1}(m)$ on (s_{n-2}, s_{n-1}) and $y_k(m) < y_{k+1}(m)$ on $[s_n, \infty)$. Now, $r_0 \notin S_2$ and there exist unique solutions of $(m_1, 1, \ldots, 1, m_{n-1} - 1, 1)$ problems. Theorem 4.1.3 implies $y_k(s_{n-1} - 1) \downarrow -\infty$, as $k \to \infty$, and Theorem 4.1.1 implies there exists $m_0 \in (s_n, s_n + n - 1]$ such that $y_k(m_0) \uparrow \infty$, as $k \to \infty$.

This time, we let $u(m)$ be the solution at the points $s_2, \ldots, s_{n-2}, s_{n-2} + 1, s_{n-1}, s_n$ of the $(m_1+1, 1, \ldots, 1, m_{n-1}-1, 1)$ conjugate problem for (4.0.1) satisfying

$$u(s_i) = v_2(s_i), \ 2 \le i \le n - 2,$$
$$u(s_{n-2} + 1) = 0,$$
$$u(s_{n-1}) = v_2(s_{n-1}),$$
$$u(s_n) = r_0.$$

Now, $y_k(s_{n-1}-1) \downarrow -\infty$, while $u(s_{n-1})-y_k(s_{n-1}) = 0$ and $u(s_n)-y_k(s_n) > 0$, for each $k \ge 1$. Also, $y_k(m_0) \uparrow \infty$, and so there exists $K \ge 1$, such that $u(m) - y_K(m)$ has a gz at s_n and a gz (or zero) at some $n_0 \in (s_n, m_0]$. In addition, $u(s_i) - y_K(s_i) = 0$, $2 \le i \le n - 1$, and then (DC) implies $u(m) = y_K(m)$ on $[a, \infty)$. This is again a contradiction.

Hence, S_2 is closed and $S_2 = \mathbb{R}$. We choose $y_n \in S_2$, and then there exists a solution $y(m)$ of (4.0.1) satisfying

$$y(s_i) = y_i, \ 1 \le i \le n.$$

We have proved thus far that, given $m_1 \in [a, \infty), m_2 = \cdots = m_{n-2} = m_n = 1$, each $(m_1, 1, \ldots, 1, m_{n-1}, 1)$ conjugate boundary value problem (4.0.1)-(4.1.1) has a unique solution on $[a, \infty)$.

We maintain the induction hypotheses associated with $m_{n-1} > 1$, and in addition, we assume that $m_n > 1$, that given $m_1 \in [a, \infty)$ and $m_2 = \cdots = m_{n-2} = 1$, there exists, for each $1 \le h < m_n$, a unique solution of each $(m_1, 1, \ldots, 1, m_{n-1}, h)$ conjugate problem for (4.0.1) on $[a, \infty)$.

Under this latter assumption, let $m_1 \in [a, \infty)$ be given, let $m_2 = \cdots = m_{n-2} = 1$, let $s_1, \ldots, s_n$ be as usual, and let $v_3(m)$ denote the solution of (4.0.1) at the points $s_1, \ldots, s_{n-2}, s_{n-1}, s_n - 1$ of the $(m_1, 1, \ldots, 1, m_{n-1}, m_n - 1)$ conjugate boundary value problem satisfying

$$v_3(s_i) = y_i, \ 1 \le i \le n - 1,$$
$$v_3(s_n - 1) = 0.$$

Define $S_3 := \{r \in \mathbb{R} \mid \text{there is a solution } y(m) \text{ of (4.0.1) satisfying } y(s_i) = v_3(s_i), 1 \le i \le n - 1, \text{ and } y(s_n) = r\}$. S_3 is a nonempty open subset of $\mathbb{R}$, and we claim that S_3 is also closed. We assume the claim to be false, and we let $r_0 \in \overline{S_3} \setminus S_3$ and a $\{r_k\} \subset S_3$, with $r_k \uparrow r_0$, be as in the previous considerations, and we let $y_k(m)$ be the corresponding solution of (4.0.1) satisfying

$$y_k(s_i) = v_3(s_i), \ 1 \le i \le n - 1,$$
$$y_k(s_n) = r_k.$$

By condition (DC), for each $k \geq 1$, $y_k(m) < y_{k+1}(m)$ on (s_{n-1}, ∞), and also since there are unique solutions of $(m_1, 1, \ldots, 1, m_{n-1}, m_n - 1)$ problems, along with $r_0 \notin S_3$, it follows from Theorem 4.1.3 that $y_k(s_n - 1) \uparrow \infty$, as $k \to \infty$, and Theorem 4.1.1 implies that, for some $m_0 \in (s_n, s_n + n - 1]$, $y_k(m_0) \uparrow \infty$, as $k \to \infty$.

Next, let $u(m)$ be the solution of the $(m_1 + 1, 1, \ldots, 1, 1, m_{n-1} - 1, m_n)$ conjugate problem for (4.0.1) with boundary conditions at the points $s_2, \ldots, s_{n-2}, s_{n-2} + 1, s_{n-1}, s_n$ and satisfying

$$u(s_i) = v_3(s_i), \ 2 \leq i \leq n - 2,$$
$$u(s_{n-2} + 1) = 0,$$
$$u(s_{n-1}) = v_3(s_{n-1}),$$
$$u(s_n) = r_0.$$

Such a solution $u(m)$ exists by our primary induction hypothesis on m_{n-1} in this section of the proof. Due to the unbounded conditions on $\{y_k(s_n - 1)\}$ and $\{y_k(m_0)\}$, while $u(s_n) > y_k(s_n)$, for each $k \geq 1$, it follows that there exists $K \geq 1$ such that $u(m) - y_K(m)$ has a gz at s_n and a gz at some $n_0 \in (s_n, m_0]$. Further, $u(s_i) - y_K(s_i) = 0$, $2 \leq i \leq n - 1$, from which it follows that $u(m) = y_K(m)$ on $[a, \infty)$. This is a contradiction.

Therefore, S_3 is closed, $S_3 = \mathbb{R}$, and choosing $y_n \in S_3$, we have a corresponding solution $y(m)$ of (4.0.1) satisfying $y(s_n) = y_n$. In particular, given $m_1 \in [a, \infty)$, $m_2 = \cdots = m_{n-2} = 1$, and $m_n \geq 1$, each $(m_1, 1, \ldots, 1, m_{n-1}, m_n)$ conjugate boundary value problem (4.0.1)-(4.1.1) has a unique solution on $[a, \infty)$.

This completes the induction on m_{n-1}. That is, given $m_1 \in [a, \infty)$, $m_2 = \cdots = m_{n-2} = 1, m_{n-1} \geq 1$ and $m_n \geq 1$, each $(m_1, 1, \ldots, 1, m_{n-1}, m_n)$ conjugate boundary value problem (4.0.1)-(4.1.1) has a unique solution on $[a, \infty)$.

Our above arguments exhibit the pattern for the induction scheme in obtaining solutions of the boundary value problems, but for completeness, we will include some of the details involved in the general induction step. To that end, we assume $1 \leq p \leq n-2$, that $m_p > 1$, and that given $m_1 \in [a, \infty)$, $m_2 = \cdots = m_{p-1} = 1$, $m_{p+1} \geq 1, \ldots, m_{n-1} \geq 1, m_n \geq 1$, there exists a unique solution of each $(m_1, 1, \ldots, 1, k, m_{p+1}, \ldots, m_n)$ conjugate boundary value problem, where $1 \leq k < m_p$, for (4.0.1) on $[a, \infty)$.

Under that assumption, of course, we will be concerned with establishing the existence of solutions of $(m_1, 1, \ldots, 1, m_p, m_{p+1}, \ldots, m_n)$ conjugate problems by proceeding through 2^{n-p} inductive steps, in which we will induct on the indices $m_n, m_{n-1}, \ldots, m_{p+1}$, along the pattern in previous parts of the proof. For each one of the 2^{n-p} steps, there are natural numbers, $1 \le j_1 < j_2 < \cdots < j_s \le n - p$, such that we are concerned, either with the $(2^{j_s} + 2^{j_{s-1}} + \cdots + 2^{j_2} + 2^{j_1})$st inductive step, or with the $(2^{j_s} + 2^{j_{s-1}} + \cdots + 2^{j_2} + 2^{j_1} + 1)$st inductive step.

(a) With the case of the $(2^{j_s} + 2^{j_{s-1}} + \cdots + 2^{j_2} + 2^{j_1})$st inductive step, the concern is with showing the existence of solutions of $(m_1, 1, \ldots, 1, m_p, 1, \ldots, 1, m_{n-j}, 1, \ldots, 1, m_{n-j_{s-1}}, 1, \ldots, 1, m_{n-j_2}, 1, \ldots, 1, m_{n-j_1+1}, m_{n-j_1+2}, \ldots, m_n)$ problems for (4.0.1), where the significance of the entries in the n-tuple are understood from previous arguments.

For this case, in addition to appropriate assumptions from preceding steps on $m_p, m_{n-j_s}, \ldots, m_{n-j_2}, m_{n-j_1+1}, \ldots, m_{n-1}$, our assumptions are that $m_n > 1$ and that, given $m_1 \in [a, \infty)$, each $(m_1, 1, \ldots, 1, m_p, 1, \ldots, 1, m_{n-j_s}, 1, \ldots, 1, m_{n-j_{s-1}}, 1, \ldots, 1, m_{n-j_2}, 1, \ldots, 1, \quad m_{n-j_1+1}, \ldots, m_{n-1}, h)$ conjugate boundary value problem, where $1 \le h < m_n$, for (4.0.1) has a unique solution on $[a, \infty)$.

Now, given $m_1 \in [a, \infty)$, let $v(m)$ be the solution of the $(m_1, 1, \ldots, 1, m_p, 1, \ldots, 1, m_{n-j_s}, 1, \ldots, 1, m_{n-j_{s-1}}, 1, \ldots, 1, m_{n-j_2}, 1, \ldots, 1, m_{n-j_1+1}, \ldots, m_{n-1}, m_n - 1)$ conjugate boundary value problem for (4.0.1) on $[a, \infty)$ with conditions at the points $s_1, \ldots, s_{n-1}, s_n - 1$, satisfying

$$v(s_i) = y_i, \ 1 \le i \le n - 1,$$
$$v(s_n - 1) = 0.$$

The set $S := \{r \in \mathbb{R} \mid \text{there is a solution } y(m) \text{ of (4.0.1) satisfying } y(s_i) = v_3(s_i), 1 \le i \le n - 1, \text{ and } y(s_n) = r\}$ is nonempty and open.

If we follow the pattern of assuming S is not closed, then with r_0, $\{r_k\}$, with $r_k \uparrow r_0$, and $y_k(m)$ as in the previous arguments, condition (DC) implies $y_k(m) < y_{k+1}(m)$ on (s_{n-1}, ∞), for each $k \ge 1$. And in complete analogy, we have also, $y_k(s_n - 1) \uparrow \infty$, as $k \to \infty$, and for some $m_0 \in (s_n, s_n + n - 1]$, $y_k(m_0) \uparrow \infty$, as $k \to \infty$.

With $u(m)$ the solution of (4.0.1) with conditions at the points, $s_2, s_3, \ldots, s_{p-1}, s_{p-1} + 1, s_p, s_{p+1}, \ldots, s_{n-1}, s_n$, of the $(m_1 + 1, 1, \ldots, 1, 1, m_p - 1, 1, \ldots, 1, m_{n-j_s}, 1, \ldots, 1, m_{n-j_{s-1}}, 1, \ldots, 1, m_{n-j_2}, 1, \ldots, 1, m_{n-j_1+1}, \ldots, m_{n-1}, m_n)$ conjugate problem satisfying

$$u(s_i) = v(s_i), \ 2 \le i \le p - 1,$$
$$u(s_{p-1} + 1) = 0,$$
$$u(s_i) = v(s_i), \ p \le i \le n - 1,$$
$$u(s_n) = r_0,$$

it then follows that, for some $K \geq 1$, $u(m) - y_K(m)$ has a gz at s_n, a gz at some $n_0 \in (s_n, m_0]$, and zeros at s_i, $2 \leq i \leq n - 1$. This contradicts (DC), and so S is closed and $S = \mathbb{R}$. For $y_n \in S$, the corresponding solution $y(m)$ is the desired solution. This completes this case.

(b) With the case of the $(2^{j_s} + 2^{j_s - 1} + \cdots + 2^{j_2} + 2^{j_1} + 1)$st inductive step, the concern is with showing existence of solutions of $(m_1, 1, \ldots, 1, m_p, 1, \ldots, 1, m_{n-j_s}, 1, \ldots, 1, m_{n-j_{s-1}}, 1, \ldots, 1, m_{n-j_2}, 1, \ldots, 1, m_{n-j_1}, 1, \ldots, 1)$ problems for (4.0.1).

As in case (a), in addition to appropriate assumptions from preceding steps on $m_p, m_{n-j_s}, \ldots, m_{n-j_2}$, our assumptions are that $m_{n-j_1} > 1$ and that, given $m_1 \in [a, \infty)$, each $(m_1, 1, \ldots, 1, m_p, 1, \ldots, 1, m_{n-j_s}, 1, \ldots, 1, m_{n-j_{s-1}}, 1, \ldots, 1, m_{n-j_2}, 1, \ldots, 1, l, m_{n-j_1+1}, \quad m_{n-j_1+2}, \ldots, m_n)$ conjugate boundary value problem, where $1 \leq l \leq m_{n-j_1}$, for (4.0.1) has a unique solution on $[a, \infty)$.

This time, given $m_1 \in [a, \infty)$, let $v(m)$ be the solution of the $(m_1, 1, \ldots, 1, m_p, 1, \ldots, 1, m_{n-j_s}, 1, \ldots, 1, m_{n-j_{s-1}}, 1, \ldots, 1, m_{n-j_2}, 1, \ldots, 1, m_{n-j_1} - 1, 1, \ldots, 1)$ conjugate problem for (4.0.1) on $[a, \infty)$, with boundary conditions at the points $s_1, \ldots, s_{n-j_1-1}, s_{n-j_1} - 1, s_{n-j_1}, s_{n-j_1+1}, \ldots, s_{n-2}, s_{n-1}$, satisfying

$$v(s_i) = y_i, \ 1 \leq i \leq n - j_1 - 1,$$
$$v(s_{n-j_1} - 1) = 0,$$
$$v(s_i) = y_i, \ n - j_1 \leq i \leq n - 1.$$

If $S := \{r \in \mathbb{R} \mid$ there is a solution $y(m)$ of (4.0.1) satisfying $y(s_i) = v(s_i), 1 \leq i \leq n - 1$, and $y(s_n) = r\}$, then S is nonempty and an open subset of $\mathbb{R}$.

If we assume S is not closed, and if r_0, $\{r_k\}$, with $r_k \uparrow r_0$, and $y_k(m)$ are as usual for these arguments, then $y_k(m) < y_{k+1}(m)$ on $[s_n, \infty)$, for all $k \geq 1$. Further, since $y_k(s_i) = v(s_i)$, $n - j_1 \leq i \leq n - 1$, it follows from (DC) that, either

(i) $y_k(s_{n-j_1} - 1) < y_{k+1}(s_{n-j_1} - 1)$, for each $k \geq 1$, if j_1 is even,

or

(i) $y_k(s_{n-j_1} - 1) > y_{k+1}(s_{n-j_1} - 1)$, for each $k \geq 1$, if j_1 is odd.

The cases are analogous, and so we will assume case (i) to be the situation. From $r_0 \notin S$, Theorem 4.1.3 implies that $y_k(s_{n-j_1} - 1) \uparrow \infty$, as $k \to \infty$. Moreover, it is again the situation that, for some $m_0 \in (s_n, s_n + n - 1]$, $y_k(m_0) \uparrow \infty$, as $k \to \infty$. Now, we let $u(m)$ be the solution of (4.0.1) with boundary conditions at the points $s_2, \ldots, s_{p-1}, s_{p-1} + 1, s_p, s_{p+1}, \ldots, s_{n-1}, s_n$ of the $(m_1 + 1, 1, \ldots, 1, 1, m_p - 1, 1, \ldots, 1, m_{n-j_s}, 1, \ldots, 1, m_{n-j_{s-1}}, 1, \ldots, 1, m_{n-j_2}, 1, \ldots, 1, m_{n-j_1}, 1, \ldots, 1)$ conjugate problem satisfying

$$u(s_i) = v(s_i), \ 2 \le i \le p - 1,$$
$$u(s_{p-1} + 1) = 0,$$
$$u(s_i) = v(s_i), \ p \le i \le n - 1,$$
$$u(s_n) = r_0.$$

As in each of the previous cases, there exists a $K \ge 1$, such that $u(m) - y_K(m)$ has a gz at s_n, a gz at some $n_0 \in (s_n, m_0]$, and zeros at s_i, $2 \le i \le n - 1$. This is a contradiction to (DC), and so S is closed and $S = \mathbb{R}$. Choosing $y_n \in S$, the corresponding solution $y(m)$ is the unique solution of (4.0.1)-(4.1.1). This completes this case.

The proof is complete. $\qquad\qquad\qquad\qquad\qquad\qquad\qquad\qquad\qquad\quad$ □

Remark 4.1.1. *As a final comment for this section, we observe that the discrete half-line* $[a, \infty)$ *is not necessary. The principal result, Theorem 4.1.4, could be stated in terms of the finite interval* $[a, b + n]$*, where b is the rightmost point at which boundary conditions are specified, so that our application of Theorem 4.1.1 can still be made in the arguments of the proof of Theorem 4.1.4.*

4.2 Focal boundary value problems: uniqueness implies existence

The results of this section were published by Henderson [54,55] in 1989, and the types of boundary value problems for (4.0.1), for which we address the question of uniqueness of solutions implying their existence, are analogous in some sense to the right focal boundary value problems (1.1.1)-(1.1.3) as they were dealt with in Section 2.5 of Chapter 2 in Theorem 2.5.1, but here with differences defined by $\Delta u(t) := u(t + 1) - u(t)$, and for $i \ge 2, \Delta^i u(t) := \Delta(\Delta^{i-1} u(t))$. Condition (DA), which was introduced in the first part of this chapter, will be assumed for most of this section.

Definition 4.2.1. *Let* $2 \le k \le n$ *and let* $m_1, \ldots, m_k$ *be positive integers such that* $\sum_{i=1}^{k} m_i = n$. *Let* $s_0 = 0$ *and for* $1 \le j \le k$, $s_j = \sum_{i=1}^{j} m_i$.

For points $t_k, t_{k-1}, \ldots, t_1$ in $[a, \infty)$, such that $t_k < t_{k-1} < \cdots < t_1$, where $t_j + m_j + 1 \leq t_{j-1}, 2 \leq j \leq k$, a boundary value problem for (4.0.1) satisfying the conditions

$$\Delta^i u(t_j) = y_{i+1}, \quad s_{j-1} \leq i \leq s_j - 1, \ 1 \leq j \leq k, \qquad (4.2.1)$$

where $y_i \in \mathbb{R}, 1 \leq i \leq n$, is called an $(m_k, \ldots, m_1)$ focal boundary value problem for (4.0.1).

In terms of generalized zeros (that is "gz"), in this section, our uniqueness assumption on solutions of (4.0.1)-(4.2.1) takes the form:

(FC) Given $2 \leq k \leq n$, positive integers $m_1, \ldots, m_k$ such that $\sum_{i=1}^{k} m_i = n$, and points $t_k < t_{k-1} < \cdots < t_1 \in [a, \infty)$, where $t_j + m_j + 1 \leq t_{j-1}, 2 \leq j \leq k$, if $u(t)$ and $v(t)$ are solutions of (4.0.1) such that $\Delta^i(u(t) - v(t)), s_{j-1} \leq i \leq s_j - 1$, (where $s_0 = 0, s_j = \sum_{i=1}^{j} m_i$), has a gz at $t_j, 1 \leq j \leq k$, then it follows that $u(t) = v(t)$ on $[t_k, t_1 + m_1 - 1]$ (hence on $[a, \infty)$).

Remark 4.2.1. (a) *Condition (FC) implies that, given $2 \leq k \leq n$, each $(m_k, \ldots, m_1)$ focal boundary value problem for (4.0.1) has at most one solution on $[a, \infty)$.*

(b) *Hartman [44] proved a discrete Rolle's Theorem in terms of generalized zeros, (in particular, Hartman proved, "if $u(t) : [a, \infty) \to \mathbb{R}$, has gz's at points $b < c$, then $\Delta u(t)$ has a gz in $[b, c)$."). It follows from this discrete Rolle's Theorem that condition (FC) implies the uniqueness of solutions of conjugate boundary value problems (4.0.1)-(4.1.1) of Section 4.1; that is condition (FC) implies condition (DC) of Section 4.1.*

(c) *Under conditions (DA) and (FC), it follows from Theorem 4.1.4 that all conjugate boundary value problems (4.0.1)-(4.1.1) have unique solutions on $[a, \infty)$.*

As a consequence of Remark 4.2.1(a) and (b), it follows that Theorems 4.1.1-4.1.3 of Section 4.1 remain valid, yet we will state a stronger version of Theorem 4.1.2 which is more suitable for application in the context of our focal boundary value problems.

Theorem 4.2.1. *Assume that with respect (4.0.1), conditions (DA) and (FC) are satisfied. Let $2 \leq k \leq n$ and positive integers $m_1, \ldots, m_k$, such that $\sum_{i=1}^{k} m_i = n$, be given, and let $s_j, 0 \leq j \leq k$, be the corresponding partial sums. Given a solution $u(t)$ of (4.0.1) on $[a, \infty)$, points $t_k < t_{k-1} < \cdots < t_1$ belonging to $[a, \infty)$, where $t_j + m_j + 1 \leq t_{j-1}, 2 \leq j \leq k$, an interval*

$[a, b]$, *where* $b \geq t_1 + m_1 - 1$, *and an* $\epsilon > 0$, *there exists a* $\delta(\epsilon, [a, b]) > 0$ *such that, if* $|\Delta^i u(t_j) - y_{i+1}| < \delta$, $s_{j-1} \leq i \leq s_j - 1$. $1 \leq j \leq k$, *then there exists a solution* $v(t)$ *of* (4.0.1) *satisfying* $\Delta^i v(t_j) = y_{i+1}$, $s_{j-1} \leq i \leq s_j - 1, 1 \leq j \leq k$, *and* $|\Delta^i v(t) - \Delta^i u(t)| < \epsilon$, $0 \leq i \leq n - 1$, *for all* $t \in [a, b]$.

This section contains two major uniqueness implies existence results for solutions of boundary value problems. In our next theorem, we prove that conditions (DA) and (FC) are sufficient for the existence of solutions for a large class of 2-point boundary value problems for (4.0.1) of which both the (m_1, m_2) conjugate problems and the (m_2, m_1) focal problems are special subclasses. Solutions of each 2-point problem in the large class are unique by condition (FC) and the discrete Rolle's Theorem. For the existence, we use shooting methods.

Theorem 4.2.2. [[54], Henderson, Thm. 3] *Assume that with respect* (4.0.1), *conditions* (DA) *and* (FC) *are satisfied. Then, given* $1 \leq k \leq n - 1$ *and* $0 \leq m \leq n - k$, *and points* $t_2 < t_1$ *in* $[a, \infty)$, *with* $t_2 + k + 1 \leq t_1$, *there exists a unique solution of* (4.0.1) *satisfying*

$$\begin{aligned} \Delta^i u(t_2) &= y_{i+(n-k)-m+1}, & m \leq i \leq k + m - 1, \\ \Delta^i u(t_1) &= y_{i+1}, & 0 \leq i \leq n - k - 1, \end{aligned} \qquad (4.2.2)$$

on $[a, \infty)$, *for every choice of* $y_i \in [a, \infty)$, $1 \leq i \leq n$.

Proof. We begin by making some observations. First, if $m = 0$, then the problem (4.0.1)-(4.2.2) is a conjugate problem, and so from Remark 4.2.1 (c), such problems have unique solutions on $[a, \infty)$. Next, if it were the case that $t_1 - t_2 = k$, then the problem (4.0.1)-(4.2.2) is an initial value problem, and hence also has a unique solution on $[a, \infty)$. Also, as stated just prior to this Theorem, solutions of each problem in (4.0.1)-(4.2.2) are unique by (FC) and the discrete Rolle's Theorem.

Let $y_i \in \mathbb{R}$, $1 \leq i \leq n$, be given throughout.

For the remainder of the proof, we employ the shooting method along with inductions on k, m, and the spacing $t_1 - t_2$.

So, first we let $k = n - 1$. At this juncture, our goal is to exhibit the existence of solutions of (4.0.1) satisfying (4.2.2), for $0 \leq m \leq n - k = 1$, and $t_1 - t_2 \geq k + 1 = n$. The observation in the first paragraph of the proof resolves the case when $m = 0$, and moreover, since the spacing $t_1 - t_2 = n - 1$ corresponds to an initial value problem, we assume that $m = 1$, that $t_1 \geq t_2 + n$, and that for each $\tau_1 \in [a, \infty)$, with $t_2 + n - 1 \leq \tau_1 < t_1$, there exists a unique solution of each boundary value problem (4.0.1)-(4.2.2) at the points t_2 and τ_1.

Now, let $v_1(t)$ be the solution of the initial value problem for (4.0.1) which satisfies

$$\Delta^i v_1(t_2) = y_{i+1}, \ 1 \le i \le n-1,$$
$$v_1(t_2) = 0.$$

Define

$$S_1 := \{r \in \mathbb{R} \,|\, \text{there is a solution } y(t) \text{ of } (4.0.1) \text{ satisfying}$$
$$\Delta^i y(t_2) = \Delta^i v_1(t_2), \ 1 \le i \le n-1, \ \text{and } y(t_1) = r\}.$$

S_1 is nonempty, since $v_1(t_1) \in S_1$, and also by Theorem 4.2.1, S_1 is an open subset of $\mathbb{R}$.

We claim that S_1 is also a closed subset of $\mathbb{R}$. We assume the claim to be false, and so, there exist an $r_0 \in \overline{S}_1 \setminus S_1$ and a strictly monotone sequence $\{r_l\} \subset S_1$ such that $\lim_{l \to \infty} r_l = r_0$. Without loss of generality, we may assume that $r_l \uparrow r_0$. For each $l \ge 1$, let $y_l(t)$ denote the corresponding solution of (4.0.1) satisfying

$$\Delta^i y_l(t_2) = \Delta^i v_1(t_2), \ 1 \le i \le n-1,$$
$$y_l(t_1) = r_l.$$

Conditions (DA) and (FC) imply that $y_l(t) < y_{l+1}(t)$ on $[t_2 + n - 1, \infty)$, for all $l \ge 1$. Now, the induction hypotheses imply the existence of unique solutions at the points t_2 and $t_1 - 1$ of (4.0.1)-(4.2.2), when $m = 1$ and $k = n-1$, and when this is coupled with Theorem 4.2.1 along with $r_0 \notin S_1$, it follows that $y_l(t_1 - 1) \uparrow \infty$, as $l \to \infty$. Moreover, by Theorem 4.1.1, there exists $t_0 \in (t_1, t_1 + n - 1]$ such that $y_l(t_0) \uparrow \infty$, as $l \to \infty$.

Now, we let $u(t)$ be the solution of (4.0.1)-(4.2.2), for $m = 0$ and $k = n - 1$, satisfying

$$\Delta^i u(t_2) = y_{i+1}, \ 1 \le i \le n-2,$$
$$u(t_2) = 0,$$
$$u(t_1) = r_0.$$

From $y_l(t_1 - 1) \uparrow \infty$ and $y_l(t_0) \uparrow \infty$, while $y_l(t_1) = r_l < r_0 = u(t_1)$, for all $l \ge 1$, it follows that, for some $L \ge 1$, $u(t) - y_L(t)$ has a gz at t_1 and also a gz (or zero) at some $\tau_0 \in (t_1, t_0]$. Further, we have $\Delta^i(u(t_2) - y_L(t_2)) = 0, 1 \le i \le n-2$, and so by repeated applications of the discrete Rolle's Theorem, there are points $t_2 \le \sigma_n < \sigma_{n-1} < \cdots < \sigma_1 \le \tau_0$ in $[a, \infty)$ such

$\Delta^{i-1}(u(t) - y_L(t))$ has a gz at σ_i, for $1 \le i \le n$. Condition (FC) implies $u(t) = y_L(t)$ on $[a, \infty)$, but this is a contradiction.

Therefore, S_1 is closed and $S_1 = \mathbb{R}$. We choose $y_1 \in S_1$, and then there exists a solution $y(t)$ of (4.0.1) satisfying

$$\Delta^i y(t_2) = y_{i+1}, \ 1 \le i \le n - 1,$$
$$y(t_1) = u_1.$$

Specifically, given $t_2 < t_1$ in $[a, \infty)$, with $t_2 + n \le t_1$, each boundary value problem (4.0.1)-(4.2.2), for $m = 1$ and $k = n - 1$, has a unique solution.

Inducting on k, we assume now that $k < n-1$ and that for each $k < h \le n-1$, each boundary value problem (4.0.1)-(4.2.2), for $0 \le m \le n-h$, has a unique solution. We proceed to establish the existence of unique solutions of (4.0.1)-(4.2.2) for this value of "k" and for $0 \le m \le n - k$. In addition to our assumption on k, since $m = 0$ corresponds to a conjugate problem, we may also assume that $1 \le m \le n - k$ and that each boundary value problem (4.0.1)-(4.2.2) has a unique solution, for each value of an index $0 \le l < m$. Also, given points $t_2 < t_1$ in $[a, \infty)$, since any problem (4.0.1)-(4.2.2) is an initial value problem if $t_1 - t_2 = k$, we assume in addition to the hypotheses on k and m that $t_1 \ge t_2 + k + 1$ and that, for each $\tau_1 \in [a, \infty)$, with $t_2 + k \le \tau_1 < t_1$, there exists a unique solution of (4.0.1)-(4.2.2) with boundary conditions specified at t_2 and τ_1.

This time, let $v_2(t)$ be the solution of (4.0.1)-(4.2.2), for the indices $m - 1$ and $k + 1$, which satisfies

$$\Delta^i v_2(t_2) = y_{i+(n-k)-m+1}, \ m \le i \le k + m - 1,$$
$$\Delta^{m-1} v_2(t_2) = 0,$$
$$\Delta^i v_2(t_1) = y_{i+1}, \ 0 \le i \le n - (k+1) - 1.$$

Define

$$S_2 := \{ r \in \mathbb{R} \, | \, \text{there is a solution } y(t) \text{ of (4.0.1) satisfying}$$
$$\Delta^i y(t_2) = \Delta^i v_2(t_2), \ m \le i \le k + m - 1, \ \Delta^i y(t_1) = \Delta^i v_2(t_1),$$
$$0 \le i \le n - k - 2, \text{ and } \Delta^{n-k-1} y(t_1) = r \}.$$

S_2 is nonempty, since $\Delta^{n-k-1} v_2(t_1) \in S_1$, and also by Theorem 4.2.1, S_2 is an open subset of $\mathbb{R}$.

We claim, as before, that S_2 is also a closed subset of $\mathbb{R}$. We assume for purpose of contradiction that S_2 is not closed. Then, there exist $r_0 \in \overline{S}_2 \backslash S_2$ and a strictly monotone sequence $\{r_l\} \subset S_2$ such that $\lim_{l \to \infty} r_l = r_0$. We

may assume that $r_l \uparrow r_0$. For each $l \geq 1$, let $y_l(t)$ denote the associated solution of (4.0.1) satisfying
$$\Delta^i y_l(t_2) = \Delta^i v_2(t_2), \ m \leq i \leq k + m - 1,$$
$$\Delta^i y_l(t_1) = \Delta^i v_2(t_1), \ 0 \leq i \leq n - k - 2,$$
$$\Delta^{n-k-1} y_l(t_1) = r_l.$$
From the boundary conditions, we note that $y_l(t) = y_{l+1}(t)$, for $t = t_1, \ldots, t_1 + n - k - 2$, and that $y_l(t_1 + n - k - 1) < y_{l+1}(t_1 + n - k - 1)$, for each $l \geq 1$. It follows from condition (FC) and repeated applications of the discrete Rolle's Theorem that $y_l(t) < y_{l+1}(t)$ on $[t_1 + n - k - 1, \infty)$, for each $l \geq 1$. Also, $r_0 \notin S_2$ and Theorem 4.1.1 imply that, for some $t_0 \in (t_1 + n - k - 1, t_1 + n - 1]$, $y_l(t_0) \uparrow \infty$, as $l \to \infty$.

Similarly, from condition (FC) and repeated applications of the discrete Rolle's Theorem, if $n - k - 1$ is even, then $y_l(t_1 - 1) < y_{l+1}(t_1 - 1)$, whereas, if $n - k - 1$ is odd, then $y_l(t_1 - 1) > y_{l+1}(t_1 - 1)$, for each $l \geq 1$. We assume that $n - k - 1$ is even and then $y_l(t_1 - 1) < y_{l+1}(t_1 - 1)$, for each $l \geq 1$.

We now claim that the sequence $\{y_l(t_1 - 1)\}$ is not bounded above. We assume this claim to be false, and so there is an $M \in \mathbb{R}$ such that $y_l(t_1 - 1) \uparrow M$, as $l \to \infty$. By the induction hypotheses on the spacing $t_1 - t_2$, there is a solution $z(t)$ of (4.0.1)-(4.2.2) at the points t_2 and $t_1 - 1$ satisfying
$$\Delta^i z(t_2) = \Delta^i v_2(t_2) = \Delta^i y_l(t_2), \ m \leq i \leq k + m - 1, \ l \geq 1,$$
$$z(t_1 - 1) = M,$$
$$\Delta^i z(t_1 - 1) = (-1)^i M + \sum_{j=1}^{i} (-1)^{j+1} \Delta^{i-j} v_2(t_1)$$
$$= (-1)^i M + \sum_{j=1}^{i} (-1)^{j+1} \Delta^{i-j} y_l(t_1), \quad 1 \leq i \leq n - k - l, l \geq 1.$$
It follows from Theorem 4.2.1 that $\{y_l(t)\}$ converges to $z(t)$ at each point of $[a, \infty)$, which in turn implies $\Delta^{n-k-1} z(t_1) = r_0$; this contradicts $r_0 \notin S_2$. We conclude that our latter claim is false, and that $y_l(t_1 - 1) \uparrow \infty$, as $l \to \infty$.

Now, we let $u(t)$ be the solution of (4.0.1)-(4.2.2), for the indices $m - 1$ and k, satisfying
$$\Delta^i u(t_2) = y_{i+(n-k)-m+1}, \ m \leq i \leq m - 2,$$
$$\Delta^{m-1} u(t_2) = 0,$$
$$\Delta^i u(t_1) = y_{i+1}, \ 0 \leq i \leq n - k - 2,$$
$$\Delta^{n-k-1} u(t_1) = r_0.$$

Since $y_l(t_1 - 1) \uparrow \infty$ and $y_l(t_0) \uparrow \infty$, where $t_0 \in (t_1 + n - k - 1, t_1 + n - 1]$ was obtained above, there exists an $L \geq 1$ such that $y_L(t_1 - 1) > u(t_1 - 1)$ and $y_L(t_0) > u(t_0)$. From $\Delta^{n-k-1} u(t_1) = r_0 > r_L = \Delta^{n-k-1} y_L(t_1)$, while $\Delta^i(u(t_1) - y_L(t_1)) = 0$, $0 \leq i \leq n - k - 2$, it follows that $u(t) - y_L(t)$ has a gz at $t_1 + n - k - 1$ and a gz at some $\tau_0 \in (t_1 + n - k - 1, t_0]$. In addition, $\Delta^i(u(t_2) - y_L(t_2)) = 0$, $m \leq i \leq k + m - 2$, and so applying the discrete Rolle's Theorem in terms of conditions satisfied by $u(t) - y_L(t)$ at $t_2, t_1, t_1 + n - k - 1$ and τ_0, we conclude that there exist points $t_2 \leq \sigma_n < \sigma_{n-1} < \cdots < \sigma_1 \leq \tau_0$ in $[a, \infty)$ such that $\Delta^{i-1}(u(t) - y_L(t))$ has a gz at σ_i, for $1 \leq i \leq n$. This is a contradiction to condition (FC).

Therefore, S_2 is also a closed subset of $\mathbb{R}$, and $S_2 = \mathbb{R}$. By choosing $y_{n-k} \in S_2$, there exists a corresponding unique solution $y(t)$ of (4.0.1)-(4.2.2), for the indices m and k, which satisfies

$$\Delta^i y(t_2) = y_{i+(n-k)-m+1}, \quad m \leq i \leq k + m - 1,$$
$$\Delta^i y(t_1) = y_{i+1}, \quad 0 \leq i \leq n - k - 1.$$

The proof is complete. $\qquad\qquad\square$

We immediately have the following concerning the existence of solutions of (m_2, m_1) focal boundary value problems for (4.0.1).

Theorem 4.2.3. [[54], Henderson, Thm. 4] *Assume that with respect* (4.0.1), *conditions* (DA) *and* (FC) *are satisfied. Then, given positive integers m_1 and m_2 such that $m_1 + m_2 = n$, each (m_2, m_1) focal boundary value problem for* (4.0.1) *has a unique solution on $[a, \infty)$.*

Proof. Given m_1 and m_2 such that $m_1 + m_2 = n$, let m_2 correspond to k and let $m_1 = n - m_2$ correspond to the case $m = n - k$ in Theorem 4.2.2. $\qquad\square$

We now deal with our second major uniqueness implies existence result for solutions of focal boundary value problems; namely, we prove that conditions (DA) and (FC) imply the existence of solutions for each $(m_k, \ldots, m_1)$ focal boundary value problem (4.0.1)-(4.2.2).

Theorem 4.2.4. [[55], Henderson, Thm. 4] *Assume that with respect* (4.0.1), *conditions* (DA) *and* (FC) *are satisfied. Then, for $2 \leq k \leq n$, each $(m_k, \ldots, m_1)$ focal boundary value problem* (4.0.1)-(4.2.2) *has a unique solution on $[a, \infty)$.*

Proof. We again begin by pointing out that, by condition (FC), solutions of each $(m_k, \ldots, m_1)$ focal boundary value problem for (4.0.1) are unique on $[a, \infty)$, when solutions exist. The proof involves shooting methods along with induction on k, induction on the tuple $(m_k, \ldots, m_1)$, and induction on the spacing $t_1 - t_2$.

Throughout, let $y_i \in \mathbb{R}$, $1 \le i \le n$, be given.

First, if $k = 2$, then by Theorem 4.2.3, each (m_2, m_1) focal boundary value problem for (4.0.1) has a unique solution on $[a, \infty)$.

For the remainder of the proof, we assume $2 < k \le n$, and that, for $2 \le s < k$, each $(i_s, \ldots, i_1)$ focal boundary value problem for (4.0.1) has a unique solution. The proof then proceeds by the following outline:

(I) For $m_1 = 1$, by inducting on m_2 and the difference $t_1 - t_2$, we obtain the existence of unique solutions of each $(m_k, \ldots, m_2, 1)$ focal boundary value problem for (4.0.1), for all positive integers $m_2, \ldots, m_k$ such that $1 + m_2 + \cdots + m_k = n$.

(II) Then, for $1 < m_1 \le n - (k - 1)$, we assume that each $(m_{kh}, \ldots, m_{2h}, h)$ focal boundary value problem, $1 \le h < m_1$, for (4.0.1) has a unique solution, and we prove, by induction on m_2 and the spacing $t_1 - t_2$, that each $(m_k, \ldots, m_2, m_1)$ focal boundary value problem for (4.0.1) has a unique solution, for all $m_2, \ldots, m_k$ such that $m_1 + m_2 + \cdots + m_k = n$.

We begin with part (I). Let $m_1 = 1$. Set $m_2 = 1$ and let $m_3, \ldots, m_k$ be positive integers such that $1+1+m_3+\cdots+m_k = n$. Let $t_k < t_{k-1} < \cdots < t_1$ in $[a, \infty)$ be given, with $t_j + m_j + 1 \le t_{j-1}$, $2 \le j \le k$. We note that in the case of $m_2 = m_1 = 1$, if $t_1 = t_2 + 1$, then the $(m_k, \ldots, m_3, 1, 1)$ focal problem can be considered as an $(i_{k-1}, \ldots, i_1)$ focal problem with conditions specified at the points, $t_k, t_{k-1}, \ldots, t_2$, where $i_1 = 2$ and $i_j = m_{j+1}, 2 \le j \le k-1$, and hence such a boundary value problem has a unique solution by the induction assumption on k. So, in addition to our foregoing assumptions, we assume that for each $t_2 + 1 \le \tau_1 < t_1$, there exists a unique solution of each $(m_k, \ldots, m_3, 1, 1)$ focal problem for (4.0.1) with boundary conditions specified at the points $t_k, t_{k-1}, \ldots, t_2, \tau_1$.

Now, let $v_1(t)$ be the solution of the $(i_{k-1}, \ldots, i_1)$ focal boundary value problem, where $i_1 = 2$ and $i_j = m_{j+1}, 2 \le j \le k - 1$, for (4.0.1) satisfying

$$v_1(t_2) = 0,$$
$$\Delta v_1(t_2) = y_2,$$
$$\Delta^i v_1(t_j) = y_{i+1}, \ s_{j-1} \le i \le s_j - 1, \ 3 \le j \le k.$$

Define

$$S_1 := \{r \in \mathbb{R} \mid \text{there is a solution } y(t) \text{ of } (4.0.1) \text{ satisfying } y(t_1) = r,$$
$$\Delta y(t_2) = \Delta v_1(t_2), \text{ and } \Delta^i y(t_j) = \Delta^i v_1(t_j), s_{j-1} \le i \le s_j - 1,$$
$$3 \le j \le k\}.$$

S_1 is nonempty, since $v_1(t_1) \in S_1$, and also by Theorem 4.2.1, S_1 is an open subset of $\mathbb{R}$.

We claim that S_1 is also a closed subset of $\mathbb{R}$. If that is not the case, then there exist an $r_0 \in \overline{S}_1 \setminus S_1$ and a strictly monotone sequence $\{r_l\} \subset S_1$ such that $\lim_{l \to \infty} r_l = r_0$. We may assume that $r_l \uparrow r_0$. Then, for each $l \ge 1$, there is an associated solution $y_l(t)$ of (4.0.1) satisfying

$$y_l(t_1) = r_l,$$
$$\Delta y_l(t_2) = \Delta v_1(t_2),$$
$$\Delta^i y_l(t_j) = \Delta^i v_1(t_j), \ s_{j-1} \le i \le s_j - 1, \ 3 \le j \le k.$$

From condition (FC), $y_l(t) < y_{l+1}(t)$ on $[t_2, \infty)$, for all $l \ge 1$. The induction hypothesis on the spacing $t_1 - t_2$ implies the existence of unique solutions of $(m_k, \ldots, m_3, 1, 1)$ focal problems, when conditions are specified at $t_k, \ldots, t_2, t_1 - 1$, which when coupled with Theorem 4.2.1 and $r_0 \notin S_1$, implies that $y_l(t_1 - 1) \uparrow \infty$, as $l \to \infty$. Moreover, from Theorem 4.1.1 and $r_0 \notin S_1$, it follows that there is a $t_0 \in (t_1, t_1 + n - 1]$ such that $y_l(t_0) \uparrow \infty$, as $l \to \infty$.

Now, we let $u(t)$ be the solution of the $(i_{k-1}, \ldots, i_1)$ focal boundary value problem for (4.0.1), where $i_1 = 1, i_2 = m_3 + 1$, and $i_j = m_{j+1}, 3 \le j \le k - 1$, with conditions at the points $t_k, t_{k-1}, \ldots, t_3, t_1$, given by

$$u(t_1) = r_0,$$
$$\Delta u(t_3) = 0,$$
$$\Delta^i u(t_j) = \Delta^i v_1(t_j), \ s_{j-1} \le i \le s_j - 1, \ 3 \le j \le k.$$

Now, $y_l(t_1 - 1) \uparrow \infty$ and $y_l(t_0) \uparrow \infty$, whereas, $y_l(t_1) = r_l < r_0 = u(t_1)$, and so there exists an $L \ge 1$ such that $u(t) - y_L(t)$ has a gz at t_1 and a gz at some $\tau_0 \in (t_1, t_0]$. By the discrete Rolle's Theorem, there exists a $\tau_1 \in [t_1, \tau_0)$ such that $\Delta(u(t) - y_L(t))$ has a gz at τ_1. We also have $\Delta^i(u(t_j) - y_L(t_j)) = 0, s_{j-1} \le i \le s_j - 1, 3 \le j \le k$, and so condition (FC) implies $u(t) = y_l(t)$ on $[a, \infty)$. This is a contradiction.

Therefore, S_1 is closed, and $S_1 = \mathbb{R}$. By choosing $y_1 \in S_1$, the corresponding solution $y(t)$ of (4.0.1) satisfies

$$y(t_1) = y_1,$$
$$\Delta^i y(t_j) = y_{i+1}, \ s_{j-1} \le i \le s_j - 1, \ 2 \le j \le k.$$

Specifically, we conclude that there exists a unique solution of each $(m_k, \ldots, m_3, 1, 1)$ focal problem for (4.0.1), for all $m_3, \ldots, m_k$ such that $1 + 1 + m_3 + \cdots + m_k = n$.

We now assume that $m_2 > 1$ and that for all $1 \leq h < m_2$ and all $m_{3h}, \ldots, m_{kh}$ such that $1 + h + m_{3h} + \cdots + m_{kh} = n$, each $(m_{kh}, \ldots, m_{3h}, h, 1)$ focal problem has a unique solution on $[a, \infty)$.

Let $m_3, \ldots, m_k$ be positive integers such that $1 + m_2 + m_3 + \cdots + m_k = n$, and let $t_k < t_{k-1} < \cdots < t_1$ in $[a, \infty)$, where $t_j + m_j + 1 \leq t_{j-1}, 2 \leq j \leq k$, be given. (We note that, if $t_2 + m_2 = t_1$, then our $(m_k, \ldots, m_2, 1)$ focal problem can be considered as an $(i_{k-1}, \ldots, i_1)$ focal problem, where $i_1 = m_2 + 1, i_j = m_{j+1}, 2 \leq j \leq k - 1$, and hence has a solution by the induction assumption on k.) So, we also assume that, for each $t_2 + m_2 \leq \tau_1 < t_1$, there exists a unique solution of the $(m_k, \ldots, m_3, m_2, 1)$ focal problem for (4.0.1) with boundary conditions at the points $t_k, t_{k-1}, \ldots, t_2, \tau_1$.

For this step, we let $v_2(t)$ be the solution of the $(i_{k-1}, \ldots, i_1)$ focal problem, where $i_1 = m_2 + 1$ and $i_j = m_{j+1}, 2 \leq j \leq k - 1$, for (4.0.1) satisfying

$$v_2(t_2) = 0,$$
$$\Delta^i v_2(t_j) = y_{i+1}, \ s_{j-1} \leq i \leq s_j - 1, \ 2 \leq j \leq k,$$

and we define

$$S_2 := \{r \in \mathbb{R} \,|\, \text{there is a solution } y(t) \text{ of (4.0.1) satisfying } y(t_1) = r,$$
$$\text{and } \Delta^i y(t_j) = \Delta^i v_2(t_j), s_{j-1} \leq i \leq s_j - 1, 2 \leq j \leq k\}.$$

S_2 is nonempty, and from continuous dependence on boundary conditions, S_2 is also open.

Again, we will claim that S_2 is closed. So, assuming that S_2 is not closed, then there exist $r_0 \in \overline{S_2} \setminus S_2$ and a strictly monotone sequence $\{r_l\} \subset S_2$ converging to r_0. We may assume, as before, that $r_l \uparrow r_0$, and also, let $y_l(t)$ be the corresponding solution of (4.0.1) which satisfies

$$y_l(t_1) = r_l,$$
$$\Delta^i y_l(t_j) = \Delta^i v_2(t_j), \ s_{j-1} \leq i \leq s_j - 1, \ 2 \leq j \leq k.$$

Since $r_0 \notin S_2$, it follows from (FC) that $y_l(t) < y_{l+1}(t)$ on $[t_1 - 1, \infty)$, for each $l \geq 1$. And by the induction hypotheses on the difference $t_1 - t_2$ (in particular, there exist unique solutions of $(m_k, \ldots, m_3, m_2, 1)$ focal problems with conditions at $t_k, t_{k-1}, \ldots, t_2, t_1 - 1$), we can argue that $y_l(t_1 - 1) \uparrow \infty$, as $l \to \infty$. And, since $r_0 \notin S_2$, Theorem 4.1.1 implies there exists $t_0 \in (t_1, t_1 + n - 1]$ such that $y_l(t_0) \uparrow \infty$, as $l \to \infty$.

From the induction assumptions on m_2, there is a unique solution $u(t)$ of the $(m_k, \ldots, m_3 + 1, m_2 - 1, 1)$ focal boundary value problem for (4.0.1) which satisfies

$$u(t_1) = r_0,$$
$$\Delta^i u(t_2) = \Delta^i v_2(t_2), \ 1 \le i \le m_2 - 1 = s_2 - 2,$$
$$\Delta^{s_2-1} u(t_3) = 0,$$
$$\Delta^i u(t_j) = \Delta^i v_2(t_j), \ s_{j-1} \le i \le s_j - 1, \ 3 \le j \le k.$$

With both sequences $\{y_l(t_1 - 1)\}$ and $\{y_l(t_0)\}$ diverging monotonically to positive infinity, while $u(t_1) = r_0 > y_l(t_1)$, for each $l \ge 1$, then there is an $L \ge 1$ such that $u(t) - y_L(t)$ has a gz at t_1 and a gz at some $\tau_0 \in (t_1, t_0]$. We also have $\Delta^i(u(t_2) - y_L(t_2)) = 0, 1 \le i \le s_2 - 2$. Repeated applications of the discrete Rolle's Theorem yields points $t_2 \le \tau_{s_2-1} < \cdots < \tau_1 < \tau_0$ such that $\Delta^i(u(t) - y_L(t))$ has a gz at τ_i, $0 \le i \le s_2 - 1$. In addition, we have $\Delta^i(u(t_j) - y_L(t_j)) = 0, s_{j-1} \le i \le s_j - 1, 3 \le j \le k$, and then by condition (FC), $u(t) = y_L(t)$ on $[a, \infty)$; a contradiction.

So, S_2 is closed and $S_2 = \mathbb{R}$. Then, choosing $y_1 \in S_2$, the corresponding solution is the desired solution. In summary, we have proved, inductively, that each $(m_k, \ldots, m_3, m_2, 1)$ focal boundary value problem for (4.0.1) has a unique solution, for all $m_2, \ldots, m_k$ such that $1 + m_2 + \cdots m_k = n$. This completes part (I).

Now, for part (II), we assume $1 < m_1 \le n - (k - 1)$ and that, for $1 \le h < m_1$ and for all positive integers $m_{2h}, \ldots, m_{kh}$ such that $h + m_{2h} + \cdots + m_{kh} = n$, each $(m_{kh}, \ldots, m_{2h}, h)$ focal boundary value problem for (4.0.1) has a unique solution.

Again, we induct on m_2, and so let $m_2 = 1$ and $m_3, \ldots, m_k$ be positive integers such that $m_1 + 1 + m_3 + \cdots + m_k = n$. We let $t_k < t_{k-1} < \cdots < t_1$ be given points in $[a, \infty)$, with $t_j + m_j + 1 \le t_{j-1}, 2 \le j \le k$. (As above, if $t_2 + 1 = t_1$, then the $(m_k, \ldots, m_3, 1, m_1)$ focal problem can be considered as an $(i_{k-1}, \ldots, i_1)$ focal problem, where $i_1 = 1 + m_1, i_j = m_{j+1}, 2 \le j \le k-1$, and hence has a unique solution by the induction assumption on k.) And so, we assume that, for each $t_2 + 1 \le \tau_1 < t_1$, there is a unique solution of the $(m_k, \ldots, m_3, 1, m_1)$ focal problem for (4.0.1) with boundary conditions specified at $t_k, t_{k-1}, \ldots, t_2, \tau_1$.

This time, let $v_3(t)$ denote the solution of the $(m_k, \ldots, m_3, 2, m_1 - 1)$

focal boundary value problem for (4.0.1) which satisfies

$$\Delta^i v_3(t_1) = y_{i+1}, \ 0 \le i \le m_1 - 2 = s_1 - 2,$$
$$\Delta^{m_1 - 1} v_3(t_2) = 0,$$
$$\Delta^{m_1} v_3(t_2) = y_{m_1 + 1},$$
$$\Delta^i v_3(t_j) = y_{i+1}, \ s_{j-1} \le i \le s_j - 1, \ 3 \le j \le k.$$

Define

$$S_3 := \{r \in \mathbb{R} \,|\, \text{there is a solution } y(t) \text{ of (4.0.1) satisfying}$$
$$\Delta^i y(t_1) = \Delta^i v_3(t_1), 0 \le i \le m_1 - 2, \ \Delta^{m_1 - 1} y(t_1) = r,$$
$$\text{and } \Delta^i y(t_j) = \Delta^i v_3(t_j), s_{j-1} \le i \le s_j - 1, 2 \le j \le k\}.$$

S_3 is a nonempty open subset of $\mathbb{R}$.

We claim that S_3 is a closed subset as well. Assuming S_3 not to be closed, let $r_0 \in \bar{S}_3 \setminus S_3$, $\{r_l\} \subset S_3$, with $r_l \uparrow r_0$, be as usual, and $y_l(t)$ denotes the corresponding solution of (4.0.1), where

$$\Delta^i y_l(t_1) = \Delta^i v_3(t_1), \ 0 \le i \le m_1 - 2,$$
$$\Delta^{m_1 - 1} y_l(t_1) = r_l,$$
$$\Delta^i y_l(t_j) = \Delta^i v_3(t_j), s_{j-1} \le i \le s_j - 1, \ 2 \le j \le k.$$

From the boundary conditions, $y_l(t) = y_{l+1}(t)$, for $t = t_1, \ldots, t_1 + m_1 - 2$, and $y_l(t_1 + m_1 - 1) < y_{l+1}(t_1 + m_1 - 1)$, for each $l \ge 1$. Condition (FC) and the discrete Rolle's Theorem give that $y_l(t) < y_{l+1}(t)$ on $[t_1 + m_1 - 1, \infty)$, for each $l \ge 1$. Since $r_0 \notin S_3$, it follows also from Theorem 4.1.1 that for some $t_0 \in (t_1 + m_1 - 1, t_1 + n - 1]$, $y_l(t_0) \uparrow \infty$, as $l \to \infty$.

From condition (FC) and again repeated applications of the discrete Rolle's Theorem, we have that, if $m_1 - 1$ is even, then $y_l(t_1 - 1) < y_{l+1}(t_1 - 1)$, and if $m_1 - 1$ is odd, then $y_l(t_1 - 1) > y_{l+1}(t_1 - 1)$, for each $l \ge 1$. We will assume the former case so that $y_l(t_1 - 1) < y_{l+1}(t_1 - 1)$, for each $l \ge 1$. By the induction on the difference $t_1 - t_2$ and $r_0 \notin S_3$, it follows that $y_l(t_1 - 1) \uparrow \infty$, as $l \to \infty$.

Now, let $u(t)$ be the solution of the $(i_{k-1}, \ldots, i_1)$ focal problem, with $i_1 = m_1, i_2 = m_3 + 1$, and $i_j = m_{j+1}, 3 \le j \le k - 1$, satisfying

$$\Delta^i u(t_1) = \Delta^i v_3(t_1), \ 0 \le i \le m_1 - 2,$$
$$\Delta^{m_1 - 1} u(t_1) = r_0,$$
$$\Delta^{m_1} u(t_3) = 0,$$
$$\Delta^i u(t_j) = \Delta^i v_3(t_j), \ s_{j-1} \le i \le s_j - 1, \ 3 \le j \le k.$$

With $y_l(t_1 - 1) \uparrow \infty$ and $y_l(t_0) \uparrow \infty$, as $l \to \infty$, and $\Delta^i u(t_1) = \Delta^i y_l(t_1)$, $0 \leq i \leq m_1 - 2$, and $\Delta^{m_1-1} u(t_1) = r_0 > \Delta^{m_1-1} y_l(t_1)$, for each $l \geq 1$, then there exists an $L \geq 1$ such that $u(t) - y_L(t)$ has a gz at $t_1 + m_1 - 1$ and a gz at some $\tau_0 \in (t_1 + m_1 - 1, t_0]$. Repeated applications of the discrete Rolle's Theorem yield points $t_1 - 1 < \tau_{m_1} < \tau_{m_1-1} < \cdots < \tau_1 < \tau_0$ in $[a, \infty)$ such that $\Delta^i(u(t) - y_L(t))$ has a gz at τ_i, $0 \leq i \leq m_1 = s_1 = s_2 - 1$. Furthermore, $\Delta^i(u(t_j) - y_L(t_j)) = 0$, $s_{j-1} \leq i \leq s_j - 1, 3 \leq j \leq k$, and so from condition (FC), $u(t) = y_L(t)$ on $[a, \infty)$. This is a contradiction.

It follows that S_3 is closed and $S_3 = \mathbb{R}$. We choose $y_{m_1} \in S_3$ and the associated solution $y(t)$ of (4.0.1) satisfying

$$\Delta^i y(t_1) = y_{i+1}, \ 0 \leq i \leq m_1 - 1 = s_1 - 1,$$
$$\Delta^{m_1} y(t_2) = y_{s_1+1},$$
$$\Delta^i y(t_1) = y_{i+1}, \ s_{j-1} \leq i \leq s_j - 1, \ 3 \leq j \leq k,$$

is the desired solution. Summarily, each $(m_k, \ldots, m_3, 1, m_1)$ focal boundary value problem for (4.0.1) has a unique solution for all positive integers $m_3, \ldots, m_k$ such that $m_1 + 1 + m_3 + \cdots + m_k = n$.

For the final step in part (II), we assume $m_2 > 1$ and that, for each $1 \leq h < m_2$, and all $m_{3h}, \ldots, m_{kh}$ such that $m_1 + h + m_{3h} + \cdots + m_{kh} = n$, each $(m_{kh}, \ldots, m_{3h}, h, m_1)$ focal boundary value problem for (4.0.1) has a unique solution.

Now, we let $m_3, \ldots, m_k$ be positive integers such that $m_1 + m_2 + m_3 + \cdots + m_k = n$, and we let $t_k < t_{k-1} < \cdots < t_1$ belong to $[a, \infty)$, where $t_j + m_j + 1 \leq t_{j-1}$, $2 \leq j \leq k$. As in the above cases, we assume also that, for each $t_2 + m_2 \leq \tau_1 < t_1$, there exists a unique solution of each $(m_k, \ldots, m_3, m_2, m_1)$ focal problem with boundary data specified at the points $t_k, \ldots, t_2, \tau_1$.

Let $v_4(t)$ be the solution of the $(m_k, \ldots, m_3, m_2+1, m_1-1)$ focal boundary value problem for (4.0.1) that satisfies

$$\Delta^i v_4(t_1) = y_{i+1}, \ 0 \leq i \leq m_1 - 2,$$
$$\Delta^{m_1-1} v_4(t_2) = 0,$$
$$\Delta^i v_4(t_j) = y_{i+1}, \ s_{j-1} \leq i \leq s_j - 1, \ 2 \leq j \leq k,$$

and define

$$S_4 := \{r \in \mathbb{R} \,|\, \text{there is a solution } y(t) \text{ of (4.0.1) satisfying}$$
$$\Delta^i y(t_1) = \Delta^i v_4(t_1), 0 \leq i \leq m_1 - 2, \ \Delta^{m_1-1} y(t_1) = r,$$
$$\text{and } \Delta^i y(t_j) = \Delta^i v_4(t_j), s_{j-1} \leq i \leq s_j - 1, 2 \leq j \leq k\}.$$

S_4 is a nonempty and open subset of $\mathbb{R}$.

We claim S_4 is also closed, and assuming this not to be the case, with $r_0 \in \overline{S}_4 \setminus S_4$, $\{r_l\} \subset S_4$ such that $r_l \uparrow r_0$, and $y_l(t)$ the corresponding solutions of (4.0.1), then as in the last argument above, $y_l(t) = y_{l+1}(t)$, for $t = t_1, \ldots, t_1 + m_1 - 2$, and $y_l(t_1 + m_1 - 1) < y_{l+1}(t_1 + m_1 - 1)$, for each $l \geq 1$. Moreover, it also is the case that $y_l(t) < y_{l+1}(t)$ on $[t_1 + m_1 - 1, \infty)$, for each $l \geq 1$, so that $y_l(t_0) \uparrow \infty$, as $l \to \infty$, for some $t_0 \in (t_1 + m_1 - 1, t_1 + n - 1]$.

As in the previous case, if $m_1 - 1$ is even, then $y_l(t_1 - 1) < y_{l+1}(t_1 - 1)$, and if $m_1 - 1$ is odd, then $y_l(t_1 - 1) > y_{l+1}(t_1 - 1)$, for each $l \geq 1$. We assume $m_1 - 1$ is even, and then from induction on the spacing $t_1 - t_2$ and $r_0 \notin S_4$, we have that $y_l(t_1 - 1) \uparrow \infty$, as $l \to \infty$.

Let $u(t)$ be the solution of the $(m_k, \ldots, m_3 + 1, m_2 - 1, m_1)$ focal problem for (4.0.1) that satisfies

$$\Delta^i u(t_1) = \Delta^i v_4(t_1),\ 0 \leq i \leq m_1 - 2,$$
$$\Delta^{m_1 - 1} u(t_1) = r_0,$$
$$\Delta^i u(t_2) = \Delta^i v_4(t_2),\ m_1 = s_1 \leq i \leq s_2 - 2,$$
$$\Delta^{s_2 - 1} u(t_3) = 0,$$
$$\Delta^i u(t_j) = \Delta^i v_4(t_j),\ s_{j-1} \leq i \leq s_j - 1,\ 3 \leq j \leq k.$$

From the conditions on the sequences, $\{y_l(t_1 - 1)\}$ and $\{y_l(t_0)\}$, coupled with $\Delta^{m_1 - 1} u(t_1) > \Delta^{m_1 - 1} y_l(t_1)$, for all $l \geq 1$, and the other boundary conditions at t_1, there exists $L \geq 1$ such that $u(t) - y_L(t)$ has a gz at $t_1 + m_1 - 1$ and a gz at some $\tau_0 \in (t_1 + m_1 - 1, t_0]$. From the boundary conditions and repeated applications of the discrete Rolle's Theorem, there exist points $t_2 \leq \tau_{s_2 - 1} < \tau_{s_2 - 2} < \cdots < \tau_1 < \tau_0$ such that $\Delta^i(u(t) - y_L(t))$ has a gz at τ_i, $0 \leq i \leq s_2 - 1$. Since $\Delta^i(u(t_j) - y_L(t_j)) = 0$, $s_{j-1} \leq i \leq s_j - 1$, $3 \leq j \leq k$, condition (ΓC) implies $u(t) - y_L(t)$ on $[a, \infty)$; a contradiction.

So, S_4 is closed and $S_4 = \mathbb{R}$. For $y_{m_1} \in S_4$, the corresponding solution $y(t)$ is the desired solution. In summary, each $(m_k, \ldots, m_3, m_2, m_1)$ focal boundary value problem for (4.0.1) has a unique solution on $[a, \infty)$, for all positive integers $m_1, \ldots, m_k$ such that $m_1 + \cdots + m_k = n$. The proof of part (II) is complete, and the proof of the theorem is complete. $\qquad\square$

Remark 4.2.2. (a) *If we replace in* (DA) *the discrete interval* $[a, \infty)$ *with the integers* $\mathbb{Z}$, *then we can establish a uniqueness implies existence result for boundary value problems for* (4.0.1) *of the (discrete) "right focal" type* *which satisfy*

$$\Delta^i u(t_j) = y_{i+1}, \ s_{j-1} \leq i \leq s_j - 1, \ 1 \leq j \leq k, \qquad (4.2.3)$$

where $2 \leq k \leq n$, $m_1, \ldots, m_k$ *are positive integers such that* $\sum_{i=1}^{k} m_i = n$, s_j *are corresponding partial sums,* $0 \leq j \leq k$, $t_1 < \cdots < t_k$ *belong to* $\mathbb{Z}$ *with* $t_j + 1 \leq t_{j+1}$, $1 \leq j \leq k - 1$, *and* $y_i \in \mathbb{R}$, $1 \leq i \leq n$.

In particular, under (DA) *(in the context of the set* $\mathbb{Z}$*), if for each* $2 \leq k \leq n$, *solutions of* (4.0.1)-(4.2.3) *are unique, when solutions exist, it can be argued in complete analogy to Theorem 4.2.4 that there exist unique solutions of* (4.0.1)-(4.2.3), $2 \leq k \leq n$, *on* $\mathbb{Z}$.

(b) *The two main results of this section, Theorem 4.2.2 and Theorem 4.2.4, can be established with respect to the finite discrete interval* $[a, b + n]$ *where* b *is the rightmost point at which boundary conditions are specified, so that our applications of Theorem 4.1.1 can be made in the arguments.*

4.3 "Between" boundary value problems: uniqueness implies existence

The results of this section were published by Henderson [56] in 1991, and the types of boundary value problems for the difference equation (4.0.1), for which we address questions of uniqueness of solutions implying their existence, are devoted to problems whose boundary conditions we will term as "between" the conjugate conditions (4.1.1) of Section 4.1 and the right focal conditions (4.2.3) in Remark 4.2.2(a) at the end of Section 4.2. In view of that, the discrete interval $[a, \infty)$ from those sections will be replaced by the integers $\mathbb{Z}$, so that condition (DA) is replaced here by the condition:

(DAZ) $f(m, r_1, \ldots, r_n) : \mathbb{Z} \times \mathbb{R}^n \to \mathbb{R}$ is continuous. and the equation $r_n = f(m, r_0, \ldots, r_{n-1})$ can be solved for r_0 as a continuous function of $r_1, \ldots, r_n$, for each $m \in \mathbb{Z}$.

Then Remark 4.0.1 and Theorem 4.1.1 remain valid with $[a, \infty)$ replaced by $\mathbb{Z}$, and both will be referred to in the development of this section, as well as to Remark 4.2.1 and Remark 4.2.2. And many of the other theorems of Sections 4.1 and 4.2 will be referred to as being valid with $[a, \infty)$ replaced by $\mathbb{Z}$. Also, when an interval, say J, is used in the discussion, the meaning will be the discrete interval of integers $J \cap \mathbb{Z}$.

Definition 4.3.1. *Let* $1 \leq \ell \leq n$, *let* $m_0 = 0$, *and let* $m_1, \ldots, m_\ell \in \mathbb{N}$ *be such that* $\sum_{i=1}^{\ell} m_i = n$. *For points in* $\mathbb{Z}$, $s_1 < \cdots < s_{m_1}$ *and* $s_{m_1 + \cdots + m_{i-1}} \leq$

$s_{m_1+\cdots+m_{i-1}+1} < \cdots < s_{m_1+\cdots+m_i}$, $2 \leq i \leq \ell$, *a boundary value problem for* (4.0.1) *satisfying*

$$\Delta^{i-1} u(s_{m_0+\cdots+m_{i-1}+j}) = y_{ij}, \ 1 \leq j \leq m_i, \ 1 \leq i \leq \ell, \qquad (4.3.1)$$

where $y_{ij} \in \mathbb{R}$, *will be called a* $B(m_1; \ldots; m_\ell)$ *boundary value problem for* (4.0.1).

Remark 4.3.1. *When* $\ell = 1$ *and* $m_1 = n$, *then* (4.3.1) *becomes* $u(s_j) = y_{1j}$, $1 \leq j \leq m_1 = n$, *so that the* $B(n)$ *problems are of the conjugate type* (4.1.1), *whereas, when* $\ell = n$ *and* $m_1 = \cdots = m_n = 1$, *the* $B(1; \ldots; 1)$ *problems are of the right focal type* (4.2.3).

We recall here for local reference sake, that for a function $u : [a, \infty) \to \mathbb{R}$, Hartman [44] defined $m_0 \in [a, \infty)$, in the case that $m_0 = a$, to be a *generalized zero* (denoted by "gz") of u if $u(a) = 0$, and $m_0 > a$ to be a *generalized zero* of u if $u(m_0) = 0$ or there is an integer $j \geq 1$ such that $(-1)^j u(m_0 - j) u(m_0) > 0$ and if $j > 1$, $u(m_0 - j + 1) = \cdots = u(m_0 - 1) = 0$.

Given $1 \leq \ell \leq n$ and $m_1, \ldots, m_\ell \in \mathbb{N}$ such that $\sum_{i=1}^{\ell} m_i = n$, our uniqueness condition on $B(m_1; \ldots; m_\ell)$ boundary value problems for (4.0.1) will be stated in terms of generalized zeros:

$(B_{m_1,\ldots,m_\ell})$ Given points in $\mathbb{Z}$, $s_1 < \cdots < s_{m_1}$ and $s_{m_1+\cdots+m_{i-1}} \leq s_{m_1+\cdots+m_{i-1}+1} < \cdots < s_{m_1+\cdots+m_i}$, for $2 \leq i \leq \ell$, if $u(m)$ and $v(m)$ are solutions of (4.0.1) such that, for each $1 \leq r \leq \ell$, $\Delta^{r-1}(u(m) - v(m))$ has m_r gz's at $s_{m_0+\cdots m_{r-1}+j}$, $1 \leq j \leq m_r$, then it follows that $u(m) = v(m)$ on $[s_1, s_n]$, (hence on $\mathbb{Z}$).

Remark 4.3.2. (a) *Given* $1 \leq \ell \leq n$ *and* $m_1, \ldots, m_\ell \in \mathbb{N}$, *condition* $(B_{m_1,\ldots,m_\ell})$ *implies solutions of* $B(m_1; \ldots; m_\ell)$ *boundary value problems* (4.0.1)-(4.3.1) *are unique, when solutions exist.*

(b) *Also, condition* (FC) *of Section 4.2 and the discrete Rolle's Theorem of Remark 4.2.1 of Section 4.2 imply condition* $(B_{m_1,\ldots,m_\ell})$, *for all* $m_1, \ldots, m_\ell$.

(c) *Further, if condition* $(B_{m_1,\ldots,m_\ell})$ *holds, for some* $m_1, \ldots, m_\ell$, *then by repeated applications of the discrete Rolle's Theorem, condition* (DC) *of Section 4.1 also holds.*

Just as with Theorem 4.1.2 of Section 4.1 and Theorem 4.2.1 of Section 4.2, our first theorem of this section concerns the continuous dependence of solutions of (4.0.1)-(4.3.1) on $B(m_1; \ldots; m_\ell)$ boundary conditions. And, like each of those, its proof involves a standard application of the Brouwer Theorem on Invariance of Domain, and so is omitted.

Theorem 4.3.1. [[56], Henderson, Thm. 5] *Let* $1 \leq \ell \leq n$ *and* $m_1, \ldots,$ $m_\ell \in \mathbb{N}$ *such that* $\sum_{i=1}^{\ell} m_i = n$ *be given. Assume that with respect to* (4.0.1), *conditions* (DAZ) *and* $(B_{m_1, \ldots, m_\ell})$ *are satisfied. Given a solution* $u(m)$ *of* (4.0.1) *on* $\mathbb{Z}$, *points in* $\mathbb{Z}$, $s_1 < \cdots < s_{m_1}$ *and* $s_{m_1 + \cdots + m_{i-1}} \leq s_{m_1 + \cdots + m_{i-1}+1} < \cdots < s_{m_1 + \cdots + m_i}$, $2 \leq i \leq \ell$, *an interval* $[a,b] \supseteq [s_1, s_n]$, *and an* $\epsilon > 0$, *there exists a* $\delta(\epsilon, [a,b]) > 0$ *such that, if* $|\Delta^{i-1} u(s_{m_1 + \cdots + s_{m-i}+j}) - y_{ij}| < \delta$, $1 \leq j \leq m_i$, $1 \leq i \leq \ell$, *then there exists a solution* $v(m)$ *of* (4.0.1) *satisfying* $\Delta^{i-1} v(s_{m_1 + \cdots + s_{m-i}+j}) = y_{ij}$, $1 \leq j \leq m_i$, $1 \leq i \leq \ell$, *and* $|\Delta^i v(m) - \Delta^i u(m)| < \epsilon$, $0 \leq i \leq n-1$, *for all* $m \in [a,b]$.

In what follows, we will establish uniqueness implies existence results for these $B(m_1; \ldots; m_\ell)$ boundary value problems for the cases of when (4.0.1) is a third order or a fourth order difference equation.

First, when $n = 3$, we consider

$$u(m+3) = f(m, u(m), u(m+1), u(m+2)), \tag{4.3.2}$$

where f satisfies condition (DAZ). Relative to (4.3.2)-(4.3.1), the problems that have not been dealt with previously in Sections 4.1 and 4.2 are the $B(1;2)$ and $B(2;1)$ boundary value problems.

The $B(1;2)$ boundary value problems for (4.3.2) on $\mathbb{Z}$ are any of the following:

$$u(s_1) = y_1, \ \Delta u(s_2) = y_2, \ \Delta^2 u(s_2) = y_3, \ s_1 < s_2, \tag{4.3.3}$$

$$u(s_1) = y_1, \ \Delta u(s_1) = y_2, \ \Delta u(s_2) = y_3, \ s_1 < s_2, \tag{4.3.4}$$

$$u(s_1) = y_1, \ \Delta u(s_2) = y_2, \ \Delta u(s_3) = y_3, \ s_1 < s_2 < s_3, \tag{4.3.5}$$

whereas, the $B(2;1)$ boundary value problems for (4.3.2) on $\mathbb{Z}$ are any of the following:

$$u(s_1) = y_1, \ u(s_2) = y_2, \ \Delta u(s_2) = y_3, \ s_1 < s_2, \tag{4.3.6}$$

$$u(s_1) = y_1, \ \Delta u(s_1) = y_2, \ \Delta u(s_2) = y_3, \ s_1 < s_2, \tag{4.3.7}$$

$$u(s_1) = y_1, \ u(s_2) = y_2, \ \Delta u(s_3) = y_3, \ s_1 < s_2 < s_3. \tag{4.3.8}$$

Theorem 4.3.2. [[56], Henderson, Thm. 6] *Assume that with respect* (4.3.2), *conditions* (DAZ) *and* $(B_{1,2})$ *are satisfied. Then, each* $B(1;2)$ *boundary value problem for* (4.3.2) *has a unique solution on* $\mathbb{Z}$.

Proof. The path of our proof will be to show existence of solutions for first (4.3.2)-(4.3.3), then (4.3.2)-(4.3.4), and finally (4.3.2)-(4.3.5). Each problem requires application of shooting methods and induction on the spacing between the boundary points.

Let $y_1, y_2, y_2 \in R$ be given throughout.

(4.3.2)-(4.3.3): Inducting on the spacing $s_2 - s_1$, we consider first the case $s_2 = s_1 + 1$, and let $z_1(m)$ be the solution of the initial value problem for (4.3.2) satisfying

$$z_1(s_2) = 0, \ \Delta z_1(s_2) = y_2, \ \Delta^2 z_1(s_2) = y_3.$$

Define

$$S_1 := \{r \in \mathbb{R} \,|\, \text{there exists a solution } y(m) \text{ of (4.3.2) satisfying}$$
$$y(s_1) = r, \Delta y(s_2) = \Delta z_1(s_2), \text{ and } \Delta^2 y(s_2) = \Delta^2 z_1(s_2)\}.$$

$z_1(s_2) \in S_1$ and so $S_1 \neq \emptyset$. Moreover, it follows from Theorem 4.3.1 that S_1 is an open subset of $\mathbb{R}$.

We claim that S_1 is also a closed subset of $\mathbb{R}$. Assume the claim to be false. Then, there exist $r_0 \in \overline{S}_1 \setminus S_1$ and a strictly monotone sequence $\{r_\ell\} \subset S_1$ such that $\lim_\ell r_\ell = r_0$. We may assume without loss of generality that $r_\ell \uparrow r_0$. For each $\ell \geq 1$, let $y_\ell(m)$ denote the associated solution of (4.3.2) satisfying

$$y_\ell(s_1) = r_\ell, \ \Delta y_\ell(s_2) = \Delta z_1(s_2), \ \Delta^2 y_\ell(s_2) = \Delta^2 z_1(s_2).$$

It follows from condition $(B_{1,2})$ that $y_\ell(m) < y_{\ell+1}(m)$ on $(-\infty, s_2]$, for all $\ell \geq 1$. Furthermore, from Theorem 4.1.1 and $r_0 \notin S_1$, $y_\ell(s_1 + 1) = y_\ell(s_2) \uparrow \infty$, as $\ell \to \infty$, and there exists $m_0 \in [s_1 - 2, s_1 - 1]$ such that $y_\ell(m_0) \uparrow \infty$, as $\ell \to \infty$.

Let $u(m)$ be the solution of (4.3.2) satisfying the initial conditions (or conjugate conditions),

$$u(s_1) = r_0, \ u(s_2) = 0, \ \Delta u(s_2) = \Delta z_1(s_2).$$

Then, for sufficiently large $K \geq 1$, $u - y_K$ has a gz in $(m_0, s_1]$ and a gz at s_2, and $\Delta(u - y_K)(s_2) = 0$. Thus, from condition (B_3) (equivalent to condition (DC)), which holds here, we have $u(m) = y_K(m)$ on $\mathbb{Z}$. This is a contradiction. So S_1 is also closed, and $S_1 = \mathbb{R}$. We choose $y_1 \in S_1$, and the corresponding solution $y(m)$ of (4.3.2) is the desired solution.

Assume now that $s_1 < s_2$ are points such that $s_2 - s_1 > 1$ and that, for points $t_1 < t_2$ with $t_2 - t_1 < s_2 - s_1$, there exist unique solutions of

(4.3.2)-(4.3.3) at t_1 and t_2. And this time, let $z_2(m)$ be the solution of the initial value problem for (4.3.2) which satisfies

$$z_2(s_2) = 0, \ \Delta z_2(s_2) = y_2, \ \Delta^2 z_2(s_2) = y_3,$$

and define

$S_2 := \{r \in \mathbb{R} \, | \, \text{there exists a solution } y(m) \text{ of } (4.3.2) \text{ satisfying}$
$$y(s_1) = r, \Delta y(s_2) = \Delta z_2(s_2), \ \text{and } \Delta^2 y(s_2) = \Delta^2 z_2(s_2)\}.$$

S_2 is a nonempty open subset of $\mathbb{R}$.

Our claim is that S_2 is also closed. Assuming that S_2 is not closed, there exist $r_0 \in \overline{S}_2 \setminus S_2$ and a strictly monotone sequence $\{r_\ell\} \subset S_2$ such that $\lim_\ell r_\ell = r_0$. Again, we may assume $r_\ell \uparrow r_0$, and for each $\ell \geq 1$, let $y_\ell(m)$ be the corresponding solution of (4.3.2) satisfying

$$y_\ell(s_1) = r_\ell, \ \Delta y_\ell(s_2) = \Delta z_2(s_2), \ \Delta^2 y_\ell(s_2) = \Delta^2 z_2(s_2).$$

From condition $(B_{1,2})$, we have $y_\ell(m) < y_{\ell+1}(m)$ on $(-\infty, s_2]$, for all $\ell \geq 1$. Furthermore, from Theorem 4.1.1 and $r_0 \notin S_2$, there exists $m_0 \in [s_1 - 2, s_1 - 1]$ such that $y_\ell(m_0) \uparrow \infty$, as $\ell \to \infty$, and from Theorem 4.3.1 and the induction assumption, $y_\ell(s_1 + 1) = \uparrow \infty$, as $\ell \to \infty$.

Let $u(m)$ be the solution of (4.3.2) satisfying the conjugate conditions

$$u(s_1) = r_0, \ u(s_2) = 0, \ \Delta u(s_2) = \Delta z21(s_2).$$

For $K \geq 1$ sufficiently large, $u - y_K$ has a gz in $(m_0, s_1]$ and a gz at $s_1 + 1$. Also, $\Delta(u - y_K)(s_2) = 0$. So by the discrete Rolle's Theorem, $\Delta(u - y_K)$ has a gz in $(m_0, s_1 + 1)$; this contradicts $(B_{1,2})$ when applied to the problems (4.3.2)-(4.3.5).

Therefore, S_2 is closed, $S_2 = \mathbb{R}$, and choosing $y_1 \in S_2$, the corresponding solution is the desired solution. We conclude that there exist unique solutions on $\mathbb{Z}$ of the problem (4.3.2)-(4.3.3).

(4.3.2)-(4.3.4): For this boundary value problem, if $s_2 = s_1 + 1$, then the problem is an initial value problem, and hence has a unique solution. Thus, we assume that $s_2 - s_1 > 1$ and that, for any points $t_1 < t_2$ with $t_2 - t_1 < s_2 - s_1$, there is a unique solution of (4.3.2)-(4.3.4) at t_1 and t_2. This time, let $z_3(m)$ be the solution of (4.3.2) satisfying conditions of the type (4.3.3)

$$z_3(s_1) = y_1, \ \Delta z_3(s_2) = y_3, \ \Delta^2 z_3(s_2) = 0.$$

Define

$S_3 := \{r \in \mathbb{R} \, | \, \text{there exists a solution } y(m) \text{ of } (4.3.2) \text{ satisfying}$
$$y(s_1) = z_3(s_1), \Delta y(s_1) = r, \ \text{and } \Delta y(s_2) = \Delta z_3(s_2)\}.$$

S_3 is a nonempty open subset of $\mathbb{R}$.

Our claim is that S_3 is closed also. If we assume S_3 is not closed, then let $r_0 \in \overline{S_3} \setminus S_3$ and $\{r_\ell\} \subset S_3$, with $r_\ell \uparrow r_0$, be selected as in the previous manner. And let $y_\ell(m)$ be the associated solution of (4.3.2) such that

$$y_\ell(s_1) = z_3(s_1), \ \Delta y_\ell(s_1) = r_\ell, \ \Delta y_\ell(s_2) = \Delta z_3(s_2).$$

In this situation, $(B_{1,2})$ implies $y_\ell(m) > y_{\ell+1}(m)$ on $(-\infty, s_1)$ and $y_\ell(m) < y_{\ell+1}(m)$ on $(s_1, s_2]$. $r_0 \notin S_3$ and so by Theorem 4.1.1, there exists $m_0 \in [s_1 - 2, s_1 - 1]$ so that $y_\ell(m_0) \downarrow -\infty$, as $\ell \to \infty$, while the induction assumption and Theorem 4.3.1 imply $y_\ell(s_1+2) \uparrow \infty$ (i.e., $\Delta y_\ell(s_1+1) \uparrow \infty$), as $\ell \to \infty$.

If we choose $u(m)$ to be the solution of (4.3.2) which satisfies the initial conditions,

$$u(s_1) = z_3(s_1), \ \Delta u(s_1) = r_0, \ \Delta^2 u(s_1) = 0,$$

then, for sufficiently large $K \geq 1$ and depending on $m_0 = s_1 - 2$ or $m_0 = s_1 - 1$, $u - y_K$ has either gz's at $s_1 - 1, s_1$, and $s_1 + 2$, or gz's at $s_1, s_1 + 1$, and $s_1 + 2$. Both scenarios contradict $(B_{1,2})$ after applications of the discrete Rolle's Theorem.

Thus, S_3 is closed and $S_3 = \mathbb{R}$. Choosing $y_2 \in S_3$ and the corresponding solution is the desired solution. Thus, there exist unique solutions of (4.3.2)-(4.3.4) on $\mathbb{Z}$.

(4.3.2)-(4.3.5): For these problems note that, if $s_1 = s_2 < s_3$, then (4.3.5) reduces to the boundary condition (4.3.4). Thus we assume $s_2 - s_1 \geq 1$ and that, for any points $t_1 \leq t_2 < t_3$ with $t_2 - t_1 < s_2 - s_1$, there exists a unique solution of (4.3.2)-(4.3.5) at t_1, t_2, and t_3. Let $z_4(m)$ be the solution of (4.3.2) satisfying the conditions of type (4.3.4) given by

$$z_4(s_2) = 0, \ \Delta z_4(s_2) = y_2, \ \Delta z_4(s_3) = y_3,$$

and we now define

$$S_4 := \{r \in \mathbb{R} \,|\, \text{there exists a solution } y(m) \text{ of (4.3.2) satisfying}$$
$$y(s_1) = r, \Delta y(s_2) = \Delta z_4(s_2), \text{ and } \Delta y(s_3) = \Delta z_4(s_3)\}.$$

As with the previous arguments, S_4 is a nonempty open subset of $\mathbb{R}$.

We claim that S_4 is closed. Again, assume this not to be the case, and let $r_0 \in \overline{S_3} \setminus S_3$ and $\{r_\ell\} \subset S_3$, with $r_\ell \uparrow r_0$, be as usual. Also, let $y_\ell(m)$ be the corresponding solution of (4.3.2) satisfying

$$y_\ell(s_1) = r_\ell, \ \Delta y_\ell(s_2) = \Delta z_4(s_2), \ \Delta y_\ell(s_3) = \Delta z_4(s_3).$$

From $(B_{1,2})$, we have $y_\ell(m) < y_{\ell+1}(m)$ on $(-\infty, s_2)$, and reasoning as we have with the other problems, there exists $m_0 \in [s_1 - 2, s_1 - 1]$ such that $y_\ell(m_0) \uparrow \infty$, and $y_\ell(s_1 + 1) \uparrow \infty$.

If $u(m)$ denotes the solution of (4.3.2) satisfying conditions of type (4.3.3) given by

$$u(s_1) = r_0, \ \Delta u(s_2) = \Delta z_4(s_2), \ \Delta^2 u(s_2) = 0,$$

then, for $K \geq 1$ sufficiently large, $u - y_K$ has a gz in $(m_0, s_1]$ and a gz at $s_1 + 1$. Moreover, $\Delta(u - y_K)(s_2) = 0$. Applying the discrete Rolle's Theorem to $u - y_K$, we reach a contradiction to $(B_{1,2})$.

Hence, again, we have $S_4 = \mathbb{R}$ and obtain the desired solution of (4.3.2)-(4.3.5). This completes the proof.

$\square$

Theorem 4.3.3. [[56], Henderson, Thm. 7] *Assume that with respect (4.3.2), conditions* (DAZ) *and* $(B_{2,1})$ *are satisfied. Then, each* $B(2;1)$ *boundary value problem for (4.3.2) has a unique solution on* $\mathbb{Z}$.

Proof. The path of our proof is similar to the one taken in Theorem 4.3.2. We establish the existence of solutions for first (4.3.2)-(4.3.6), then (4.3.2)-(4.3.7), and finally (4.3.2)-(4.3.8). Also, shooting methods and induction on the spacing between the boundary points are employed.

Let $y_1, y_2, y_2 \in R$ be given throughout.

(4.3.2)-(4.3.6): In terms of notation of Section 4.1, this is an $(m_1, m_2, 1)$ conjugate boundary value problem (4.3.2). Since $(B_{2,1})$ implies condition (B_3) (equivalent to condition (DC)), it follows from Theorem 4.1.4 that (4.3.2)-(4.3.6) has a unique solution on $\mathbb{Z}$.

(4.3.2)-(4.3.7): This boundary value problem is the same problem as (4.3.2)-(4.3.4) in Theorem 4.3.2. We proceed here as we did in that situation, except that we choose $z_2(m)$ to be the solution of (4.3.2) satisfying the conjugate conditions

$$z_2(s_1) = y_1, \ z_2(s_2) = 0, \ \Delta z_2(s_2) = y_3.$$

From this point, the argument is exactly as the one in Theorem 4.3.2, except condition $(B_{2,1})$ is employed at several steps.

(4.3.2)-(4.3.8): First, if the points $s_1 < s_2 < s_3$ are such that $s_2 - s_1 = 1$, then the problem reduces to the boundary value problem (4.3.2)-(4.3.7). Hence, we have unique solutions for that case. So, we assume $s_2 - s_1 > 1$

and that, for all points $t_1 < t_2 < t_3$ with $t_2 - t_1 < s_2 - s_1$, there exist unique solutions of (4.3.2)-(4.3.8) at the points t_1, t_2, and t_3.

We let $z(t)$ denote the solution of (4.3.2) which satisfies conditions of type (4.3.7),

$$z(s_2) = y_2, \ \Delta z(s_2) = 0, \ \Delta z(s_3) = y_3.$$

Set

$$S := \{r \in \mathbb{R} \mid \text{there exists a solution } y(m) \text{ of (4.3.2) satisfying}$$
$$y(s_1) = r, y(s_2) = z(s_2), \text{ and } \Delta y(s_3) = \Delta z(s_3)\}.$$

S is a nonempty open subset of $\mathbb{R}$.

If we assume that S is not closed, then we again have $r_0 \in \overline{S} \setminus S$ and $\{r_\ell\} \subset S$, with $r_\ell \uparrow r_0$ as in the arguments of the previous theorem . With $y_\ell(m)$ the solution of (4.3.2) satisfying

$$y_\ell(s_1) = r_\ell, \ y_\ell(s_2) = z(s_2), \ \Delta y_\ell(s_3) = \Delta z(s_3),$$

$(B_{2,1})$ implies $y_\ell(m) < y_{\ell+1}(m)$ on $(-\infty, s_2)$, for all $\ell \geq 1$. Again, $r_0 \notin S$ and Theorems 4.1.1 and 4.3.1 imply there exists $m_0 \in [s_1 - 2, s_1 - 1]$ such that $y_\ell(m_0) \uparrow \infty$ and $y_\ell(s_1 + 1) \uparrow \infty$, as $\ell \to \infty$.

If $u(m)$ is the solution of (4.3.2) satisfying the conjugate boundary conditions,

$$u(s_1) = r_0, \ u(s_3) = 0, \ \Delta u(s_3) = \Delta z(s_3),$$

then for $K \geq 1$ sufficiently large, $u - y_K$ has a gz in $(m_0, s_1]$ and a gz at $s_1 + 1$, and $\Delta(u - y_K)(s_3) = 0$. This contradicts $(B_{2,1})$, when applied to problems satisfying conditions of the form (4.3.8).

Thus, S is closed and $S = \mathbb{R}$. Choose $y_1 \in S$, and the corresponding solution is the solution we sought. We conclude that (4.3.2)-(4.3.8) has unique solutions on $\mathbb{Z}$. The theorem is proved. $\qquad\qquad\square$

The proofs that are given above for Theorems 4.3.2 and 4.3.3 are not patterned and hence the proofs themselves most likely do not lead to generalizations. We will conclude this section by considering some of the $B(m_1; \ldots, m_\ell)$ boundary value problems for the fourth order equation

$$u(m + 4) = f(m, u(m), u(m + 1), u(m + 2), u(m + 3)), \qquad (4.3.9)$$

where f satisfies condition (DAZ). Relative to (4.3.9)-(4.3.1), the problems not dealt with in Sections 4.1 and 4.2 correspond to $\ell = 2$ and $\ell = 3$. For $\ell = 2$, we would consider $B(3; 1), B(2; 2)$, and $B(1; 3)$ boundary value problems for (4.3.9), and for $\ell = 3$, there are $B(2;1;1), B(1;2;1)$, and $B(1;1;2)$ boundary value problems for (4.3.9).

For example, the $B(2;2)$ boundary value problems for (4.3.9) on $\mathbb{Z}$ are any of the following:

$$u(s_1) = y_1, u(s_2) = y_2, \Delta u(s_2) = y_3, \Delta^2 u(s_2) = y_4, s_1 < s_2, \quad (4.3.10)$$

$$u(s_1) = y_1, \Delta u(s_1) = y_2, \Delta u(s_2) = y_3, \Delta^2 u(s_2) = y_4, s_1 < s_2, \quad (4.3.11)$$

$$u(s_1) = y_1, u(s_2) = y_2, \Delta u(s_3) = y_3, \Delta^2 u(s_3) = y_4, s_1 < s_2 < s_3, \quad (4.3.12)$$

$$u(s_1) = y_1, u(s_2) = y_2, \Delta u(s_2) = y_3, \Delta u(s_3) = y_4, s_1 < s_2 < s_3, \quad (4.3.13)$$

$$u(s_1) = y_1, \Delta u(s_1) = y_2, \Delta u(s_2) = y_3, \Delta u(s_3) = y_4, s_1 < s_2 < s_3, \quad (4.3.14)$$

$$u(s_1) = y_1, u(s_2) = y_2, \Delta u(s_3) = y_3, \Delta u(s_4) = y_4, s_1 < \cdots < s_4. \quad (4.3.15)$$

For the family of $B(2;2)$ boundary value problems we will outline a proof for uniqueness implies existence of solutions. These arguments are not patterned, but they are typical of a process that can be successfully taken for the other families of problems for (4.3.9).

Theorem 4.3.4. [[56], Henderson, Thm. 8] *Assume that with respect* (4.3.9), *conditions* (DAZ) *and* $(B_{2,2})$ *are satisfied. Then, each* $B(2;2)$ *boundary value problem for* (4.3.9) *has a unique solution on* $\mathbb{Z}$.

Proof. Let $y_1, y_2, y_3, y_4 \in \mathbb{R}$ be given.

(4.3.9)-(4.3.10): In the notation of Section 4.1, this is an $(m_1, m_2, 1, 1)$ conjugate boundary value problem for (4.3.9). Since $(B_{2,2})$ implies condition (B_4) (equivalent to condition (DC)), it follows from Theorem 4.1.4 that (4.3.9)-(4.3.10) has a unique solution on $\mathbb{Z}$.

(4.3.9)-(4.3.11): In conjunction with the usual induction argument, if we let $z(m)$ be the solution of (4.3.9) satisfying conditions of type (4.3.10)

$$z(s_1) = y_1, \ z(s_2) = 0, \ \Delta z(s_2) = y_3, \ \Delta^2 z(s_2) = y_4,$$

then we can argue that the set,

$$S_1 := \{\Delta y(s_1) \,|\, y(m) \text{ is a solution of (4.3.9) and}$$

$$y(s_1) = z(s_1), \Delta y(s_2) = \Delta z(s_2), \text{ and } \Delta^2 y(s_2) = \Delta^2 z(s_2)\},$$

consists of all the real numbers. So, for $y_2 \in S_2$, the corresponding solution $y(m)$ of (4.3.9) satisfies (4.3.11).

(4.3.9)-(4.3.12): For this problem, let $z(m)$ be the solution of (4.3.9) satisfying conditions of type (4.3.11),

$$z(s_2) = y_2, \ \Delta z(s_2) = 0, \ \Delta z(s_3) = y_3, \ \Delta^2 z(s_3) = y_4.$$

Defining

$S_2 := \{y(s_1) \mid y(m)$ is a solution of $(4.3.9)$ and
$$y(s_2) = z(s_2), \Delta y(s_3) = \Delta z(s_3), \text{ and } \Delta^2 y(s_3) = \Delta^2 z(s_3)\},$$

it can be shown that $S_2 = \mathbb{R}$. Thus, $(4.3.9)$-$(4.3.12)$ has a unique solution $\mathbb{Z}$.

$(4.3.9)$-$(4.3.13)$: Continuing, let $z(m)$ denote the solution of $(4.3.9)$ satisfying conditions of type $(4.3.12)$,

$$z(s_1) = y_1, \; z(s_2) = y_2, \; \Delta z(s_3) = y_4, \; \Delta^2 z(s_3) = 0.$$

We can then argue that

$S_3 := \{\Delta y(s_2) \mid y(m)$ is a solution of $(4.3.9)$ and
$$y(s_1) = z(s_1), y(s_2) = z(s_2), \text{ and } \Delta y(s_3) = \Delta z(s_3)\}$$

consists of the entire real line. Hence, we conclude that $(4.3.9)$-$(4.3.13)$ has a unique solution on $\mathbb{Z}$.

$(4.3.9)$-$(4.3.14)$: In this case, we take $z(m)$ to be the solution of $(4.3.9)$ satisfying conditions of type $(4.3.13)$,

$$z(s_1) = y_1, \; z(s_2) = 0, \; \Delta z(s_2) = y_3, \; \Delta z(s_3) = y_4,$$

and we define

$S_4 := \{\Delta y(s_1) \mid y(m)$ is a solution of $(4.3.9)$ and
$$y(s_1) = z(s_1), \Delta y(s_2) = \Delta z(s_2), \text{ and } \Delta y(s_3) = \Delta z(s_3)\}.$$

Here, we also have $S_4 = \mathbb{R}$, and so for $y_2 \in S_4$, $(4.3.9)$-$(4.3.14)$ has a unique solution.

$(4.3.9)$-$(4.3.15)$: Finally, we let $z(m)$ be the solution of $(4.3.9)$ satisfying conditions of the form $(4.3.14)$,

$$z(s_2) = y_2, \; \Delta z(s_2) = 0, \; \Delta z(s_3) = y_3, \; \Delta z(s_4) = y_4,$$

and set

$S_5 := \{y(s_1) \mid y(m)$ is a solution of $(4.3.9)$ and
$$y(s_2) = z(s_2), \Delta y(s_3) = \Delta z(s_3), \text{ and } \Delta y(s_4) = \Delta z(s_4)\}.$$

As with the other cases, $S_5 = \mathbb{R}$, and so choosing $y_1 \in S_5$, the boundary value problem $(4.3.9)$-$(4.3.14)$ has a unique solution. $\square$

4.4 Lidstone boundary value problems: uniqueness implies existence

This section is devoted to establishing the existence of unique solutions of (4.0.1), when $n = 4$, satisfying *Lidstone* boundary conditions. Such a difference equation was labeled in the previous section, but because of the distinctiveness of the boundary conditions here, we will relabel our difference equation and pair it with the boundary conditions now under consideration. In particular, our concern is with unique solutions of the difference equation,

$$u(m + 4) = f(m, u(m), u(m + 1), u(m + 2), u(m + 3)), \qquad (4.4.1)$$

satisfying the boundary conditions,

$$u(m_1) = u_1, \ \Delta^2 u(m_2) = u_2, \ \Delta^2 u(m_3) = u_3, u(m_4) = u_4, \qquad (4.4.2)$$

where f satisfies condition (DAZ) of the previous Section 4.3, $m_i \in \mathbb{Z}$ and $u_i \in \mathbb{R}$, for $1 \le i \le 4$, and $m_1 \le m_2 < m_3 - 2 \le m_4 - 4$.

Such boundary value problems are called *Lidstone boundary value problems*, because of their analogy to Lidstone problems for ordinary differential equations [16]. The results of this section were published by Henderson and Johnson [63] in 2000.

Remark 4.0.1 and Theorem 4.1.1 remain valid with $[a, \infty)$ replaced by $\mathbb{Z}$, and both will be referred to in the development of this section, as well as to Remark 4.2.1 and Remark 4.2.2. And many of the other theorems of Sections 4.1 and 4.2 will be referred to as being valid with $[a, \infty)$ replaced by $\mathbb{Z}$. Also, when an interval, say J, is used in the discussion, the meaning will be the discrete interval of integers $J \cap \mathbb{Z}$.

Also, we recall again for local reference sake, that for a function $u : [a, \infty) \to \mathbb{R}$, Hartman [44] defined $m_0 \in [a, \infty)$, in the case that $m_0 = a$, to be a *generalized zero* (denoted by "gz") of u if $u(a) = 0$, and $m_0 > a$ to be a *generalized zero* of u if $u(m_0) = 0$ or there is an integer $j \ge 1$ such that $(-1)^j u(m_0 - j) u(m_0) > 0$ and if $j > 1$, $u(m_0 - j + 1) = \cdots = u(m_0 - 1) = 0$.

The problems which will be considered will be classified as two-point, three-point, and four-point Lidstone problems. Unique solutions are to be obtained by using shooting methods. We now state our fundamental uniqueness condition on solutions of boundary value problems for (4.4.1)

(LC) (a) For any $m_1, m_2, m_3, m_4 \in \mathbb{Z}$, if $\phi(m)$ and $\psi(m)$ are solutions of (4.4.1) such that $\phi(m) - \psi(m)$ has a gz at m_1, $\Delta(\phi(m) - \psi(m))$ has a gz at m_2, $\Delta^2(\phi(m) - \psi(m))$ has a gz at m_3, and $\Delta^3(\phi(m) - \psi(m))$ has a gz at m_4, then $\phi(m) = \psi(m)$ on $\mathbb{Z}$.

Remark 4.4.1. (a) *If* (LC) *holds, then repeated applications of the discrete Rolle's Theorem* (*see Remark* 4.2.1) *imply that solutions of any conjugate* ((4.1.1) *of Section* 4.1), *any right focal* ((4.2.3) *in Remark* 4.2.2(a) *at the end of Section* 4.2), *or any Lidstone* ((4.4.1)-(4.4.2)) *boundary value problems are unique when such solutions exist on $\mathbb{Z}$, (that is, conditions* (DC) *and* (FC) *respectively hold).*

(b) *And in particular, under conditions* (DAZ) *and* (LC), *there exist unique solutions of conjugate boundary value problems* (*from Section* 4.1), *and there exist unique solutions of right focal boundary value problems* (*from Section* 4.2). *And so, solutions of conjugate, right focal, and Lidstone boundary value problems depend continuously on boundary conditions.*

We now consider the two-point Lidstone problem for (4.4.1) satisfying

$$u(m_1) = u_1, \ \Delta^2 u(m_1) = u_2, \ \Delta^2 u(m_2) = u_3, \ u(m_2) = u_4, \qquad (4.4.3)$$

where $m_1 < m_2 - 2$.

Theorem 4.4.1. [[63], Henderson and Johnson, Thm. 3.1] *Suppose that with respect to* (4.4.1), *conditions* (DAZ) *and* (LC) *are satisfied. Then, given $m_1 < m_2 - 2$ in $\mathbb{Z}$ and given $u_i \in \mathbb{R}$, $1 \le i \le 4$, there exists a unique solution of* (4.4.1)-(4.4.3) *on $\mathbb{Z}$.*

Proof. Let $m_1 < m_2 - 2$ in $\mathbb{Z}$ and $u_1, u_2, u_3, u_4 \in \mathbb{R}$ be given. By Remark 4.4.1 (b), there exists a unique solution $y(m)$ of (4.4.1) subject to the conjugate boundary conditions,

$$y(m_1) = u_1,$$
$$y(m_2) = u_4,$$
$$\Delta y(m_2) = 0,$$
$$\Delta^2 y(m_2) = u_3.$$

Define $S \subseteq \mathbb{R}$ by

$$S := \{\Delta^2 v(m_1) | v(m) \text{ is a solution of } (4.4.1), \text{ and } v(m_1) = y(m_1) = u_1,$$
$$\Delta^2 v(m_2) = \Delta^2 y(m_2) = u_3, v(m_2) = y(m_2) = u_4\}.$$

Note that $S \neq \emptyset$, since $\Delta^2 y(m_1) \in S$. Also, by the continuous dependence statement in Remark 4.4.1 (b), S is an open subset or $\mathbb{R}$.

We next claim that S is also a closed subset of $\mathbb{R}$. Assume otherwise. Then we have the existence of a limit point $n_0 \in \overline{S} \setminus S$ and a strictly monotone sequence $\{n_k\}_{k=1}^{\infty} \subset S$ such that $n_k \to n_0$. Assume without loss of generality that $n_k \uparrow n_0$. By the definition of S, for each $k \geq 1$, there exists a solution $y_k(m)$ of (4.4.1) such that

$$y_k(m_1) = y(m_1) = u_1,$$
$$\Delta^2 y_k(m_1) = n_k,$$
$$\Delta^2 y_k(m_2) = \Delta^2 y(m_2) = u_3,$$
$$y_k(m_2) = y(m_2) = u_4.$$

By the discrete Rolle's Theorem and condition (LC), it follows that for all $k \geq 1$, we have $\Delta^2 y_{k+1}(m) > \Delta^2 y_k(m)$ on $(-\infty, m_2)$. Since $y_k(m_1) = y_{k+1}(m_1)$ and $y_k(m_2) = y_{k+1}(m_2)$, we must have that $y_k(m) < y_{k+1}(m)$ on $(-\infty, m_2] \setminus \{m_1\}$, for all $k \geq 1$. By Theorem 4.1.1, $\{y_k(m)\}_{k=1}^{\infty}$ is not bounded on any finite subset of $\mathbb{Z}$ having at least four points.

Now, we let w be the unique solution of the conjugate boundary value problem consisting of (4.4.1) with boundary conditions

$$w(m_1) = y_k(m_1),$$
$$\Delta w(m_1) = 0,$$
$$\Delta^2 w(m_1) = n_0,$$
$$w(m_2) = y_k(m_2).$$

(We note that $\Delta^2 w(m_1) > \Delta^2 y_k(m_1)$, for all $k \geq 1$.) It follows that for every $k \geq 1$, $w(m) > y_k(m)$ at $m = m_1 + 1$ or $m = m_1 + 2$. Since $\{y_k(m)\}_{k=1}^{\infty}$ is unbounded above on each finite subset of $(-\infty, m_2) \setminus \{m_1\}$, so also is $\{\Delta^2 y_k(m)\}_{k=1}^{\infty}$ at points of (m_1, m_2). Then there exist $K \geq 1$ and integers α and β where $m_1 < \alpha < \beta < m_2$ such that $(w - y_K)(m)$ has a gz at β and $\Delta^2(w - y_K)(m)$ has a gz at α. Moreover, it follows, from $\Delta^2 w(m_1) > \Delta^2 y_K(m_1)$ and the discrete Rolle's Theorem, that there exists $\gamma \in [\alpha, \beta)$ such that $\Delta(w - y_K)(m)$ has a gz at γ. Also, there exists a $\rho \in [\gamma, m_2)$ such that $\Delta^2(w - y_K)(m)$ has a gz at ρ. Thus, we have that $(w - y_K)(m)$ has gz's at m_1 and m_2 and that $\Delta^2(w - y_K)(m)$ has gz's at α and ρ. So it must be that $w(m) = y_K(m)$, for all $m \in \mathbb{Z}$, by the uniqueness of Lidstone boundary value problems. But $\Delta^2 w(m_1) = n_0 > \Delta^2 y_k(m_1)$ for all $k \geq 1$. This is a contradiction. Hence S is a closed set.

Since $\emptyset \neq S \subseteq \mathbb{R}$ is both open and closed, we must have that $S = \mathbb{R}$. Thus, choosing $u_2 \in S$, the corresponding solution is the desired unique solution of (4.4.1)-(4.4.3) on $\mathbb{Z}$. $\qquad \square$

Next we establish the existence of solutions of the three-point Lidstone problems for (4.4.1). There are two such problems for (4.4.1), those satisfying, either

$$u(m_1) = u_1, \ \Delta^2 u(m_1) = u_2, \ \Delta^2 u(m_2) = u_3, \ u(m_3) = u_4, \qquad (4.4.4)$$

where $m_1 < m_2 - 2 < m_3 - 4$, or its dual

$$u(m_1) = u_1, \ \Delta^2 u(m_2) = u_2, \ \Delta^2 u(m_3) = u_3, \ u(m_3) = u_4. \qquad (4.4.5)$$

where $m_1 < m_2 < m_3 - 2$.

We will make the argument for solutions of only (4.4.1)-(4.4.4).

Theorem 4.4.2. [[63], Henderson and Johnson, Thm. 4.1] *Suppose that with respect to (4.4.1), conditions (DAZ) and (LC) are satisfied. Then, given $m_1 < m_2 - 2 < m_3 - 4$ in $\mathbb{Z}$ and given $u_i \in \mathbb{R}$, $1 \le i \le 4$, there exists a unique solution of (4.4.1)-(4.4.4) on $\mathbb{Z}$.*

Proof. Let integers $m_1 < m_2 - 2 < m_3 - 4$ be given, and let $u_1, u_2, u_3, u_4 \in \mathbb{R}$. Let $y(m)$ be the solution of (4.4.1)-(4.4.3) given by Theorem 4.4.1 which satisfies

$$y(m_1) = u_1,$$
$$\Delta^2 y(m_1) = u_2,$$
$$\Delta^2 y(m_2) = u_3,$$
$$y(m_2) = 0.$$

Define a set $S \subseteq \mathbb{R}$ by

$$S := \{v(m_3) \mid v \text{ is a solution of } (4.4.1), \text{ and } v(m_1) = y(m_1) = u_1,$$
$$\Delta^2 v(m_1) = \Delta^2 y(m_1) = u_2, \text{ and } \Delta^2 v(m_2) = \Delta^2 y(m_2) = u_3\}.$$

$S \ne \emptyset$, since $y(m_3) \in S$. Again, by the continuous dependence statement in Remark 4.4.1 (b), S is an open subset or $\mathbb{R}$.

Again, we claim that S is also a closed subset of $\mathbb{R}$. We assume otherwise. Then there exist a limit point $n_0 \in \overline{S} \setminus S$ and a monotone sequence $\{n_k\}_{k=1}^{\infty} \subset S$ such that $n_k \to n_0$. We may assume, with no loss of generality that $\{n_k\}_{k=1}^{\infty}$ is increasing. By definition of S, for $k \ge 1$, there exists a solution $y_k(m)$ of (4.4.1) such that

$$y_k(m_1) = y(m_1),$$
$$\Delta^2 y_k(m_1) = \Delta^2 y(m_1),$$
$$\Delta^2 y_k(m_2) = \Delta^2 y(m_2),$$
$$y_k(m_3) = n_k.$$

It then follows from the discrete Rolle's Theorem and condition (LC) that for every $k \geq 1$, $y_k(m) < y_{k+1}(m)$ on (m_1, ∞). Also, $\{y_k(m)\}_{k=1}^{\infty}$ is unbounded above on each finite subinterval of $[m_2, \infty)\backslash\{m_3\}$. Let w denote the unique solution of the conjugate boundary value problem for (4.4.1) satisfying the boundary conditions

$$
\begin{aligned}
w(m_1) &= y_k(m_1), \\
\Delta w(m_1) &= 0, \\
\Delta^2 w(m_1) &= \Delta^2 y_k(m_1), \\
w(m_3) &= n_0.
\end{aligned}
$$

(Note that $w(m_3) > y_k(m_3)$, for all $k \geq 1$.) Since $\{y_k(m_3 - 1)\}_{k=1}^{\infty}$ and $\{y_k(m_3 + 1)\}_{k=1}^{\infty}$ are not bounded above, but $y_k(m_3) < y(m_3)$ for all k, there exists a $K \geq 1$ such that $(w - y_K)(m)$ has a gz at m_3 and a gz at $m_3 + 1$. We also have that $\Delta(w - y_K)(m)$ has a gz at m_3. Thus, there exists a $v \in [m_1, m_3)$ such that $\Delta(w - y_K)(m)$ has a gz at v. So there must be an $r \in [m_1, v - 1]$ such that $\Delta^2(w - y_K)(m)$ has a gz at r. But $w(m_1) = y_K(m_1)$ and $\Delta^2 w(m_1) = \Delta^2 y_K(m_1)$. Thus, utilizing the points $m_1, m_3 + 1$ and r, it follows by the uniqueness of three-point Lidstone boundary value problems that we must have $w(m) = y_K(m)$ for all $m \in \mathbb{Z}$. But $w(m_3) = n_0 > y_k(m_3)$ for all $k \geq 1$. This is a contradiction. Thus S must be a closed subset of $\mathbb{R}$.

Since $S \subseteq \mathbb{R}$ is both open and closed, we must have that $S = \mathbb{R}$. Thus choosing $u_4 \in S$, the corresponding solution is the desired unique solution of (4.4.1)-(4.4.4) on $\mathbb{Z}$. $\qquad\square$

We state the theorem for the dual three-point Lidstone problem without proof.

Theorem 4.4.3. [[63], Henderson and Johnson, Thm. 4.2] *Suppose that with respect to* (4.4.1), *conditions* (DAZ) *and* (LC) *are satisfied. Then, given* $m_1 < m_2 < m_3 - 2$ *in* $\mathbb{Z}$ *and given* $u_i \in \mathbb{R}$, $1 \leq i \leq 4$, *there exists a unique solution of* (4.4.1)-(4.4.5) *on* $\mathbb{Z}$.

For the final result of this section, we establish the existence of solutions of the four-point Lidstone problem for (4.4.1) which satisfies

$$u(m_1) = u_1, \quad \Delta^2 u(m_2) = u_2, \quad \Delta^2 u(m_3) = u_3, \quad u(m_4) = u_4, \qquad (4.4.6)$$

where $m_1 < m_2 < m_3 - 2 < m_4 - 4$.

Theorem 4.4.4. [[63], Henderson and Johnson, Thm. 5.1] *Suppose that with respect to* (4.4.1), *conditions* (DAZ) *and* (LC) *are satisfied. Then, given* $m_1 < m_2 < m_3 - 2 < m_4 - 4$ *in* $\mathbb{Z}$ *and given* $u_i \in \mathbb{R}$, $1 \le i \le 4$, *there exists a unique solution of* (4.4.1)-(4.4.6) *on* $\mathbb{Z}$.

Proof. Let integers $m_1 < m_2 < m_3 - 2 < m_4 - 4$ be given and let $u_1, u_2, u_3, u_4 \in \mathbb{R}$. Let $y(m)$ be the solution of (4.4.1)-(4.4.4) given by Theorem 4.4.2 which satisfies

$$y(m_2) = 0,$$
$$\Delta^2 y(m_2) = u_2,$$
$$\Delta^2 y(m_3) = u_3,$$
$$y(m_4) = u_4.$$

Define a set $S \subseteq \mathbb{R}$ by

$$S := \{v(m_1) \mid v \text{ is a solution of (4.4.1), and } \Delta^2 v(m_2) = \Delta^2 y(m_2) = u_2,$$
$$\Delta^2 v(m_3) = \Delta^2 y(m_3) = u_3, v(m_4) = y(m_4) = u_4\}.$$

Again S is a nonempty open subset of $\mathbb{R}$.

We show that S is also a closed subset of $\mathbb{R}$. We assume otherwise. Then there exist a limit point $n_0 \in \overline{S} \setminus S$ and a monotone sequence $\{n_k\}_{k=1}^{\infty} \subset S$ such that $n_k \to n_0$. Assume $\{n_k\}_{k=1}^{\infty} \uparrow n_0$. By the definition of S, for $k \ge 1$, there exists a solution $y_k(m)$ of (4.4.1) such that

$$y_k(m_1) = n_k,$$
$$\Delta^2 y_k(m_2) = \Delta^2 y(m_2),$$
$$\Delta^2 y_k(m_3) = \Delta^2 y(m_3),$$
$$y_k(m_4) = y(m_4).$$

It then follows from the discrete Rolle's Theorem and condition (LC) that for all $k \ge 1$, $y_k(m) < y_{k+1}(m)$ on $(-\infty, m_4)$. Also, $\{y_k(m)\}_{k=1}^{\infty}$ is unbounded above on each finite subset of $(-\infty, m_2] \setminus \{m_1\}$. Let w denote the unique solution of the three-point Lidstone boundary-value problem for (4.4.1) which satisfies

$$w(m_1) = n_0,$$
$$\Delta^2 w(m_2) = \Delta^2 y(m_2) = u_2,$$
$$\Delta^2 w(m_3) = \Delta^2 y(m_3) = u_3,$$
$$w(m_3) = 0.$$

(Here $w(m_1) > y_k(m_1)$, for all $k \geq 1$.) Since $\{y_k(m_1-1)\}_{k=1}^{\infty}$ and $\{y_k(m_1+1)\}_{k=1}^{\infty}$ are unbounded above, but $y_k(m_1) < w(m_1)$ for all k, there exists a $K \geq 1$ such that $(w-y_K)(m)$ has a gz at m_1 and has a gz at m_1+1. Then $\Delta(w-y_K)(m)$ has a gz at m_1. Also, $\Delta^2(w-y_K)$ has gz's at m_2 and m_3, so there exists an $\omega \in (m_2, m_3)$ such that $\Delta^3(w-y_K)(m)$ has a gz at ω. Hence, utilizing the points m_1, m_2 and ω, it follows by condition (LC) that $w(m) = y_K(m)$ for all $m \in \mathbb{Z}$. But $w(m_1) = n_0 > y_k(m_1)$ for all $k \geq 1$; again, a contradiction. Thus S must be a closed subset of $\mathbb{R}$.

We conclude that $S = \mathbb{R}$. Thus, choosing $u_1 \in S$, the corresponding solution is the sought after unique solution of (4.4.1)-(4.4.6) on $\mathbb{Z}$. □

Chapter 5

Boundary Value Problems for Dynamic Equations on Time Scales

In 1990, Hilger published a pioneering paper [74] on the topic of *measure chains* in which he developed a theory that both unified and extended the theories between the continuous calculus and the discrete calculus. Given $\emptyset \neq \mathbb{T} \subseteq \mathbb{R}$, if $\mathbb{T}$ is closed and equipped with the subspace topology inherited from the Euclidean topology on $\mathbb{R}$, then $\mathbb{T}$ is called a *measure chain*; Hilger [74]. Within the context of measure chains, such a subset $\mathbb{T}$ of $\mathbb{R}$ came to be called a *time scale* [6]. Now, almost exclusively, if $\mathbb{T}$ is a nonempty, closed subset of $\mathbb{R}$ and is endowed with the subspace Euclidean topology, then $\mathbb{T}$ is called a *time scale*.

In this chapter, we are concerned with uniqueness of solutions implying the existence of solutions for boundary value problems for nonlinear dynamic equations on a time scale. Before introducing the problems of interest for this chapter, we will present some definitions and notation, mostly from Hilger's work [74], which are now common in the literature and are sufficient for the presentation of this chapter's results.

Let $\emptyset \neq \mathbb{T} \subseteq \mathbb{R}$, be a time scale. For $t < \sup \mathbb{T}$ and $r > \inf \mathbb{T}$, we define the *forward jump operator*, σ, and the *backward jump operator*, ρ, respectively, by

$$\sigma(t) := \inf\{\tau \in \mathbb{T} \,|\, \tau > t\} \in \mathbb{T},$$

and

$$\rho(t) := \sup\{\tau \in \mathbb{T} \,|\, \tau < t\} \in \mathbb{T}.$$

If $\sigma(t) > t$, then t is said to be *right scattered*, while if $\rho(r) < r$, then r is said to be *left scattered*. If $\sigma(t) = t$, then t is said to be *right dense*, and if $\rho(r) = r$, then r is said to be *left dense*.

Definition 5.0.1. *For $x : \mathbb{T} \to \mathbb{R}$ and $t \in \mathbb{T}$ (if $t = \sup \mathbb{T}$, assume t is not left scattered), define the "delta derivative" of $x(t)$, denoted $x^{\Delta}(t)$, to be the number (when it exists) with the property that, for any $\epsilon > 0$, there is a neighborhood, U, of t such that*

$$\left| [x(\sigma(t)) - x(s)] - x^{\Delta}(t)[\sigma(t) - s] \right| \leq \epsilon \left| \sigma(t) - s \right|,$$

for all $s \in U$. For $n \geq 2$, define the "nth order delta derivative" of $x(t)$ by $x^{\Delta^n}(t) := (x^{\Delta^{n-1}})^{\Delta}(t)$. If $F^{\Delta}(t) = h(t)$, then define the "integral" by $\int_a^t h(s)ds := F(t) - F(a)$.

Remark 5.0.1. *Suppose $x : \mathbb{T} \to \mathbb{R}$ and that $x^{\Delta}(t)$ exists for all $t \in \mathbb{T}$. We then observe that, if $\mathbb{T} = \mathbb{Z}$, then $x^{\Delta}(t) = \Delta x(t) = x(t+1) - x(t)$, for all $t \in \mathbb{Z}$, whereas, if $\mathbb{T} = \mathbb{R}$, then $x^{\Delta}(t) = x'(t)$, for all $t \in \mathbb{R}$.*

Definition 5.0.2. *Let $\mathbb{T}$ be a time scale and let $y : \mathbb{T} \to \mathbb{R}$. A point $t \in \mathbb{T}$ is said to be a "generalized zero" (denoted GZ) of y if $y(t) = 0$, or $y(\rho(t))y(t) < 0$.*

For functions of higher order delta differentiability, Bohner and Eloe [9] defined GZ's of higher orders.

Definition 5.0.3. *For $y : \mathbb{T} \to \mathbb{R}$, a point $a \in \mathbb{T}$ is a "generalized zero (GZ) of order greater than or equal to k," if either*

$$y^{\Delta^j}(a) = 0, \quad j = 0, \ldots, k-1,$$

or

$$y^{\Delta^j}(a) = 0, \; j = 0, \ldots, k-2, \; \text{and} \; y^{\Delta^{k-1}}(\rho(a))y^{\Delta^{k-1}}(a) < 0. \qquad (5.0.1)$$

Note that (5.0.1) is equivalent to

$$y^{\Delta^j}(a) = 0, \; j = 0, \ldots, k-2, \; \text{and} \; (-1)^{k-1}y(\rho(a))y^{\Delta^{k-1}}(a) < 0. \qquad (5.0.2)$$

In this chapter, we let $\emptyset \neq \mathbb{T} \subseteq \mathbb{R}$ be a time scale with $\inf \mathbb{T} = -\infty$ and $\sup \mathbb{T} = +\infty$. Throughout, for notation, we shall use the convention that, for each $O \subseteq \mathbb{R}$,

$$O_{\mathbb{T}} := O \cap \mathbb{T}.$$

For $n \geq 2$, we will be concerned with uniqueness of solutions implying the existence of solutions for certain boundary value problems for the nonlinear equation,

$$y^{\Delta^n}(t) = f(t, y(t), y^{\Delta}(t), \ldots, y^{\Delta^{n-1}}(t)), \quad t \in \mathbb{T}. \qquad (5.0.3)$$

Equation (5.0.3) is called *an nth order dynamic equation on the time scale* $\mathbb{T}$.

We assume throughout the chapter the conditions:

(TSA) $f(t, r_1, \ldots, r_n) : \mathbb{T} \times \mathbb{R}^n \to \mathbb{R}$ is continuous.
(TSB) Solutions of initial value problems for (5.0.3) exist and are unique on all of $\mathbb{T}$.

In particular, our uniqueness implies existence results will be restricted to (5.0.3) of orders $n = 2, 3, 4$. Moreover, the boundary value problems will be aptly referred to as conjugate, right focal and nonlocal boundary value problems.

Fundamental for our main results, we will often rely on a "compactness condition" assumption with respect to solutions of (5.0.3).

(TSCP) If there exists a sequence of solutions $\{y_k(t)\}_{k=1}^{\infty}$ of (5.0.3), a compact interval $[c, d]_{\mathbb{T}}$, with $Card[c, d]_{\mathbb{T}} \geq n$, and a number $M > 0$ such that $|y_k(t)| \leq M$, for all $k \geq 1$ and each $t \in [c, d]_{\mathbb{T}}$, then there is a subsequence $\{y_{k_j}(t)\}_{j=1}^{\infty}$ such that $\{y_{k_j}^{\Delta^i}(t)\}_{j=1}^{\infty}$ converges uniformly on each compact subinterval $[a, b]_{\mathbb{T}} \subseteq \mathbb{T}$, for each $i = 0, \ldots, n - 1$.

Chyan [10] proved a Kamke-type theorem concerning solutions of initial value problems for first order systems of dynamic equations on time scales.

Theorem 5.0.1. [[10], Chyan, Thm. 2.1] *Let* $g(t, u) : \mathbb{T} \times \mathbb{R}^m \to \mathbb{R}^m$ *be continuous. Assume that solutions of initial value problems for* $y^{\Delta} = g(t, y)$ *are unique on* $\mathbb{T}$. *Given any* $(t_0, \alpha_0) \in \mathbb{T} \times \mathbb{R}^m$, *let* $y(t, t_0, \alpha_0)$ *denote the solution of the initial value problem*

$$y^{\Delta}(t) = g(t, y), \tag{5.0.4}$$

$$y(t_0) = \alpha_0. \tag{5.0.5}$$

Assume that $t_n \to t_0$ *in* $\mathbb{T}$ *and* $\alpha_n \to \alpha_0$ *in* $\mathbb{R}^m$. *Then the solutions* $y_n(t, t_n, \alpha_n)$ *of (5.0.4) with*

$$y(t_n) = \alpha_n \tag{5.0.6}$$

converge uniformly to $y(t, t_0, \alpha_0)$ *on each compact subset of* $\mathbb{T}$.

Remark 5.0.2. *We note that under conditions* (TSA) *and* (TSB), *a Kamke-type theorem for solutions of initial value problems for the higher order dynamic equation* (5.0.3) *on a time scale would be proved in the same way as Theorem 5.0.1.*

5.1   Conjugate boundary value problems: uniqueness implies existence

In this section, we will concentrate on uniqueness of solutions implying their existence for conjugate type boundary value problems for (5.0.3) of orders $n = 2, 3, 4$.

Definition 5.1.1. *Given* $k \in \{2, \ldots, n\}$, $m_1, \ldots, m_k \in \mathbb{N}$ *such that* $\sum_{\ell=1}^{k} m_\ell = n$, *and* $t_1 < \cdots < t_k$ *in* $\mathbb{T}$, *with* $\sigma^{m_\ell - 1}(t_\ell) < t_{\ell+1}$, $1 \le \ell \le k-1$, *a boundary value problem for* (5.0.3) *satisfying the conditions*

$$y^{\Delta^j}(t_i) = y_{ij}, \quad 0 \le j \le m_i - 1, \ 1 \le i \le k, \tag{5.1.1}$$

where $y_{ij} \in \mathbb{R}$, *is called a* k-*point conjugate or an* $(m_1, \ldots, m_k)$ *conjugate boundary value problem for* (5.0.3).

The uniqueness condition of which we will make extensive use is stated as follows:

(TSC) Given $k \in \{2, \ldots, n\}$, $m_1, \ldots, m_k \in \mathbb{N}$ such that $\sum_{\ell=1}^{k} m_\ell = n$, and $t_1 < \cdots < t_k$ in $\mathbb{T}$, with $\sigma^{m_\ell - 1}(t_\ell) < t_{\ell+1}$, $1 \le \ell \le k - 1$, if $y(t)$ and $z(t)$ are solutions of (5.0.3) such that $y(t) - z(t)$ has a GZ at t_j of order m_j, $1 \le j \le k$, then $y(t) = z(t)$ on $[t_1, t_k]_{\mathbb{T}}$ (hence on $\mathbb{T}$).

As a consequence of Theorem 5.0.1, Chyan and Yin [12] proved the following concerning the continuous dependence of solutions of (5.0.3) with respect to initial conditions.

Theorem 5.1.1. [[12], Chyan and Yin] *Assume that conditions* (TSA) *and* (TSB) *are satisfied. Given a solution* $y(t)$ *of* (5.0.3) *on* $\mathbb{T}$, *an interval* $[a, b]_{\mathbb{T}}$, *a point* $t_0 \in [a, b]_{\mathbb{T}}$, *and* $\epsilon > 0$, *there exists a* $\delta(\epsilon, [a, b]_{\mathbb{T}}) > 0$ *such that, if* $|y^{\Delta^i}(t_0) - z_{i+1}| < \delta$, $i = 0, \ldots, n - 1$, *then there exists a solution* $z(t)$ *of* (5.0.3) *satisfying* $z^{\Delta^i}(t_0) = z_{i+1}$, $i = 0, \ldots, n - 1$, *and* $|y^{\Delta^i}(t) - z^{\Delta^i}(t)| < \epsilon$ *on* $[a, b]_{\mathbb{T}}$, $i = 0, \ldots, n - 1$.

In turn, if (TSC) is also assumed, then Theorem 5.1.1 and an application of the Brouwer Theorem on Invariance of Domain imply solutions of k-point conjugate problems, (5.0.3)-(5.1.1) depend continuously on boundary conditions.

Theorem 5.1.2. [[73], Henderson and Yin, Thm. 2.2] *Assume that with respect to (5.0.3), conditions (TSA), (TSB) and (TSC) are satisfied. Given a solution $y(t)$ of (5.0.3) on $\mathbb{T}$, an interval $[a,b]_\mathbb{T}$, $k \in \{2,\ldots,n\}$, $m_1,\ldots,m_k$, such that $\sum_{i=1}^k m_i = n$, points $t_1 < \cdots < t_k$ in $[a,b]_\mathbb{T}$, with $\sigma^{m_\ell - 1}(t_\ell) < t_{\ell+1}$, $1 \leq \ell \leq k-1$, and an $\epsilon > 0$, there exists a $\delta(\epsilon,[a,b]_\mathbb{T}) > 0$ such that, if $|y^{\Delta^j}(t_i) - z_{i,j}| < \delta$, $0 \leq j \leq m_i - 1$, $1 \leq i \leq k$, then there exists a solution $z(t)$ of (5.0.3) satisfying $z^{\Delta^j}(t_i) = z_{i,j}$, $0 \leq j \leq m_i - 1$, $1 \leq i \leq k$, and $|y^{\Delta^i}(t) - z^{\Delta^i}(t)| < \epsilon$ on $[a,b]_\mathbb{T}$, $i = 0,\ldots,n$.*

The first uniqueness of solutions implies their existence result for this section was obtained by Chyan [10] for 2-point conjugate problems for (5.0.3)-(5.1.1), when $n = 2$; in particular, for the boundary value problem,

$$y^{\Delta\Delta}(t) = f(t,y(t),y^{\Delta}(t)), \quad t \in \mathbb{T}, \tag{5.1.2}$$

$$y(t_1) = y_1, \quad y(t_2) = y_2. \tag{5.1.3}$$

Theorem 5.1.3. [[10], Chyan, Thm. 2.3] *Assume that with respect to (5.1.2), conditions (TSA), (TSB), (TSC) and (TSCP) are satisfied. Then, for any $t_1 < t_2$ in $\mathbb{T}$ and any $y_1, y_2 \in \mathbb{R}$, the boundary value problem (5.1.2)-(5.1.3) has a unique solution on $\mathbb{T}$.*

Proof. The uniqueness follows from assumption (TSC). Let $t_1 < t_2$ belong to $\mathbb{T}$ and choose $y_1, y_2 \in \mathbb{R}$. Then let $y(t)$ denote the solution of (5.1.2) which satisfies the initial conditions $y(t_1) = y_1$, $y^{\Delta}(t_1) = 0$. Now define the set

$$S := \{w(t_2) \mid w(t) \text{ is a solution of (5.1.2) and } w(t_1) = y(t_1)\}.$$

First, $S \neq \emptyset$ since $y(t_2) \in S$. We claim first that $S \subset \mathbb{R}$ is an open set. To show this, choose $s \in S$. By definition of S, there exists a solution $y_s(t)$ of (5.1.2) such that

$$y_s(t_1) = y(t_1) \quad \text{and} \quad y_s(t_2) = s.$$

By Theorem 5.1.2, there is an $\varepsilon > 0$, such that if $|s - r| < \varepsilon$, then there is a solution $z(t)$ of (5.1.2) satisfying $z(t_1) = y_s(t_1) = y(t_1)$, and $z(t_2) = r$.

That is, $r \in S$, and therefore, $(s - \varepsilon, s + \varepsilon) \subseteq S$. And S is an open subset of $\mathbb{R}$.

Next we claim $S \subset \mathbb{R}$ is a closed set. To show this, suppose the contrary. Then there is a limit point $r_0 \in \overline{S} \setminus S$ and a strictly monotone sequence $\{r_k\}_{k=1}^{\infty} \subset S$ such that $r_k \to r_0$. We may assume without loss of generality that $r_k \uparrow r_0$. Now let $z(t)$ be the solution of (5.1.2) satisfying $z(t_2) = r_0$, and $z^\Delta(t_2) = 0$. Since $r_0 \notin S$, then $z(t_1) \neq y_k(t_1)$, for all $k \geq 1$. We have two subcases.

Case (i). Suppose $z(t_1) > y_1(t_1)$. Then, by assumption (TSB), $y_1(t) \leq y_k(t) \leq z(t)$ on $[t_1, t_2]_{\mathbb{T}}$, for all $k \geq 1$. Therefore there exists $M > 0$ such that $|y_k(t)| \leq M$ for every $t \in [t_1, t_2]_{\mathbb{T}}$. By compactness condition (TSCP) there is a subsequence $\{y_{k_j}(t)\}_{j=1}^{\infty}$ such that $\{y_{k_j}(t)\}_{j=1}^{\infty}$ and $\{y_{k_j}^\Delta(t)\}_{j=1}^{\infty}$ converge uniformly on each compact subinterval of $\mathbb{T}$. Theorem 5.0.1 now implies that this subsequence converges uniformly on $[t_1, t_2]_{\mathbb{T}}$ to a solution $Y(t)$ of (5.1.2). It is immediate that $Y(t)$ must satisfy

$$Y(t_1) = y(t_1) \quad \text{and} \quad Y(t_2) = r_0.$$

But this implies $r_0 \in S$, a contradiction.

Case (ii). Suppose $z(t_1) < y(t_1)$. Then we consider some $c \in \mathbb{T}$ such that $t_2 < c$. Such a c exists, since $\sup T = +\infty$. We see that $y_k(t) \leq z(t)$ on $[t_2, c]_{\mathbb{T}}$. We can argue as in Case (i), except we will take the interval $[t_2, c]_{\mathbb{T}}$. We again obtain a contradiction.

So, S is also a closed subset of $\mathbb{R}$. And then, $S = \mathbb{R}$. If we choose $y_2 \in \mathbb{R}$, then there is a corresponding solution $w(t)$ of (5.1.2) satisfying $w(t_1) = y_1$ and $w(t_2) = y_2$. The proof is complete.   □

The next uniqueness of solutions implies their existence results for this section were obtained by Henderson and Yin [71] for 3-point conjugate problems and 2-point conjugate problems for (5.0.3)-(5.1.1), when $n = 3$; in particular, for solutions of the third order dynamic equation,

$$y^{\Delta^3}(t) = f(t, y(t), y^\Delta(t), y^{\Delta\Delta}(t)), \quad t \in \mathbb{T}, \tag{5.1.4}$$

satisfying in one case the 3-point conjugate conditions,

$$y(t_1) = y_1, \quad y(t_2) = y_2, \quad y(t_3) = y_3, \tag{5.1.5}$$

where $t_1 < t_2 < t_3$ in $\mathbb{T}$, and $y_1, y_2, y_3 \in \mathbb{R}$, and in other cases, either the 2-point conjugate conditions,

$$y(t_1) = y_1, \quad y^\Delta(t_1) = y_2, \quad y(t_2) = y_3, \tag{5.1.6}$$

where $t_1 \leq \sigma(t_1) < t_2$ in $\mathbb{T}$, and $y_1, y_2, y_3 \in \mathbb{R}$, or the 2-point conjugate conditions,

$$y(t_1) = y_1, \quad y(t_2) = y_2, \quad y^\Delta(t_2) = y_3, \tag{5.1.7}$$

where $t_1 < t_2$ in $\mathbb{T}$, and $y_1, y_2, y_3 \in \mathbb{R}$. We note that with respect to (5.1.4)-(5.1.5), if $t_2 = \sigma(t_1)$ and $t_3 = \sigma(t_2)$, then the boundary value problem is actually an initial value problem.

Theorem 5.1.4. [[71], Henderson and Yin, Thm. 3.1] *Assume that with respect to (5.1.4), conditions* (TSA), (TSB), (TSC) *and* (TSCP) *are satisfied. Then, for any $t_1 < t_2 < t_3$ in $\mathbb{T}$ and any $y_1, y_2, y_3 \in \mathbb{R}$, the boundary value problem (5.1.4)-(5.1.5) has a unique solution on $\mathbb{T}$.*

Proof. We remark that the uniqueness follows from assumption (TSC). So, now let $t_1 < t_2 < t_3$ and $y_1, y_2, y_3 \in \mathbb{R}$ be given as stated. There are a number of cases and subcases.

Case (i): Assume t_1 and t_2 are right dense.
Let $z(t)$ be the solution of (5.1.4) satisfying the initial conditions,

$$z(t_1) = y_1, \quad z^\Delta(t_1) = 0, \quad z^{\Delta^2}(t_1) = 0.$$

Define

$$S_1 = \{r \in \mathbb{R} \,|\, \text{there is a solution } x(t) \text{ of (5.1.4) with } x(t_1) = z(t_1),$$
$$x(t_2) = z(t_2), x(t_3) = r\}.$$

Since $z(t_3) \in S_1$, $S_1 \neq \emptyset$. We argue that S_1 is both open and closed.

To see that S_1 is open, choose $s \in S_1$. Then there is a solution $x_s(t)$ such that $x_s(t_1) = z(t_1), x_s(t_2) = z(t_2)$, and $x_s(t_3) = s$. By Theorem 5.1.2, there exists an $\varepsilon > 0$ such that for any $u_1, u_2, u_3 \in \mathbb{R}$ with $|x_s(t_i) - u_i| < \varepsilon$, (5.1.4) has a solution $u(t)$ with $u(t_i) = u_i$, $i = 1, 2, 3$. In particular, $(s - \varepsilon, s + \varepsilon) \subseteq S_1$, and S_1 is open.

For contradiction purposes, assume that S_1 is not closed. Then there exists $r_0 \in \overline{S}_1 \setminus S_1$ and a strictly monotone sequence $\{r_k\} \subset S_1$ such that $\lim_{k \to \infty} r_k = r_0$. We may assume without loss of generality that $r_k \uparrow r_0$. For each $k \in \mathbb{N}$, let $x_k(t)$ to be the solution of (5.1.4) with

$$x_k(t_1) = z(t_1), \quad x_k(t_2) = z(t_2), \quad x_k(t_3) = r_k.$$

If follows from (TSC) that, for each $k \in \mathbb{N}$, $x_k^\Delta(t_i) \neq x_{k+1}^\Delta(t_i)$, $i = 1, 2$. Since $r_{k+1} > r_k$, we have from (TSC) that, for $k \in \mathbb{N}$,

$$x_k(t) < x_{k+1}(t) \text{ on } (-\infty, t_1)_\mathbb{T} \cup (t_2, \infty)_\mathbb{T},$$

and

$$x_k(t) > x_{k+1}(t) \text{ on } (t_1, t_2)_{\mathbb{T}}.$$

By the compactness condition (TSCP) and the right density of t_1 and t_2, there exist points $\tau_1 < t_1 < \tau_2 < t_2 < \tau_3 < t_3 < \tau_4$ in $\mathbb{T}$ such that

$$x(\tau_i) \uparrow +\infty, \ i = 1, 3, 4, \text{ and } x_k(\tau_2) \downarrow -\infty.$$

Now let $w(t)$ be the solution (5.1.4) satisfying the initial conditions,

$$w(t_3) = r_0, \ w^\Delta(t_3) = 0, w^{\Delta^2}(t_3) = 0.$$

Since $w(t_3) = r_0 > r_k = x_k(t_3)$, for all $k \in \mathbb{N}$, it follows that, for sufficiently large K, $x_K(t) - w(t)$ has GZ's in $(\tau_1, \tau_2]_{\mathbb{T}}$, $(\tau_2, \tau_3]_{\mathbb{T}}$, $(\tau_3, t_3]_{\mathbb{T}}$, and $(t_3, \tau_4]_{\mathbb{T}}$. It follows from (TSC) that $w(t) = y_K(t)$ on $\mathbb{T}$, a contradiction. Therefore, S_1 is closed. Consequently, $S_1 = \mathbb{R}$. Choosing $y_3 \in S_1$, we have a solution $\bar{x}(t)$ of (5.1.4) satisfying

$$\bar{x}(t_1) = z(t_1) = y_1, \ \bar{x}(t_2) = z(t_2), \ \bar{x}(t_3) = y_3.$$

Next define

$$S_2 = \{r \in \mathbb{R} \,|\, \text{there is a solution } x(t) \text{ of (5.1.4) with } x(t_1) = z(t_1),$$
$$x(t_2) = r, x(t_3) = y_3\}.$$

For the solution just produced, $\bar{x}(t_2) \in S_2$, and so $S_2 \neq \emptyset$. Again, by Theorem 5.1.2, S_2 is an open subset of $\mathbb{R}$.

We now claim that S_2 is also closed. Assuming S_2 is not closed, there exist $r_0 \in \bar{S}_2 \setminus S_2$ and a strictly monotone sequence $\{r_k\} \subset S_2$ such that $\lim_{k\to\infty} r_k = r_0$. We may assume again that $r_k \uparrow r_0$, and as before, let $x_k(t)$ denote the corresponding solution of (5.1.4) with

$$x_k(t_1) = z(t_1), \ x_k(t_2) = r_k, \ x_k(t_3) = y_3.$$

Again, by (TSC), for $k \in \mathbb{N}$, $x_k^\Delta(t_i) \neq x_{k+1}^\Delta(t_i)$, $i = 1, 3$. Using $r_{k+1} > r_k$ and (TSC), we have that, for $k \in \mathbb{N}$,

$$x_k(t) > x_{k+1}(t) \text{ on } (-\infty, t_1)_{\mathbb{T}} \cup (t_3, \infty)_{\mathbb{T}},$$

and

$$x_k(t) < x_{k+1}(t) \text{ on } (t_1, t_3)_{\mathbb{T}}.$$

By (TSCP) and the right density of t_1 and t_2, there exist points $\tau_1 < t_1 < \tau_2 < t_2 < \tau_3 < t_3 < \tau_4$ in $\mathbb{T}$ such that

$$x_k(\tau_i) \downarrow -\infty, \ i = 1, 4, \text{ and } x_k(\tau_j) \uparrow +\infty, \ j = 2, 3.$$

This time, let $w(t)$ be the solution of (5.1.4) satisfying the initial conditions,

$$w(t_2) = r_0, \ w^\Delta(t_2) = 0, \ w^{\Delta^2}(t_2) = 0.$$

Since $w(t_2) = r_0 > r_k = x_k(t_2)$, for all k, it follow that, for sufficiently large K, $x_K(t) - w(t)$ has GZ's in $(\tau_1, \tau_2]_\mathbb{T}$, $(\tau_2, t_2]_\mathbb{T}$, $(t_2, \tau_3]_\mathbb{T}$, and $(\tau_3, \tau_4]_\mathbb{T}$, and (TSC) implies $w(t) = y_K(t)$ on $\mathbb{T}$; again, a contradiction. Therefore S_2 is also closed. So $S_2 = \mathbb{R}$, and if we choose $y_2 \in S_2$, we have a solution $y(t)$ of (5.1.4) satisfying

$$y(t_1) = z(t_1) = y_1, \ y(t_2) = y_2, \ y(t_3) = y_3.$$

This concludes case (i).

Case (ii): Assume t_1 is right dense and t_2 is right scattered. There are a couple of subcases.

Subcase (ii.1): Assume $\sigma(t_2) = t_3$.
Our arguments are much like the those before. Let $z(t)$ be the solution (5.1.4) satisfying the initial conditions,

$$z(t_1) = y_1, \ z^\Delta(t_1) = 0, \ z^{\Delta^2}(t_1) = 0.$$

Define

$$S_1 = \{r \in \mathbb{R} \,|\, \text{there is a solution } x(t) \text{ of } (5.1.4) \text{ with } x(t_1) = z(t_1),$$
$$x(t_2) = z(t_2), x(t_3) = r\}.$$

S_1 is nonempty and open.

We wish to show S_1 is also closed, and so we assume not. As above, let $r_0 \in \overline{S}_1 \setminus S_1$, $\{r_k\} \subset S_1$, with $r_k \uparrow r_0$, and let $x_k(t)$ denote the corresponding solution of (5.1.4) such that

$$x_k(t_1) = z(t_1), \ x_k(t_2) = z(t_2), \ x_k(t_3) = r_k.$$

In this subcase,

$$x_k(t) < x_{k+1}(t) \text{ on } (-\infty, t_1)_\mathbb{T} \cup (t_2, +\infty)_\mathbb{T},$$

and

$$x_k(t) > x_{k+1}(t) \text{ on } (t_1, t_2)_\mathbb{T}.$$

By (TSCP), there are points $\tau_1 < t_1 < \tau_2 < t_2 < t_3 < \tau_3$ in $\mathbb{T}$ such that

$$x_k(\tau_i) \uparrow +\infty, \ i = 1, 3, \ x_k(\tau_2) \downarrow -\infty.$$

Recalling that $\sigma(t_2) = t_3$ in this subcase, we let $w(t)$ be the solution of the initial value problem for (5.1.4) such that

$$w(t_2) = z(t_2) = x_k(t_2),$$
$$w(t_3) = w(\sigma(t_2)) = r_0,$$
$$w^\Delta(t_3) = 0.$$

Then, for sufficiently large K, $x_K(t) - w(t)$ has a GZ in $(\tau_1, \tau_2]$, a zero at t_2, and a GZ in $(l_3, \imath_3]$. Again, we contradict (TSC), and S_1 is closed.

So, $S_1 = \mathbb{R}$ and choosing $y_3 \in S_1$, there is solution $\overline{x}(t)$ of (5.1.4) for which

$$\overline{x}(t_1) = z(t_1) = y_1, \ \overline{x}(t_2) = z(t_2), \ \overline{x}(t_3) = y_3.$$

Remaining in Subcase (ii.1), we now define

$$S_2 = \{r \in \mathbb{R} \,|\, \text{there is a solution } x(t) \text{ of (5.1.4) with } x(t_1) = z(t_1),$$
$$x(t_2) = r, x(t_3) = y_3\}.$$

S_2 is a nonempty open subset of $\mathbb{R}$.

If we assume S_2 is not closed, let $r_0 \in \overline{S}_2 \setminus S_2$, $\{r_k\} \subset S_2$ with $r_k \uparrow r_0$ be as usual, and let $x_k(t)$ be the solution of (5.1.4) such that

$$x_k(t_1) = z(t_1), \ x_k(t_2) = r_k, \ x_k(t_3) = y_3.$$

Now,

$$x_k(t) > x_{k+1}(t) \text{ on } (-\infty, t_1)_\mathbb{T} \cup (t_3, \infty)_\mathbb{T},$$

and

$$x_k(t) < x_{k+1}(t) \text{ on } (t_1, t_3)_\mathbb{T}.$$

In this subcase, there are points $\tau_1 < t_1 < \tau_2 < t_2 < t_3 < \tau_3$ such that

$$x_k(\tau_i) \downarrow -\infty, \ i = 1, 3, \ x_k(\tau_2) \uparrow +\infty.$$

Recalling again that $\sigma(t_2) = t_3$, we choose $w(t)$ to be the solution of the initial value problem for (5.1.4) such that

$$w(t_2) = r_0,$$
$$w(t_3) = w(\sigma(t_2)) = y_3 = x_k(t_3),$$
$$w^\Delta(t_3) = 0.$$

For K large, $x_K(t) - w(t)$ this time has GZ's in $(\tau_1, \tau_2]$ and $(\tau_2, t_2]$, and has a zero at t_3, which contradicts (TSC). Thus S_2 is closed.

$S_2 = \mathbb{R}$ and we may choose $y_2 \in S_2$ so that we have a solution $y(t)$ of (5.1.4) satisfying

$$y(t_1) = z(t_1) = y_1, \ y(t_2) = y_2, \ y(t_3) = y_3.$$

This concludes Subcase (ii.1).

Subcase (ii.2): Assume $\sigma(t_2) < t_3$.
Recalling that in this case, Case (ii), t_2 is right scattered, and by using solutions guaranteed by Subcase (ii.1), we let $z(t)$ be the solution of the three-point problem for (5.1.4) satisfying,

$$z(t_1) = y_1, \ z(t_2) = y_2, \ z(\sigma(t_2)) = 0,$$

and define the nonempty open set,

$$S_1 = \{r \in \mathbb{R} \, | \text{ there is a solution } x(t) \text{ of } (5.1.4) \text{ with } x(t_1) = z(t_1),$$
$$x(t_2) = z(t_2), \ x(t_3) = r\}.$$

If we assume S_1 is not closed, then let r_0 and $\{r_k\}$ with $r_k \uparrow r_0$, be as usual, and let $x_k(t)$ be the solution of (5.1.4) such that

$$x_k(t_1) = z(t_1), \ x_k(t_2) = z(t_2), \ x_k(t_3) = r_k.$$

In this case,

$$x_k(t) < x_{k+1}(t) \text{ on } (-\infty, t_1)_{\mathbb{T}} \cup (t_2, \infty)_{\mathbb{T}},$$

and

$$x_k(t) > x_{k+1}(t) \text{ on } (t_1, t_2)_{\mathbb{T}}.$$

It follows that there are points $\tau_1 < t_1 < \tau_2 < t_2 < \sigma(t_2) \leq \tau_3 < t_3 < \tau_4$ in $\mathbb{T}$ such that

$$x_k(\tau_i) \uparrow +\infty, \ i = 1, 3, 4, \ x_k(\tau_2) \downarrow -\infty.$$

Within this scenario, we argue by two additional sub-subcases.

Subcase (ii.2.1): Assume t_3 is right scattered.
In this setting, let $w(t)$ be the solution of the three-point problem for (5.1.4) guaranteed by Subcase (ii.1) which satisfies

$$w(t_1) = z(t_1) = x_k(t_1),$$
$$w(t_3) = r_0,$$
$$w(\sigma(t_3)) = 0.$$

Then, for K large, $x_K(t) - w(t)$ has a zero at t_1 and GZ's in $(\tau_3, t_3]$ and $(t_3, \tau_4]$, which contradicts (TSC).

Subcase (ii.2.2): Assume t_3 is right dense.
This time, let $w(t)$ be the solution of the three-point problem for (5.1.4) guaranteed by Case (i) which satisfies,

$$w(t_1) = z(t_1) = x_k(t_1),$$
$$w(t_3) = r_0,$$
$$w(\tau_4) = 0.$$

Again, for K large, $x_K(t) - w(t)$ has a zero at t_1 and GZ's in $(\tau_3, t_3]$ and $(t_3, \tau_4]$, the same contradiction.

From the two subcases, we conclude S_1 is closed so that $S_1 = \mathbb{R}$. We choose $y_3 \in S_1$, and so there is a solution $y(t)$ of (5.1.4) for which

$$y(t_1) = z(t_1) = y_1, \ y(t_2) = z(t_2) = y_2, \ y(t_3) = y_3.$$

This concludes Subcase (ii.2), which also concludes Case (ii).

Case (iii): Assume t_1 is right scattered. As above there are various subcases with some of them involving further subcases.

Subcase (iii.1): Assume $\sigma(t_1) = t_2$.
If $\sigma(t_2) = t_3$, then (5.1.4)-(5.1.5) is an initial value problem which has a unique solution on $\mathbb{T}$. Thus, in this subcase, we also assume that $\sigma(t_2) < t_3$.
Let $z(t)$ be the solution of the initial value problem for (5.1.4) satisfying

$$z(t_3) = y_3, \ z^{\Delta}(t_3) = 0, \ z^{\Delta^2}(t_3) = 0,$$

and define the nonempty open subset

$$\begin{aligned} S_1 = \ & \{r \in \mathbb{R} \,|\, \text{there is a solution } x(t) \text{ of } (5.1.4) \text{ with } x(t_1) = r, \\ & x(t_2) = z(t_2), x(t_3) = z(t_3)\}. \end{aligned}$$

Assuming S_1 is not closed, let r_0 and $\{r_k\}$, with $r_k \uparrow r_0$, be the usual, and let $x_k(t)$ be the appropriate solution of (5.1.4). Then, by (TSC),

$$x_k(t) < x_{k+1}(t) \text{ on } (-\infty, t_2)_\mathbb{T} \cup (t_3, \infty)_\mathbb{T},$$

and

$$x_k(t) > x_{k+1}(t) \text{ on } (t_2, t_3)_\mathbb{T}.$$

In this setting, there are points $\tau_1 < t_1 < t_2 = \sigma(t_1) < \tau_2 < t_3 < \tau_3$ in $\mathbb{T}$ for which

$$x_k(\tau_i) \uparrow +\infty, \ i = 1, 3, \ x_k(\tau_2) \downarrow -\infty.$$

We choose $w(t)$ as the solution of the initial value problem for (5.1.4) satisfying

$$\begin{aligned} w(t_1) &= r_0, \\ w(t_2) &= w(\sigma(t_1)) = z(t_2) = x_k(t_2), \\ w^{\Delta}(t_2) &= 0. \end{aligned}$$

With K large enough, $x_K(t) - w(t)$ has GZ's in $(\tau_1, t_1]$ and $(\tau_2, \tau_3]$ and a zero at t_2. This is again a contradiction.

S_1 is closed, $S_1 = \mathbb{R}$ and choosing $y_1 \in S_1$, there is a solution $\overline{x}(t)$ of (5.1.4) satisfying

$$\overline{x}(t_1) = y_1, \ \overline{x}(t_2) = z(t_2), \ \overline{x}(t_3) = z(t_3) = y_3.$$

Still remaining in Subcase (iii.1), with $\sigma(t_1) = t_2$ and $\sigma(t_2) < t_3$, we define a nonempty open set

$$S_2 = \{r \in \mathbb{R} \mid \text{there is a solution } x(t) \text{ of } (5.1.4) \text{ satisfying } x(t_1) = y_1,$$
$$x(t_2) = r, x(t_3) = y_3\}.$$

As in the pattern, if S_2 is not closed, let r_0, $r_k \uparrow r_0$ be as usual, and $x_k(t)$ the associated solution of (5.1.4). Here

$$x_k(t) > x_{k+1}(t) \text{ on } (-\infty, t_1)_{\mathbb{T}} \cup (t_3, \infty)_{\mathbb{T}},$$

and

$$x_k(t) < x_{k+1}(t) \text{ on } (t_1, t_3)_{\mathbb{T}},$$

and so there are points $\tau_1 < t_1 < t_2 = \sigma(t_1) < \tau_2 < t_3 < \tau_3$ such that

$$x_k(\tau_i) \downarrow -\infty, \ i = 1, 3, \ x_k(\tau_2) \uparrow +\infty.$$

Then, for K large, $x_K(t) - w(t)$ has a zero at t_1, and GZ's in $(t_2, \tau_2]_{\mathbb{T}}$ and $(\tau_2, \tau_3]_{\mathbb{T}}$, a contradiction. Therefore S_2 is closed, $S_2 = \mathbb{R}$, and so choosing $y_2 \in S_2$, there is a solution $y(t)$ of (5.1.4) for which

$$y(t_1) = y_1, \ y(t_2) = y_2, \ y(t_3) = y_3.$$

This concludes Subcase (iii.1).

Subcase (iii.2): Assume that $\sigma(t_1) < t_2 < t_3$.
Recall that Case (iii) involves t_1 is right scattered.
    This time let $z(t)$ be the solution of the 3-point problem dealt with in Subcase (iii.1) satisfying

$$z(t_1) - y_1, \ z(\sigma(t_1)) = 0, \ z(t_2) - y_2,$$

and define

$$S_1 = \{r \in \mathbb{R} \mid \text{there is a solution } x(t) \text{ of } (5.1.4) \text{ such that } x(t_1) = z(t_1),$$
$$x(t_2) = z(t_2), x(t_3) = r\}.$$

We assume the nonempty open S_1 is not closed. Let $r_0 \in \overline{S}_1 \setminus S_1$, $\{r_k\} \subset S$ with $r_k \uparrow r_0$ and associated solution $x_k(t)$ be as in the pattern. Then

$$x_k(t) < x_{k+1}(t) \text{ on } (-\infty, t_1)_{\mathbb{T}} \cup (t_2, \infty)_{\mathbb{T}},$$

and
$$x_k(t) > x_{k+1}(t) \text{ on } (t_1, t_2)_{\mathrm{T}}.$$

There are further subcases to resolve.

Subcase (iii.2.1): Assume $\sigma(t_2) = t_3$.
Then there are points $\tau_1 < t_1 < \sigma(t_1) \le \tau_2 < t_2 < t_3 = \sigma(t_2) < \tau_3$ such that

$$x_k(\tau_i) \uparrow +\infty, \ i = 1, 3, \ x(\tau_2) \downarrow -\infty.$$

Let $w(t)$ be the solution of the 3-point problem for (5.1.4) assured by Subcase (iii.1) satisfying

$$\begin{aligned}
w(t_2) &= z(t_2) = x_k(t_2), \\
w(t_3) &= w(\sigma(t_2)) = r_0, \\
w(\tau_3) &= 0.
\end{aligned}$$

With K sufficiently large, $x_K(t) - w(t)$ has a GZ in $(\tau_1, \tau_2]$, a zero at t_2, and a GZ in $(t_3, \tau_3]$, which is a contradiction.

Subcase (iii.2.2): Assume $\sigma(t_2) < t_3$.
Then there exist points $\tau_1 < t_1 < \sigma(t_1) \le \tau_2 < t_2 < \tau_3 < t_3 < \tau_4$ such that

$$x_k(\tau_i) \uparrow +\infty, \ i = 1, 3, 4, \ x_k(\tau_2) \downarrow -\infty.$$

Let $w(t)$ be the solution of the 3-point boundary value problem for (5.1.4) given by Subcase (iii.1) satisfying

$$\begin{aligned}
w(t_1) &= z(t_1) = x_k(t_1), \\
w(\sigma(t_1)) &= 0, \\
w(\tau_3) &= r_0.
\end{aligned}$$

For large enough K, $x_K(t) - w(t)$ has a zero at t_1, and GZ's in $(\tau_2, \tau_3]$, $(\tau_3, t_3]$ and $(t_3, \tau_4]$, which is again a contradiction.

As a consequence S_1 is closed, $S_1 = \mathbb{R}$, and choosing $y_3 \subset S_1$, there is a solution $y(t)$ of (5.1.4) satisfying

$$y(t_1) = z(t_1) = y_1, \ y(t_2) = z(t_2) = y_2, \ y(t_3) = y_3.$$

This concludes Subcase (iii.2), which completes in turn Case (iii). The proof is complete. $\square$

By arguments much along the lines of those involved in Theorem 5.1.4, yet very much relying on Theorem 5.1.4 itself, the 2-point conjugate boundary value problems for (5.1.4) also have solutions. We state that result in this section, and give only the briefest parts of the tedious proof.

Theorem 5.1.5. [[71], Henderson and Yin, Thm. 4.1] *Assume that with respect to (5.1.4), conditions* (TSA), (TSB), (TSC) *and* (TSCP) *are satisfied.*

(a) *Given any $t_1 \leq \sigma(t_1) < t_2$ in $\mathbb{T}$ and any $y_1, y_2, y_3 \in \mathbb{R}$, the boundary value problem (5.1.4)-(5.1.6) has a unique solution on $\mathbb{T}$.*

(b) *Given any $t_1 < t_2$ in $\mathbb{T}$ and any $y_1, y_2, y_3 \in \mathbb{R}$, the boundary value problem (5.1.4)-(5.1.7) has a unique solution on $\mathbb{T}$.*

Proof. We will give the briefest of outlines of the proofs for each of the 2-point boundary value problems.

For the boundary value problem (5.1.4)-(5.1.6), let $t_1 \leq \sigma(t_1) < t_2$ in $\mathbb{T}$ and $y_1, y_2, y_3 \in \mathbb{R}$ be given, and let $\tau \in \mathbb{T}$, with $\tau < t_1$, be arbitrary, but fixed. We let $z(t)$ be the solution, given by Theorem 5.1.4, of the 3-point conjugate boundary value problem satisfying

$$z(\tau) = 0, \; z(t_1) = y_1, \; z(t_2) = y_3,$$

and we define

$$S_1 = \{r \in \mathbb{R} \mid \text{ there is a solution } x(t) \text{ of } (5.1.4) \text{ such that } x(t_1) = y_1,$$
$$x^{\Delta}(t_1) = r, \; x(t_2) = y_3\}.$$

$z^{\Delta}(t_1) \in S_1$, and then various cases are used in the arguments that S_1 is a subset of $\mathbb{R}$ which is both open and closed. So $S_1 = \mathbb{R}$. Choose $y_2 \in S_2$, and the corresponding solution is the desired solution of (5.1.4)-(5.1.6).

For the boundary value problem (5.1.4)-(5.1.7), let $t_1 < t_2$ in $\mathbb{T}$ and $y_1, y_2, y_3 \in \mathbb{R}$ be given, and again let $\tau \in \mathbb{T}$, with $\tau < t_1$, be arbitrary, but fixed. We let $\zeta(t)$ be the solution, given by Theorem 5.1.4, of the 3-point conjugate boundary value problem satisfying

$$\zeta(\tau) = 0, \; \zeta(t_1) = y_1, \; \zeta(t_2) = y_2,$$

and we define

$$S_2 = \{r \in \mathbb{R} \mid \text{ there is a solution } x(t) \text{ of } (5.1.4) \text{ such that } x(t_1) = y_1,$$
$$x(t_2) = y_2, \; x^{\Delta}(t_2) = r\}.$$

$\zeta^{\Delta}(t_2) \in S_2$, and again various cases are used in the arguments that S_2 is a subset of $\mathbb{R}$ which is both open and closed. So $S_2 = \mathbb{R}$. Choose $y_3 \in S_2$, and the corresponding solution is the desired solution of (5.1.4)-(5.1.7). $\square$

The next two results concerning uniqueness of solutions implying their existence for this section were obtained by Henderson and Yin in [72] and [73], first for 4-point conjugate problems, then followed by 3-point conjugate problems and 2-point conjugate problems for (5.0.3)-(5.1.1), when $n = 4$; in particular, for solutions of the fourth order dynamic equation,

$$y^{\Delta^4}(t) = f(t, y(t), y^{\Delta}(t), y^{\Delta^2}(t), y^{\Delta^3}(t)), \quad t \in \mathbb{T}, \qquad (5.1.8)$$

satisfying in the first case the 4-point conjugate conditions,

$$y(t_1) = y_1, \quad y(t_2) = y_2, \quad y(t_3) = y_3, \quad y(t_4) = y_4, \qquad (5.1.9)$$

and in the second cases, either the 3-point conjugate conditions,

$$y(t_1) = y_1, \quad y^{\Delta}(t_1) = y_2, \quad y(t_2) = y_3, \quad y(t_3) = y_4, \qquad (5.1.10)$$

or the 3-point conjugate conditions,

$$y(t_1) = y_1, \quad y(t_2) = y_2, \quad y(t_3) = y_3, \quad y^{\Delta}(t_3) = y_4, \qquad (5.1.11)$$

or the 3-point conjugate conditions,

$$y(t_1) = y_1, \quad y(t_2) = y_2, \quad y^{\Delta}(t_2) = y_3, \quad y(t_3) = y_4, \qquad (5.1.12)$$

and in the last cases, either the 2-point conjugate conditions,

$$y(t_1) = y_1, \quad y^{\Delta}(t_1) = y_2, \quad y^{\Delta^2}(t_1) = y_3, \quad y(t_2) = y_4, \qquad (5.1.13)$$

or the 2-point conjugate conditions,

$$y(t_1) = y_1, \quad y(t_2) = y_2, \quad y^{\Delta}(t_2) = y_3, \quad y^{\Delta^2}(t_2) = y_4, \qquad (5.1.14)$$

or the 2-point conjugate conditions,

$$y(t_1) = y_1, \quad y^{\Delta}(t_1) = y_2, \quad y(t_2) = y_3, \quad y^{\Delta}(t_2) = y_4, \qquad (5.1.15)$$

where in each case $y_1, y_2, y_3, y_4 \in \mathbb{R}$, and where the spacing between t_j and t_{j+1} is determined by Definition 5.1.1.

Theorem 5.1.6. [[72], Henderson and Yin, Thm. 3.1] *Assume that with respect to* (5.1.8), *conditions* (TSA), (TSB), (TSC) *and* (TSCP) *are satisfied. Then, given any* $t_1 < t_2 < t_3 < t_4$ *in* $\mathbb{T}$ *and any* $y_1, y_2, y_3, y_4 \in \mathbb{R}$, *the boundary value problem* (5.1.8)-(5.1.9) *has a unique solution on* $\mathbb{T}$.

Proof. We remark as often above, uniqueness of solutions is provided by (TSC). So, let $t_1 < t_2 < t_3 < t_4$ and y_1, y_2, y_3, y_4 be given as stated. There are a number of cases and subcases.

Case (i): Assume $Card(t_i, t_{i+1})_{\mathbb{T}} \geq 2$ for $i = 1, 2, 3$.
Let $z(t)$ be the solution of (5.1.8) satisfying the initial conditions,

$$z(t_1) = y_1, \ z^{\Delta}(t_1) = 0, \ z^{\Delta^2}(t_1) = 0, \ z^{\Delta^3}(t_1) = 0.$$

Define

$$S_1 = \{r \in \mathbb{R} \mid \text{ there is a solution } x(t) \text{ of } (5.1.8) \text{ with } x(t_1) = z(t_1),$$
$$x(t_2) = z(t_2), x(t_3) = z(t_3), x(t_4) = r\}.$$

Since $z(t_4) \in S_1$, $S_1 \neq \emptyset$. We argue that S_1 is both open and closed.

Exactly as in Theorem 5.1.4, it follows from the continuous dependence by Theorem 5.1.2 that S_1 is an open subset of $\mathbb{R}$.

For contradiction purposes, assume that S_1 is not closed. Then there exists $r_0 \in \overline{S}_1 \setminus S_1$ and a strictly monotone sequence $\{r_k\} \subset S_1$ such that $\lim_{k \to \infty} r_k = r_0$. We may assume without loss of generality that $r_k \uparrow r_0$. For each $k \in \mathbb{N}$, let $x_k(t)$ to be the solution of (5.1.8) with

$$x_k(t_1) = z(t_1), \ x_k(t_2) = z(t_2), \ x_k(t_3) = z(t_3), \ x_k(t_4) = r_k.$$

If follows from (TSC) that, for each $k \in \mathbb{N}$, $x_k^{\Delta}(t_i) \neq x_{k+1}^{\Delta}(t_i)$, $i = 1, 2, 3$. Since $r_{k+1} > r_k$, we have from (TSC) that, for $k \in \mathbb{N}$,

$$x_k(t) < x_{k+1}(t) \text{ on } (t_1, t_2)_{\mathbb{T}} \cup (t_3, \infty)_{\mathbb{T}},$$

and

$$x_k(t) > x_{k+1}(t) \text{ on } (-\infty, t_1)_{\mathbb{T}} \cup (t_2, t_3)_{\mathbb{T}}.$$

By the "compactness condition" (TSCP) and the cardinality of $(t_i, t_{i+1})_{\mathbb{T}}$, $i = 1, 2, 3$, there exist points $\tau_1 < t_1 < \tau_2 < t_2 < \tau_3 < t_3 < \tau_4 < t_4 < \tau_5$ in $\mathbb{T}$ such that

$$x(\tau_i) \downarrow -\infty, \ i = 1, 3 \text{ and } x_k(\tau_j) \uparrow +\infty, \ j = 2, 4, 5.$$

Now let $w(t)$ be the solution (5.1.8) satisfying the initial conditions,

$$w(t_4) = r_0, \ w^{\Delta}(t_4) = 0, \ w^{\Delta^2}(t_4) = 0, \ w^{\Delta^3}(t_4) = 0.$$

Since $w(t_4) = r_0 > r_k = x_k(t_4)$, for all $k \in \mathbb{N}$, it follows that, for sufficiently large K, $x_K(t) - w(t)$ has GZ's in $(\tau_1, \tau_2]_{\mathbb{T}}$, $(\tau_2, \tau_3]_{\mathbb{T}}$, $(\tau_3, \tau_4]_{\mathbb{T}}$, and $(\tau_4, \tau_5]_{\mathbb{T}}$. It follows from (TSC) that $w(t) = x_K(t)$ on $\mathbb{T}$, a contradiction. Therefore, S_1 is closed. Consequently, $S_1 = \mathbb{R}$. Choosing $y_4 \in S_1$, we have a solution $\overline{x}(t)$ of (5.1.8) satisfying

$$\overline{x}(t_1) = z(t_1) = y_1, \ \overline{x}(t_2) = z(t_2), \ \overline{x}(t_3) = z(t_3), \ \overline{x}(t_4) = y_4.$$

Next define

$$S_2 = \{r \in \mathbb{R} \mid \text{there is a solution } x(t) \text{ of (5.1.8) with } x(t_1) = z(t_1) = y_1,$$
$$x(t_2) = z(t_2), x(t_3) = r, x(t_4) = y_4\}.$$

For the solution just produced, $\bar{x}(t_3) \in S_2$, and so $S_2 \neq \emptyset$. Again, by Theorem 5.1.2, S_2 is an open subset of $\mathbb{R}$.

As is the custom, we now claim that S_2 is also closed. Assuming S_2 is not closed, there exist $r_0 \in \bar{S}_2 \setminus S_2$ and a strictly monotone sequence $\{r_k\} \subset S_2$ such that $\lim_{k\to\infty} r_k = r_0$. We may assume again that $r_k \uparrow r_0$, and as before, let $x_k(t)$ denote the corresponding solution of (5.1.8) with

$$x_k(t_1) = z(t_1) = y_1, \ x_k(t_2) = z(t_2), \ x_k(t_3) = r_k, \ x_k(t_4) = y_4.$$

Again, by (TSC), for $k \in \mathbb{N}$, $x_k^\Delta(t_i) \neq x_{k+1}^\Delta(t_i)$, $i = 1, 2, 4$. Using $r_{k+1} > r_k$ and (TSC), we have that, for $k \in \mathbb{N}$,

$$x_k(t) < x_{k+1}(t) \text{ on } (-\infty, t_1)_{\mathbb{T}} \cup (t_2, t_4)_{\mathbb{T}},$$

and

$$x_k(t) > x_{k+1}(t) \text{ on } (t_1, t_2)_{\mathbb{T}} \cup (t_4, \infty)_{\mathbb{T}}.$$

By (TSCP) and the cardinality of $(t_i, t_{i+1})_{\mathbb{T}}$, $i = 1, 2, 3$, there exist points $\tau_1 < t_1 < \tau_2 < t_2 < \tau_3 < t_3 < \tau_4 < t_4 < \tau_5$ in $\mathbb{T}$ such that

$$x_k(\tau_i) \uparrow +\infty, \ i = 1, 3, 4 \text{ and } x_k(\tau_j) \downarrow -\infty, \ j = 2, 5.$$

This time, let $w(t)$ be the solution of (5.1.8) satisfying the initial conditions,

$$w(t_3) = r_0, \ w^\Delta(t_3) = 0, \ w^{\Delta^2}(t_3) = 0, \ w^{\Delta^3}(t_3) = 0.$$

Since $w(t_3) = r_0 > r_k = x_k(t_3)$, for all k, it follow that, for sufficiently large K, $x_K(t) - w(t)$ has GZ's in $(\tau_1, \tau_2]_{\mathbb{T}}$, $(\tau_2, \tau_3]_{\mathbb{T}}$, $(\tau_3, \tau_4]_{\mathbb{T}}$ and $(\tau_4, \tau_5]_{\mathbb{T}}$, and (TSC) implies $w(t) = x_K(t)$ on $\mathbb{T}$; again, a contradiction. Therefore S_2 is also closed. So $S_2 = \mathbb{R}$, and if we choose $y_3 \in S_2$, we have a solution $\bar{\bar{x}}(t)$ of (5.1.8) satisfying

$$\bar{\bar{x}}(t_1) = z(t_1) = y_1, \ \bar{\bar{x}}(t_2) = z(t_2), \ \bar{\bar{x}}(t_3) = y_3, \ \bar{\bar{x}}(t_4) = y_4.$$

Finally, in this case, we define

$$S_3 = \{r \in \mathbb{R} \mid \text{there is a solution } x(t) \text{ of (5.1.8) with } x(t_1) = z(t_1),$$
$$x(t_2) = r, x(t_3) = y_3, x(t_4) = y_4\}.$$

For the solution just produced, $\bar{\bar{x}}(t_2) \in S_3$, so $S_3 \neq \emptyset$, and by Theorem 5.1.2, S_3 is an open subset of $\mathbb{R}$.

We now claim that S_3 is also closed. Assuming S_3 is not closed, there exist $r_0 \in \overline{S}_3 \setminus S_3$ and a strictly monotone sequence $\{r_k\} \subset S_3$ such that $\lim_{k \to \infty} r_k = r_0$. We may assume again that $r_k \uparrow r_0$, and as before, let $x_k(t)$ denote the corresponding solution of (5.1.8) with

$$x_k(t_1) = z(t_1) = y_1, \; x_k(t_2) = r_k, \; x_k(t_3) = y_3, \; x_k(t_4) = y_4.$$

Again, by (TSC), for $k \in \mathbb{N}$, $x_k^\Delta(t_i) \neq x_{k+1}^\Delta(t_i)$, $i = 1, 3, 4$. Using $r_{k+1} > r_k$ and (TSC), we have that, for $k \in \mathbb{N}$,

$$x_k(t) > x_{k+1}(t) \text{ on } (-\infty, t_1)_\mathbb{T} \cup (t_3, t_4)_\mathbb{T},$$

and

$$x_k(t) < x_{k+1}(t) \text{ on } (t_1, t_3)_\mathbb{T} \cup (t_4, \infty)_\mathbb{T}.$$

By (TSCP) and the cardinality of $(t_i, t_{i+1})_\mathbb{T}$, $i = 1, 2, 3$, there exist points $\tau_1 < t_1 < \tau_2 < t_2 < \tau_3 < t_3 < \tau_4 < t_4 < \tau_5$ in $\mathbb{T}$ such that

$$x_k(\tau_i) \downarrow -\infty, \; i = 1, 4 \text{ and } x_k(\tau_j) \uparrow +\infty, \; j = 2, 3, 5.$$

This time, let $w(t)$ be the solution of (5.1.8) satisfying the initial conditions,

$$w(t_2) = r_0, \; w^\Delta(t_2) = 0, \; w^{\Delta^2}(t_2) = 0, \; w^{\Delta^3}(t_2) = 0.$$

Since $w(t_2) = r_0 > r_k = x_k(t_2)$, for all k, it follow that, for sufficiently large K, $x_K(t) - w(t)$ has GZ's in $(\tau_1, \tau_2]_\mathbb{T}$, $(\tau_2, \tau_3]_\mathbb{T}$, $(\tau_3, \tau_4]_\mathbb{T}$ and $(\tau_4, \tau_5]_\mathbb{T}$, and (TSC) implies $w(t) = x_K(t)$ on $\mathbb{T}$; again, a contradiction. Therefore S_3 is also closed. So $S_3 = \mathbb{R}$, and if we choose $y_2 \in S_3$, we have a solution $y(t)$ of (5.1.8) satisfying

$$y(t_1) = z(t_1) = y_1, \; y(t_2) = y_2, \; y(t_3) = y_3, \; y(t_4) = y_4.$$

This concludes Case (i).

Case (ii): Assume that the $Card(t_i, t_{i+1})_\mathbb{T} \geq 2$, $i = 1, 2$ and $Card(t_3, t_4)_\mathbb{T} < 2$.
Pick $\xi_1 \in (-\infty, t_1)_\mathbb{T}$ such that $Card(\xi_1, t_1)_\mathbb{T} \geq 2$. By Case (i), there exists a solution $z(t)$ of (5.1.8) satisfying the boundary conditions,

$$z(\xi_1) = 0, z(t_1) = y_1, z(t_2) = y_2, z(t_3) = y_3.$$

Define

$$S_1 = \{r \in \mathbb{R} \mid \text{ there is a solution } x(t) \text{ of (5.1.8) with } x(t_1) = z(t_1) = y_1,$$
$$x(t_2) = z(t_2) = y_2, x(t_3) = z(t_3) = y_3, x(t_4) = r\}.$$

S_1 is nonempty and open. We wish to show S_1 is also closed, and so we assume not. There are a couple of subcases.

Subcase (ii.1): Assume $Card(t_3, t_4)_{\mathbb{T}} = 1$; that is, $(t_3, t_4)_{\mathbb{T}} = \{\sigma(t_3)\}$. As above, let $r_0 \in \overline{S}_1 \setminus S_1$, $\{r_k\} \subset S_1$, with $r_k \uparrow r_0$, and $x_k(t)$ denote the corresponding solution of (5.1.8) such that

$$x_k(t_1) = z(t_1), \; x_k(t_2) = z(t_2), \; x_k(t_3) = z(t_3), \; x_k(t_4) = r_k.$$

In this case,

$$x_k(t) > x_{k+1} \text{ on } (-\infty, t_1)_{\mathbb{T}} \cup (t_2, t_3)_{\mathbb{T}},$$

and

$$x_k(t) < x_{k+1} \text{ on } (t_1, t_2)_{\mathbb{T}} \cup (t_3, \infty)_{\mathbb{T}}.$$

By (TSCP) and the cardinality of $(t_i, t_{i+1})_{\mathbb{T}}$, $i = 1, 2$, there exist points $\tau_1 < t_1 < \tau_2 < t_2 < \tau_3 < t_3$ and $t_4 < \tau_4$ in $\mathbb{T}$ such that

$$x_k(\tau_i) \downarrow -\infty, i = 1, 3, \quad x_k(\tau_i) \uparrow +\infty, i = 2, 4.$$

Recalling that $(t_3, t_4)_{\mathbb{T}} = \{\sigma(t_3)\}$ in this subcase, we let $w(t)$ be the solution of the initial value problem for (5.1.8) such that

$$\begin{aligned} w(t_3) &= z(t_3) = x_k(t_3) = y_3, \\ w(\sigma(t_3)) &= 0, \\ w(\sigma^2(t_3)) &= w(t_4) = r_0, \\ w^{\Delta^3}(t_3) &= 0. \end{aligned}$$

Then, for sufficiently large K, $x_K(t) - w(t)$ has a zero at t_3 and GZ's in $(\tau_1, \tau_2]_{\mathbb{T}}$, $(\tau_2, \tau_3]_{\mathbb{T}}$, and $(t_4, \tau_4]_{\mathbb{T}}$. This contradicts (TSC), and S_1 is closed. So, $S_1 = \mathbb{R}$ and choosing $y_4 \in S_1$, there is a solution $y(t)$ of (5.1.8) for which

$$y(t_1) = y_1, \; y(t_2) = y_2, \; y(t_3) = y_3, \; y(t_4) = y_4.$$

This concludes Subcase (ii.1).

Subcase (ii.2): Assume that the $Card(t_3, t_4)_{\mathbb{T}} = 0$; that is, $t_3 < \sigma(t_3) = t_4$. Mimicking the argument from Subcase (ii.1), we establish the existence of the same $\tau_1, \tau_2, \tau_3, \tau_4$, with the same monotonicity conditions on the x_k's. We let $w(t)$ to be the solution of the initial value problem for (5.1.8) such that

$$\begin{aligned} w(t_3) &= z(t_3) = x_k(t_3) = y_3, \\ w(\sigma(t_3)) &= w(t_4) = r_0, \\ w^{\Delta^2}(t_3) &= 0, \\ w^{\Delta^3}(t_3) &= 0. \end{aligned}$$

For K large, $x_K(t) - w(t)$, has a zero at t_3 and GZ's in $(\tau_1, \tau_2]_\mathbb{T}$, $(\tau_2, \tau_3]_\mathbb{T}$, and $(t_4, \tau_4]_\mathbb{T}$ which contradicts (TSC). Thus, S_1 is closed. So, $S_1 = \mathbb{R}$ and choosing $y_4 \in S_1$, there is a solution $y(t)$ of (5.1.8) for which

$$y(t_1) = y_1, \ y(t_2) = y_2, \ y(t_3) = y_3, \ y(t_4) = y_4.$$

This concludes Subcase (ii.2) and Case (ii).

Case (iii): Assume that the $Card(t_1, t_2)_\mathbb{T} \geq 2$ and $Card(t_2, t_3)_\mathbb{T} < 2$. Pick $\xi_1 \in (-\infty, t_1)_\mathbb{T}$ such that $Card(\xi_1, t_1)_\mathbb{T} \geq 2$. By Case (ii), there exists a solution $z(t)$ of (5.1.8) satisfying the boundary conditions,

$$z(\xi_1) = 0, \ z(t_1) = y_1, \ z(t_3) = y_3, \ z(t_4) = y_4.$$

Define

$$S_1 = \{r \in \mathbb{R} \mid \text{ there is a solution } x(t) \text{ of } (5.1.8) \text{ with } x(t_1) = z(t_1) = y_1,$$
$$x(t_2) = r, x(t_3) = z(t_3) = y_3, x(t_4) = z(t_4) = y_4\}.$$

S_1 is nonempty and open. We wish to show that that S_1 is also closed. We assume the contrary. Again, there are a couple of subcases.

Subcase (iii.1): Assume $Card(t_2, t_3)_\mathbb{T} = 1$; that is, $(t_2, t_3)_\mathbb{T} = \{\sigma(t_2)\}$. Let $r_0 \in \overline{S} \setminus S_1$, $\{r_k\} \subset S_1$, with $r_k \uparrow r_0$, and $x_k(t)$ denote the corresponding solution of (5.1.8) such that

$$x_k(t_1) = z(t_1), \ x_k(t_2) = r_k, \ x_k(t_3) = z(t_3), \ x_k(t_4) = z(t_4).$$

There are further subcases to resolve. We will divide them into the following cases:

(iii.1.1) $Card(t_3, t_4)_\mathbb{T} \geq 2$.
(iii.1.2) $Card(t_3, t_4)_\mathbb{T} = 1$.
(iii.1.3) $Card(t_3, t_4)_\mathbb{T} = 0$.

Subcase (iii.1.1): In this case, $Card(t_3, t_4)_\mathbb{T} \geq 2$. We have

$$x_k(t) > x_{k+1}(t) \text{ on } (-\infty, t_1)_\mathbb{T} \cup (t_3, t_4)_\mathbb{T}$$

and

$$x_k(t) < x_{k+1}(t) \text{ on } (t_1, t_3)_\mathbb{T} \cup (t_4, \infty)_\mathbb{T}.$$

By (TSCP) and the cardinality of $(t_1, t_2)_\mathbb{T}$ and $(t_3, t_4)_\mathbb{T}$, there exist points $\tau_1 < t_1 < \tau_2 < t_2$ and $t_3 < \tau_3 < t_4 < \tau_4$ in $\mathbb{T}$ such that

$$x_k(\tau_i) \downarrow -\infty \ i = 1, 3, \ \ x_k(\tau_i) \uparrow +\infty, \ i = 2, 4.$$

Recalling $(t_2, t_3)_T = \{\sigma(t_2)\}$ in this case, we let $w(t)$ be the solution of the initial problem for (5.1.8) such that

$$w(t_2) = r_0,$$
$$w(\sigma(t_2)) = 0,$$
$$w(\sigma^2(t_2)) = w(t_3) = y_3,$$
$$w^{\Delta^3}(t_2) = 0.$$

Then, for sufficiently large K, $x_K(t) - w(t)$ has a zero at t_3 and GZ's in $(\tau_1, \tau_2]_T$, $(\tau_2, t_2]_T$, and $(\tau_3, \tau_4]_T$, which contradicts (TSC). Thus, S_1 is closed. This concludes Subcase (iii.1.1).

Subcase (iii.1.2): In the case that $Card(t_3, t_4) = 1$, we have,

$$x_k(t) > x_{k+1}(t) \text{ on } (-\infty, t_1)_T \cup (t_3, t_4)_T$$

and

$$x_k(t) < x_{k+1}(t) \text{ on } (t_1, t_3)_T \cup (t_4, \infty)_T.$$

By (TSCP), there exist $\tau_1 < t_1 < \tau_2 < t_2$, and $t_4 < \tau_4$ such that

$$x_k(\tau_1) \downarrow -\infty \text{ and } x_k(\tau_i) \uparrow +\infty, \ i = 2, 4.$$

By (TSCP) and $Card(t_2, \sigma(t_3))_T$, either

$$x_k(\sigma(t_2)) \uparrow +\infty \text{ or } x_k(\sigma(t_3)) \downarrow -\infty.$$

If $x_k(\sigma(t_2)) \uparrow +\infty$, by Case (ii), we let $w(t)$ be the solution of the boundary value problem for (5.1.8) such that

$$w(\xi_1) = 0,$$
$$w(t_1) = y_1,$$
$$w(t_2) = r_0,$$
$$w(t_3) = y_3.$$

For large K, $x_K(t) - w(t)$ has a zero at t_3 and GZ's in $(\tau_1, \tau_2]_T$, $(\tau_2, t_2]_T$, and $(t_2, \sigma(t_2)]_T$, , which contradicts (TSC). Thus, S_1 is closed. If $x_k(\sigma(t_3)) \downarrow -\infty$, then we use the same $w(t)$. For large K, $x_K(t) - w(t)$ has a zero at t_3 and GZ's in $(\tau_1, \tau_2]_T$, $(\tau_2, t_2]_T$, and $(\sigma(t_3), \tau_4]$ which contradicts (TSC). This concludes the Subcase (iii.1.2).

Subcase (iii.1.3): In the case that $Card(t_3, t_4) = 0$ we have,

$$x_k(t) > x_{k+1}(t) \text{ on } (-\infty, t_1)_T$$

and

$$x_k(t) < x_{k+1}(t) \text{ on } (t_1, t_3)_T \cup (t_4, \infty)_T.$$

By (TSCP), there exist $\tau_1 < t_1 < \tau_2 < t_2$, and $t_4 < \tau_4$ such that

$$x_k(\tau_1) \downarrow -\infty, \quad x_k(\tau_i) \uparrow +\infty, \quad i = 2, 4.$$

Let $w(t)$ be the solution of the initial value problem for (5.1.8) such that

$$\begin{aligned} w(t_2) &= r_0, \\ w(\sigma(t_2)) &= 0, \\ w(t_3) &= y_3, \\ w(t_4) &= y_4. \end{aligned}$$

For large K, $x_K(t) - w(t)$ has two zero's at t_3, t_4 and GZ's in $(\tau_1, \tau_2]_{\mathbb{T}}$, $(\tau_2, t_2]_{\mathbb{T}}$, which contradicts (TSC). Thus, S_1 is closed. This concludes Subcase (iii.1.3).

From Subcase (iii.1.i), $i = 1, 2, 3$, we conclude that $S_1 = \mathbb{R}$. Choosing $y_2 \in S_1$, there is a solution $y(t)$ of (5.1.8) for which

$$y(t_1) = y_1, \ y(t_2) = y_2, \ y(t_3) = y_3, \ y(t_4) = y_4.$$

This concludes Subcase (iii.1)

Subcase (iii.2): Assume that the $Card(t_2, t_3)_{\mathbb{T}} = 0$; that is, $t_2 < \sigma(t_2) = t_3$. Still with S_1, $z(t)$, and the x_k's of Case (iii), there are further cases to resolve. We divide them into the following cases:

(iii.2.1) $Card(t_3, t_4)_{\mathbb{T}} \geq 2$.
(iii.2.2) $Card(t_3, t_4)_{\mathbb{T}} = 1$ and $Card(t_3, t_4)_{\mathbb{T}} = 0$.

Subcase (iii.2.1): In the case that $Card(t_3, t_4) \geq 2$, we have

$$x_k(t) > x_{k+1}(t) \text{ on } (-\infty, t_1)_{\mathbb{T}} \cup (t_3, t_4)_{\mathbb{T}}$$

and

$$x_k(t) < x_{k+1}(t) \text{ on } (t_1, t_3)_{\mathbb{T}} \cup (t_4, \infty)_{\mathbb{T}}.$$

By (TSCP) and the cardinality of $(t_1, t_2)_{\mathbb{T}}$ and $(t_3, t_4)_{\mathbb{T}}$, there exist points $\tau_1 < t_1 < \tau_2 < t_2$ and $t_3 < \tau_3 < t_4 < \tau_4$ in $\mathbb{T}$ such that

$$x_k(\tau_i) \downarrow -\infty \ i = 1, 3, \quad x_k(\tau_i) \uparrow +\infty, \ i = 2, 4.$$

Recalling $t_2 < \sigma(t_2) = t_3$, we let $w(t)$ be the solution of the initial value problem for (5.1.8) such that

$$\begin{aligned} w(t_2) &= r_0, \\ w(\sigma(t_2)) &= w(t_3) = y_3, \\ w^{\Delta^2}(t_2) &= 0, \\ w^{\Delta^3}(t_2) &= 0. \end{aligned}$$

Then, for sufficient large K, $x_K(t) - w(t)$ has a zero at t_3 and GZ's in $(\tau_1, \tau_2]_{\mathbb{T}}$, $(\tau_2, t_2]_{\mathbb{T}}$, and $(\tau_3, \tau_4]_{\mathbb{T}}$, which contradicts (TSC). Thus, S_1 is closed. This concludes the Subcase (iii.2.1).

Subcase (iii.2.2) In this case, we will consider both $Card(t_3, t_4) = 1$ and $Card(t_3, t_4) = 0$. We have

$$x_k(t) > x_{k+1}(t) \text{ on } (-\infty, t_1)_{\mathbb{T}}$$

and

$$x_k(t) < x_{k+1}(t) \text{ on } (t_1, t_3)_{\mathbb{T}} \cup (t_4, \infty)_{\mathbb{T}}.$$

By (TSCP), there exist $\tau_1 < t_1 < \tau_2 < t_2$, and $t_4 < \tau_4$ such that

$$x_k(\tau_1) \downarrow -\infty, \quad x_k(\tau_i) \uparrow +\infty, \ i = 2, 4.$$

For the case where $Card(t_3, t_4) = 1$, let $w(t)$ be the solution of the initial value problem for (5.1.8) such that

$$
\begin{aligned}
w(t_2) &= r_0, \\
w(\sigma(t_2)) &= w(t_3) = y_3, \\
w(\sigma(t_3)) &= 0, \\
w(t_4) &= y_4.
\end{aligned}
$$

For the case where $Card(t_3, t_4) = 0$, let $w(t)$ be the solution of the initial value problem for (5.1.8) such that

$$
\begin{aligned}
w(t_2) &= r_0, \\
w(\sigma(t_2)) &= w(t_3) = y_3, \\
w(t_4) &= y_4, \\
w^{\Delta^3}(t_2) &= 0.
\end{aligned}
$$

In both cases, for large K, $x_K(t) - w(t)$ has two zero's at t_3 and t_4 and GZ's in $(\tau_1, \tau_2]_{\mathbb{T}}$ and $(\tau_2, t_2]_{\mathbb{T}}$, which contradicts (TSC). Thus, S_1 is closed. This concludes Subcase (iii.2.2). From Subcase (iii.2.i), $i = 1, 2$, we conclude that $S_1 = \mathbb{R}$. Choosing $y_2 \in S_1$, there is a solution $y(t)$ of (5.1.8) for which

$$y(t_1) = y_1, \ y(t_2) = y_2, \ y(t_3) = y_3, \ y(t_4) = y_4.$$

This concludes Subcase (iii.2).
    We conclude Case (iii).

    Case (iv): Assume that the $Card(t_1, t_2)_{\mathbb{T}} < 2$.
Pick $\xi_1 \in (-\infty, t_1)_{\mathbb{T}}$ such that $Card(\xi_1, t_1) \geq 2$. By Case (iii), there exists a solution $z(t)$ of (5.1.8) satisfying the boundary conditions,

$$z(\xi_1) = 0, \ z(t_2) = y_2, \ z(t_3) = y_3, \ z(t_4) = y_4.$$

Define

$$S_1 = \{r \in \mathbb{R} \mid \text{ there is a solution } x(t) \text{ of (5.1.8) with } x(t_1) = r,$$
$$x(t_2) = z(t_2) = y_2, x(t_3) = z(t_2) = y_3, x(t_4) = z(t_4) = y_4\}.$$

S_1 is nonempty and open. We wish to show that that S_1 is also closed. We assume the contrary. Again, there are a couple of subcases.

Subcase (iv.1): Assume $Card(t_1, t_2)_\mathbb{T} = 1$; that is, $(t_1, t_2)_\mathbb{T} = \{\sigma(t_1)\}$. Let $r_0 \in \overline{S} \setminus S_1$, $\{r_k\} \subset S_1$, with $r_k \uparrow r_0$, and $x_k(t)$ denotes the corresponding solution of (5.1.8) such that

$$x_k(t_1) = r_k, \quad x_k(t_2) = z(t_2), \quad x_k(t_3) = z(t_3), \quad x_k(t_4) = z(t_4).$$

There are further subcases to resolve. We will divide them into the following cases:

(iv.1.1) $Card(t_2, t_3)_\mathbb{T} \geq 2$,
(iv.1.2) $Card(t_2, t_3)_\mathbb{T} = 1$,
(iv.1.3) $Card(t_2, t_3)_\mathbb{T} = 0$.

Subcase (iv.1.1): In this case, $Card(t_2, t_3)_\mathbb{T} \geq 2$. We have

$$x_k(t) < x_{k+1}(t) \text{ on } (-\infty, t_2)_\mathbb{T}$$

and

$$x_k(t) > x_{k+1}(t) \text{ on } (t_2, t_3)_\mathbb{T} \cup (t_4, \infty)_\mathbb{T}.$$

By (TSCP) and $Card(t_2, t_3)_\mathbb{T}$, there exist points $\tau_1 < t_1$, $t_2 < \tau_2 < t_3$ and $t_4 < \tau_4$ in $\mathbb{T}$ such that

$$x_k(\tau_1) \uparrow +\infty, \quad x_k(\tau_i) \downarrow +\infty, \ i = 2, 4.$$

By Case (iii), we let $w(t)$ be the solution of the boundary problem for (5.1.8) such that

$$w(\xi_1) = 0,$$
$$w(t_1) = r_0,$$
$$w(t_2) = y_2,$$
$$w(t_3) = y_3.$$

Then, for sufficiently large K, $x_K(t) - w(t)$ has a zero at t_3, a GZ of order 2 at t_2 and a GZ in $(\tau_1, t_1]_\mathbb{T}$, which contradicts (TSC). Thus, S_1 is closed. This concludes Subcase (iv.1.1).

Subcase (iv.1.2): We now consider the case where $Card(t_2, t_3) = 1$. We have

$$x_k(t) < x_{k+1}(t) \text{ on } (-\infty, t_2)_\mathbb{T}$$

and

$$x_k(t) > x_{k+1}(t) \text{ on } (t_2, t_3)_{\mathbb{T}} \cup (t_4, \infty)_{\mathbb{T}}.$$

By (TSCP), there exist $\tau_1 < t_1$ and $t_4 < \tau_4$ such that

$$x_k(\tau_1) \uparrow +\infty \text{ and } x_k(\tau_4) \downarrow -\infty.$$

By (TSCP) and $Card(t_1, \sigma(t_2))_{\mathbb{T}}$, either

$$x_k(\sigma(t_1)) \uparrow +\infty \text{ or } x_k(\sigma(t_2)) \downarrow -\infty.$$

If $x_k(\sigma(t_1)) \uparrow +\infty$, by Case (iii), let $w(t)$ be the solution of the boundary value problem for (5.1.8) such that

$$\begin{aligned} w(\xi_1) &= 0, \\ w(t_1) &= r_0, \\ w(t_2) &= y_2, \\ w(t_3) &= y_3. \end{aligned}$$

For sufficiently large K, $x_K(t) - w(t)$ has two zero's at t_2 and t_3 and two GZ's in $(\tau_1, t_1]_{\mathbb{T}}$ and $(t_1, \sigma(t_1)]_{\mathbb{T}}$, which contradicts (TSC). Thus S_1 is closed. If $x_k(\sigma(t_2)) \downarrow -\infty$, use the same $w(t)$. For large K, $x_K(t) - w(t)$ has a zero at t_3, a GZ of order 2 at t_2 and a GZ in $(\tau_1, t_1]$ which contradicts (TSC). S_1 is closed. This concludes the Subcase (iv.1.2).

Subcase (iv.1.3): In this case, $Card(t_2, t_3)_{\mathbb{T}} = 0$. We have

$$x_k(t) < x_{k+1}(t) \text{ on } (-\infty, t_2)_{\mathbb{T}}$$

and

$$x_k(t) > x_{k+1}(t) \text{ on } (t_4, \infty)_{\mathbb{T}}.$$

By (TSCP), there exist $\tau_1 < t_1$ and $t_4 < \tau_4$. such that

$$x_k(\tau_1) \uparrow +\infty \text{ and } x_k(\tau_4) \downarrow -\infty.$$

By (TSCP) and $Card(t_1, t_3)_{\mathbb{T}}$, we have

$$x_k(\sigma(t_1)) \uparrow +\infty.$$

By Case (iii), let $w(t)$ be the solution of the boundary value problem for (5.1.8) such that

$$\begin{aligned} w(\xi_1) &= 0, \\ w(t_1) &= r_0, \\ w(t_2) &= y_2, \\ w(t_3) &= y_3. \end{aligned}$$

Then, for sufficient large K, $x_K(t) - w(t)$ has two zero's at t_2 and t_3 and two GZ's in $(\tau_1, t_1]_{\mathbb{T}}$ and $(t_1, \sigma(t_1)]_{\mathbb{T}}$ which contradicts (TSC). Thus, S_1 is closed. This concludes the Subcase (iv.1.3). From Subcase (iv.1.i) $i = 1, 2, 3$, we conclude that $S_1 = \mathbb{R}$. Choosing $y_1 \in S_1$, there is a solution $y(t)$ of (5.1.8) for which

$$y(t_1) = y_1, \ y(t_2) = y_2, \ y(t_3) = y_3, \ y(t_4) = y_4.$$

This concludes Subcase (iv.1).

Subcase (iv.2): Assume $Card(t_1, t_2)_{\mathbb{T}} = 0$, that is $t_1 < \sigma(t_1) = t_2$. Let $r_0 \in \overline{S} \setminus S_1$, $\{r_k\} \subset S_1$, with $r_k \uparrow r_0$, and $x_k(t)$ denotes the corresponding solution of (5.1.8) such that

$$x_k(t_1) = r_k, \ x_k(t_2) = z(t_2), \ x_k(t_3) = z(t_3), \ x_k(t_4) = z(t_4).$$

There are further subcases to resolve. We will divide them into the following cases:

(iv.2.1) $Card(t_2, t_3)_{\mathbb{T}} \geq 2$.
(iv.2.2) $Card(t_2, t_3)_{\mathbb{T}} = 1$.
(iv.2.3) $Card(t_2, t_3)_{\mathbb{T}} = 0$.

Subcase (iv.2.1): In this case, $Card(t_2, t_3)_{\mathbb{T}} \geq 2$. We have

$$x_k(t) < x_{k+1}(t) \text{ on } (-\infty, t_1)_{\mathbb{T}}$$

and

$$x_k(t) > x_{k+1}(t) \text{ on } (t_2, t_3)_{\mathbb{T}} \cup (t_4, \infty)_{\mathbb{T}}.$$

By (TSCP) and $Card(t_2, t_3)_{\mathbb{T}}$, there exist points $\tau_1 < t_1$, $t_2 < \tau_2 < t_3$ and $t_4 < \tau_4$ in $\mathbb{T}$ such that

$$x_k(\tau_1) \uparrow +\infty, \ x_k(\tau_i) \downarrow -\infty, \ i = 2, 4.$$

By Case (iii), we let $w(t)$ be the solution of the boundary problem for (5.1.8) such that

$$\begin{aligned} w(\xi_1) &= 0, \\ w(t_1) &= r_0, \\ w(t_2) &= y_2, \\ w(t_3) &= y_3. \end{aligned}$$

Then, for sufficiently large K, $x_K(t) - w(t)$ has a zero at t_3, a GZ of order 2 at t_2 and a GZ in $(\tau_1, t_1]_{\mathbb{T}}$, which contradicts (TSC). Thus, S_1 is closed. This concludes Subcase (iv.2.1).

Subcase (iv.2.2): We now consider the case where $Card(t_2, t_3) = 1$. We have

$$x_k(t) < x_{k+1}(t) \text{ on } (-\infty, t_1)_\mathbb{T}$$

and

$$x_k(t) > x_{k+1}(t) \text{ on } (t_2, t_3)_\mathbb{T} \cup (t_4, \infty)_\mathbb{T}.$$

By (TSCP), there exist $\tau_1 < t_1$ and $t_4 < \tau_4$ such that

$$x_k(\tau_1) \uparrow +\infty \text{ and } x_k(\tau_4) \downarrow -\infty.$$

By (TSCP) and $Card(t_1, t_3)_\mathbb{T}$, $x_k(\sigma(t_2)) \downarrow -\infty$. By Case (iii), let $w(t)$ be the solution of the boundary value problem for (5.1.8) such that

$$
\begin{aligned}
w(\xi_1) &= 0, \\
w(t_1) &= r_0, \\
w(t_2) &= y_2, \\
w(t_3) &= y_3.
\end{aligned}
$$

For large K, $x_K(t) - w(t)$ has a zero at t_3, a GZ of order 2 at t_2 and a GZ in $(\tau_1, t_1]_\mathbb{T}$, which contradicts (TSC). Thus S_1 is closed. This concludes the Subcase (iv.2.2).

Subcase (iv.2.3): In this case, $Card(t_2, t_3) = 0$. We have

$$x_k(t) < x_{k+1}(t) \text{ on } (-\infty, t_1)_\mathbb{T}$$

and

$$x_k(t) > x_{k+1}(t) \text{ on } (t_4, \infty)_\mathbb{T}.$$

By (TSCP), there exist $\tau_1 < t_1$ and $t_4 < \tau_4$ such that

$$x_k(\tau_1) \uparrow +\infty \text{ and } x_k(\tau_4) \downarrow -\infty.$$

We now have three more further subcases:

- $Card(t_3, t_4)_\mathbb{T} \geq 2$.
- $Card(t_3, t_4)_\mathbb{T} = 1$.
- $Card(t_3, t_4)_\mathbb{T} = 0$.

For the case that $Card(t_3, t_4)_\mathbb{T} \geq 2$, there exists a $t_3 < \tau_3 < t_4$ such that $x_k(\tau_3) \uparrow +\infty$. Let $w(t)$ be the solution of the initial value problem of (5.1.8) such that

$$
\begin{aligned}
w(t_1) &= r_0, \\
w(t_2) &= y_2, \\
w(t_3) &= y_3, \\
w^{\Delta^3}(t_1) &= 0.
\end{aligned}
$$

For sufficiently large K, $x_K(t) - w(t)$ has two zero's at t_2 and t_3 and two GZ's in $(\tau_1, t_1)_\mathbb{T}$ and $(\tau_3, \tau_4)_\mathbb{T}$, which contradicts (TSC). S_1 is closed.

For the case $Card(t_3, t_4)_\mathbb{T} = 1$, by (TSCP) and the $Card(t_1, \sigma(t_3))_\mathbb{T}$, $x_k(\sigma(t_3)) \uparrow +\infty$. Using the same $w(t)$ as in the previous case, for sufficiently large K, $x_K(t) - w(t)$ has two zero's at t_2 and t_3 and two GZ's in $(\tau_1, t_1)_\mathbb{T}$ and $(\sigma(t_3), \tau_4)_\mathbb{T}$, which contradicts (TSC). S_1 is closed.

The last case we will discuss is for $Card(t_3, t_4)_\mathbb{T} = 0$. Let $w(t)$ be the solution of the initial value problem of (5.1.8) such that

$$w(t_1) = r_0,$$
$$w(t_2) = y_2,$$
$$w(t_3) = y_3,$$
$$w(t_4) = y_4.$$

For sufficiently large K, $x_K(t) - w(t)$ has three zero's t_2, t_3, and t_4 and a GZ in $(\tau_1, t_1)_\mathbb{T}$, which contradicts (TSC). This completes Subcase (iv.2.3). From Subcase (iv.2.i) $i = 1, 2, 3$, we conclude that $S_1 = \mathbb{R}$. Choosing $y_1 \in S_1$, there is a solution $y(t)$ of (5.1.8) for which

$$y(t_1) = y_1, \; y(t_2) = y_2, \; y(t_3) = y_3, \; y(t_4) = y_4.$$

This concludes Subcase (iv.2), which concludes Case (iv). The proof is complete. $\qquad\square$

We now show that under the same hypotheses as Theorem 5.1.6, there exists a unique solution of (5.1.8) satisfying each of the 3-point conjugate boundary value problems (that is, for the boundary conditions (5.1.10), (5.1.11), (5.1.12), respectively), and there exists a unique solution of (5.1.8) satisfying each of the 2-point conjugate boundary value problems (that is, for the boundary conditions (5.1.13), (5.1.14), (5.1.15), respectively).

Theorem 5.1.7. [[73], Henderson and Yin, Thm. 3.1] *Assume that with respect to (5.1.8), conditions* (TSA), (TSB), (TSC) *and* (TSCP) *are satisfied. Then, given $k \in \{2, 3\}$, each k-point conjugate boundary value problem for (5.1.8) has a unique solution on $\mathbb{T}$.*

Proof. Let the boundary values $y_1, y_2, y_3, y_4 \in \mathbb{R}$ be given throughout. The proof will involve several cases depending on the nature of the boundary conditions according to number of boundary points involved.

Case (i): For $t_1 < t_2 < t_3$ in $\mathbb{T}$, we will focus first on the 3-point problems for (5.1.8).

Subcase (i.1): Initially, we will deal with solutions satisfying (5.1.10), that is,

$$y(t_1) = y_1, \quad y^\Delta(t_1) = y_2, \quad y(t_2) = y_3, \quad y(t_3) = y_4,$$

or with solutions satisfying (5.1.11), that is,

$$y(t_1) = y_1, \quad y(t_2) = y_2, \quad y(t_3) = y_3, \quad y^\Delta(t_3) = y_4.$$

Actually, we will now give the details concerning the existence of solutions for only (5.1.8)-(5.1.10). Let $z(t)$ be the solution of (5.1.8) satisfying the initial conditions

$$z(t_1) = y_1, \ z^\Delta(t_1) = y_2, \ z^{\Delta^2}(t_1) = 0, z^{\Delta^3}(t_1) = 0.$$

Define

$$S_1 = \{r \in \mathbb{R} \mid \text{ there is a solution } x(t) \text{ of (5.1.8) with } x(t_1) = z(t_1),$$
$$x^\Delta(t_1) = z^\Delta(t_1), \ x(t_2) = z(t_2), \ x(t_3) = r\}.$$

Since $z(t_3) \in S_1$, $S_1 \neq \emptyset$. By Theorem 5.1.2, S_1 is open. We want to show that S_1 is closed.

For contradiction purposes, assume that S_1 is not closed. Then there exist $r_0 \in \overline{S}_1 \setminus S_1$ and a strictly monotone sequence $\{r_k\} \subset S_1$ such that $\lim_{k \to \infty} r_k = r_0$. We may assume, without loss of generality, that $r_k \uparrow r_0$. For each $k \in \mathbb{N}$, let $x_k(t)$ be the solution of (5.1.8) with

$$x_k(t_1) = z(t_1), \ x_k^\Delta(t_1) = z^\Delta(t_1), \ x_k(t_2) = z(t_2), \ x_k(t_3) = r_k.$$

It follows from (TSC) that, for each k, $x_k^{\Delta^2}(t_1) \neq x_{k+1}^{\Delta^2}(t_1)$ and $x_k^\Delta(t_2) \neq x_{k+1}^\Delta(t_2)$. Since $r_{k+1} > r_k$, we have from (TSC) that for $k \in \mathbb{N}$,

$$x_k(t) < x_{k+1}(t) \text{ on } (t_3, \infty)_{\mathbb{T}}.$$

By the "compactness condition" (TSCP) and the cardinality of $(t_3, \infty)_{\mathbb{T}}$, there exists at least one point $\tau_1 \in (t_3, \infty)_{\mathbb{T}}$ such that

$$x_k(\tau_1) \uparrow +\infty.$$

Now, $Card(t_1, t_2)_{\mathbb{T}} + Card(t_2, t_3)_{\mathbb{T}} \geq 1$; otherwise, this is an initial value problem. Again, by the "compactness condition" (TSCP), there exists a point $\tau_2 \in (t_1, t_2)_{\mathbb{T}} \cup (t_2, t_3)_{\mathbb{T}}$ such that either

$$x_k(\tau_2) \downarrow -\infty \text{ if } \tau_2 \in (t_1, t_2)_{\mathbb{T}} \text{ or } x_k(\tau_2) \uparrow +\infty \text{ if } \tau_2 \in (t_2, t_3)_{\mathbb{T}}.$$

Now, from Theorem 5.1.6, let $w(t)$ be the solution of (5.1.8) satisfying the boundary conditions

$$w(t_1) = y_1, \ w(t_2) = z(t_2), \ w(t_3) = r_0, \ w(\xi) = 0$$

for some fixed $\xi \in (t_3, \infty)_\mathbb{T}$ such that $Card(t_3, \xi)_\mathbb{T} \geq 2$. Since $w(t_3) = r_0 > r_k = x_k(t_3)$ for all $k \in \mathbb{N}$, it follows that, for sufficiently large K, $x_K(t) - w(t)$ has GZ's at t_1, t_2 and in $(t_3, \tau_1]_\mathbb{T}$. Also, $x_K(t) - w(t)$ has a GZ in either $(t_1, t_2)_\mathbb{T}$ or $(t_2, t_3)_\mathbb{T}$. This is a contradiction to (TSC). Therefore, S_1 is closed. Consequently, $S_1 = \mathbb{R}$. Choosing $y_4 \in S_1$, we have a solution $\overline{x}(t)$ of (5.1.8) satisfying

$$\overline{x}(t_1) = z(t_1), \ \overline{x}^\Delta(t_1) = z^\Delta(t_1), \ \overline{x}(t_2) = z(t_2), \ \overline{x}(t_3) = y_4.$$

Next, define

$$S_2 = \{r \in \mathbb{R} \mid \text{there is a solution } x(t) \text{ of (5.1.8) with } x(t_1) = z(t_1),$$
$$x^\Delta(t_1) = z^\Delta(t_1), \ x(t_2) = r, \ x(t_3) = y_4\}.$$

For the solution just produced, $\overline{x}(t_2) \in S_2$ so $S_2 \neq \emptyset$. Again, by Theorem 5.1.2, S_2 is an open subset of $\mathbb{R}$.

We now claim that S_2 is also closed. Assuming S_2 is not closed, there exist $r_0 \in \overline{S}_2 \setminus S_2$ and a strictly monotone sequence $\{r_k\} \subset S_2$ such that $\lim_{k \to \infty} r_k = r_0$. We may assume again that $r_k \uparrow r_0$, and as before, let $x_k(t)$ denote the corresponding solution (5.1.8) with

$$x_k(t_1) = z(t_1), \ x_k^\Delta(t_1) = z^\Delta(t_1), \ x_k(t_2) = r_k, \ x_k(t_3) = y_4.$$

Again, by (TSC), for $k \in \mathbb{N}$, $x_k^{\Delta^2}(t_1) \neq x_{k+1}^{\Delta^2}(t_1)$ and $x_k^\Delta(t_3) \neq x_{k+1}^\Delta(t_3)$. Using $r_{k+1} > r_k$, (TSC) and $Card(t_1, t_2)_\mathbb{T} \geq 2$ (or this is an "2-point" problem which we will discuss later), we have that for $k \in \mathbb{N}$,

$$x_k(t) < x_{k+1}(t) \text{ on } (t_1, t_2)_\mathbb{T}$$

and

$$x_k(t) > x_{k+1}(t) \text{ on } (t_3, \infty)_\mathbb{T}.$$

By (TSCP) and the cardinality of $(t_1, t_2)_\mathbb{T}$, there exist $t_1 < \tau_1 < t_2$ and $t_3 < \tau_2$ in $\mathbb{T}$ such that

$$x_k(\tau_1) \uparrow +\infty \text{ and } x_k(\tau_2) \downarrow -\infty.$$

This time, from Theorem 5.1.6, let $w(t)$ be the solution of (5.1.8) satisfying the boundary conditions,

$$w(t_1) = z(t_1), \ w(t_2) = r_0, \ w(t_3) = z(t_3) = y_4, w(\xi) = 0,$$

for some fixed $\xi \in (t_3, \infty)_\mathbb{T}$ such that $Card(t_3, \xi)_\mathbb{T} \geq 2$. Since $w(t_2) = r_0 > r_k = x_k(t_2)$, for all k, it follows that, for sufficiently large K, $x_K(t) - w(t)$ has GZ's at t_1 and t_2 and in $(t_3, \tau_2)_\mathbb{T}$. Since the $Card(t_1, t_2)_\mathbb{T} \geq 2$, either $x_K(t) - w(t)$ has a GZ in $(t_1, t_2)_\mathbb{T}$, or t_1 is a GZ of order 2. Assumption (TSC) implies that $w(t) = x_K(t)$ on $\mathbb{T}$ which is contradiction. Therefore, S_2 is also closed. If we choose $y_3 \in S_2$, we have a solution $y(t)$ of (5.1.8) satisfying

$$y(t_1) = y_1, \ y^\Delta(t_1) = y_2, \ y(t_2) = y_3, \ y(t_3) = y_4.$$

Analogous arguments yield the existence of solutions for the 3-point problem (5.1.8)-(5.1.11).

This concludes the Subcase (i.1).

Subcase (i.2): In this subcase, we consider the existence of a solution of (5.1.8) satisfying Eq. (5.1.12); that is, satisfying the boundary conditions

$$y(t_1) = y_1, \ y(t_2) = y_2, \ y^\Delta(t_2) = y_3, \ y(t_3) = y_4.$$

By Subcase (i.1), there is a unique solution $z(t)$ of (5.1.8) satisfying the boundary conditions,

$$z(\tau) = 0, \ z(t_1) = y_1, \ z(t_2) = y_2, \ z^\Delta(t_2) = y_3,$$

for a fixed $\tau \in (-\infty, t_1)_\mathbb{T}$ and the $Card(\tau, t_1)_\mathbb{T} \geq 2$. Define

$$S = \{r \in \mathbb{R} \mid \text{there is a solution } x(t) \text{ of (5.1.8) with } x(t_1) = z(t_1) = y_1,$$
$$x(t_2) = z(t_2) = y_2, \ x^\Delta(t_2) = z^\Delta(t_2) = y_3, \ x(t_3) = r\}.$$

S is nonempty and open.

Again, we wish to show that S is closed. And, again, we assume the contrary. Then, there exist an $r_0 \in \overline{S} \setminus S$ and a strictly monotone sequence $\{r_k\} \subset S$ such that $\lim_{k \to \infty} r_k = r_0$. We may assume, without loss of generality, that $r_k \uparrow r_0$. For each $k \in \mathbb{N}$, let $x_k(t)$ be the solution of (5.1.8) with

$$x_k(t_1) = z(t_1), \ x_k(t_2) = z(t_2), \ x_k^\Delta(t_2) = z^\Delta(t_2), \ x_k(t_3) = r_k.$$

It follows from (TSC), that for each k, $x_k^\Delta(t_1) \neq x_{k+1}^\Delta(t_1)$ and $x_k^{\Delta^2}(t_2) \neq x_{k+1}^{\Delta^2}(t_2)$. Since $r_{k+1} > r_k$ and $Card(t_2, t_3)_\mathbb{T} \geq 2$ (or this is a "2-point" problem which we will discuss later), we have from (TSC) that for $k \in \mathbb{N}$, there exist $\tau_1, \tau_2 \in \mathbb{T}$ such that $t_2 < \tau_1 < t_3 < \tau_2$ and

$$x_k(\tau_1) \uparrow +\infty \text{ and } x_k(\tau_2) \uparrow +\infty.$$

We know that $Card(t_1, t_2)_\mathbb{T} \geq 1$ (or this is a "2-two-point problem), and so let $\xi \in \mathbb{T}$ and $t_1 < \xi < t_2$. Again, from Theorem 5.1.6, let $w(t)$ be the solution of (5.1.8) satisfying the boundary conditions

$$w(t_1) = z(t_1), w(\xi) = z(\xi), \ w(t_2) = z(t_2), \ w(t_3) = r_0.$$

Since $w(t_3) = r_0 > r_k = x_k(t_3)$ for all $k \in \mathbb{N}$, it follows that for sufficiently large K, $x_K(t) - w(t)$ has GZ's at t_1, t_2 and in $(t_3, \tau_2)_\mathbb{T}$. Either t_1 is a double GZ or there is a GZ in $(t_1, t_2)_\mathbb{T}$. This is a contradiction to (TSC). Therefore S is closed. Consequently, $S = \mathbb{R}$. Choosing $r = y_4$, we have a solution $y(t)$ of (5.1.8) satisfying

$$y(t_1) = z(t_1) = y_1, \ y(t_2) = z(t_2) = y_2, \ y^\Delta(t_2) = z^\Delta(t_2) = y_3, \ y(t_3) = y_4.$$

This concludes Subcase (i.2), and Case (i) is also concluded.

Case (ii): For $t_1 < t_2$ in $\mathbb{T}$, we will focus now on the 2-point problems for (5.1.8).

Subcase (ii.1): Initially, we will deal with solutions satisfying (5.1.13), that is,

$$y(t_1) = y_1, \ y^\Delta(t_1) = y_2, \ y^{\Delta^2}(t_1) = y_3, \ y(t_2) = y_4,$$

or with solutions satisfying (5.1.14), that is,

$$y(t_1) = y_1, \ y(t_2) = y_2, \ y^\Delta(t_2) = y_3, \ y^{\Delta^2}(t_2) = y_4.$$

Actually, we will now give the details concerning the existence of solutions for only (5.1.8)-(5.1.13). Let $z(t)$ be the solution of (5.1.8) satisfying the initial conditions

$$z(t_1) = y_1, \ z^\Delta(t_1) = y_2, \ y^{\Delta^2}(t_1) = y_3, \ z^{\Delta^3}(t_1) = 0.$$

Define

$$S = \{ r \in \mathbb{R} \mid \text{ there is a solution of (5.1.8) with } x(t_1) = z(t_1),$$
$$x^\Delta(t_1) = z^\Delta(t_1), \ x^{\Delta^2}(t_1) = z^{\Delta^2}(t_1), \ x(t_2) = r \}.$$

Since $z(t_2) \in S$, $S \neq \emptyset$, and again, S is open.

We wish to show that S is also closed. For contradiction purposes, assume that S is not closed. Then there exist an $r_0 \in \overline{S} \setminus S$ and a strictly monotone sequence $\{r_k\} \subset S$ such that $\lim_{k \to \infty} r_k = r_0$. We may assume, without loss of generality, that $r_k \uparrow r_0$. For each $k \in \mathbb{N}$, let $x_k(t)$ be the solution of (5.1.8) with

$$x_k(t_1) = z(t_1), \ x_k^\Delta(t_1) = z^\Delta(t_1), \ x_k^{\Delta^2}(t_1) = x_k^{\Delta^2}(t_1), \ x_k(t_2) = r_k.$$

It follows form (TSC) that, for each k, $x_k^{\Delta^3}(t_1) \neq x_{k+1}^{\Delta^3}(t_1)$. Since $r_{k+1} > r_k$, and there are at least three points from $\mathbb{T}$ in $(t_1, t_2)_\mathbb{T}$ (otherwise, this is an initial value problem), we have from (TSC) that for $k \in \mathbb{N}$,

$$x_k(t) < x_{k+1}(t) \text{ on } (t_1, t_2)_\mathbb{T}.$$

By the "compactness condition" (TSCP) and the cardinality of $(t_1, t_2)_\mathbb{T}$, there exist points $\tau_1 \in (t_1, t_2)_\mathbb{T}$ and $\tau_2 \in (t_2, \infty)_\mathbb{T}$ such that

$$x_k(\tau_1) \uparrow +\infty \text{ and } x_k(\tau_2) \uparrow +\infty.$$

Now, from the previous results in Case (i), let $w(t)$ be the solution of (5.1.8) satisfying the boundary conditions

$$w(t_1) = y_1, \ w^\Delta(t_1) = y_2, \ w(t_2) = r_0, \ w(\xi) = 0,$$

for some fixed $\xi \in (t_3, \infty)_\mathbb{T}$ such that $Card(t_3, \xi)_\mathbb{T} \geq 2$. Since $w(t_2) = r_0 > r_k = x_k(t_2)$ for all $k \in \mathbb{N}$, it follows that, for sufficiently large K, $x_K(t) - w(t)$ has GZ's at t_1 of order 2, a GZ in $(t_1, \tau_1]_\mathbb{T}$ and a GZ in $(t_2, \tau_2]_\mathbb{T}$. This is a contradiction to (TSC). Therefore, S is closed. Consequently, $S = \mathbb{R}$. Choosing $y_4 \in S$, we have a solution $y(t)$ of (5.1.8) satisfying

$$y(t_1) = y_1, \ y^\Delta(t_1) = y_2, \ y^{\mathcal{D}^2}(t_1) = y_3, \ y(t_2) = y_4.$$

Analogous arguments yield the existence of solutions for the 3-point problem (5.1.8)-(5.1.14).

This concludes Subcase (ii.1).

Subcase (ii.2): In this subcase, we consider the existence of a solution of (5.1.8) satisfying (5.1.15); that is satisfying the boundary conditions,

$$y(t_1) = y_1, \ y^\Delta(t_1) = y_2, \ y(t_2) = y_3, \ y^\Delta(t_2) = y_4.$$

First, we note that $Card(t_1, t_2)_\mathbb{T} \geq 2$ (otherwise the problem is an initial value problem). Using the result from Subcase (i.1), let $z(t)$ be the solution of (5.1.8) satisfying the boundary conditions

$$z(t_1) = y_1, \ z^\Delta(t_1) = y_2, \ z(t_2) = y_3, \ z(\tau)) = 0,$$

for fixed $\tau \in (t_2, \infty)_\mathbb{T}$ and $Card(t_2, \tau) > 2$. Define

$$S = \{r \in S \mid \text{ there is a solution of (5.1.8) with } x(t_1) = z(t_1),$$
$$x^\Delta(t_1) = z^\Delta(t_1), \ x(t_2) = z(t_2), \ x^\Delta(t_2) = r\}.$$

Since $z^\Delta(t_2) \in S$, $S \neq \emptyset$, and S is also open.

We now show that S is closed. For contradiction purposes, assume that S is not closed. Then there exist $r_0 \in \overline{S} \setminus S$ and a strictly monotone sequence $\{r_k\} \subset S$ such that $\lim_{k \to \infty} r_k = r_0$. We may assume, without loss of generality, that $r_k \uparrow r_0$. For each $k \in \mathbb{N}$, let $x_k(t)$ be the solution of (5.1.8) with

$$x_k(t_1) = z(t_1), \ x_k^\Delta(t_1) = z^\Delta(t_1), \ x(t_2) = z(t_2), \ x_k^\Delta(t_2) = r_k.$$

It follows from (TSC) that, for each k, $x_k^{\Delta^2}(t_1) \neq x_{k+1}^{\Delta^2}(t_1)$. Since $r_{k+1} > r_k$, we have from (TSC) that for $k \in \mathbb{N}$,

$$x_k(t) > x_{k+1}(t) \text{ on } (t_1, t_2)_{\mathbb{T}} \text{ and } x_k(t) < x_{k+1}(t) \text{ on } (t_2, \infty)_{\mathbb{T}}.$$

By (TSCP), there exist $\tau_1 \in (t_1, t_2)_{\mathbb{T}}$ and $\tau_2 \in (t_2, \infty)_{\mathbb{T}}$ such that

$$x_k(\tau_1) \downarrow -\infty \text{ and } x_k(\tau_2) \uparrow +\infty.$$

Now, using the results from Case (i), let $w(t)$ be the solution of (5.1.8) satisfying the boundary conditions

$$w(\xi) = 0, \ w(t_1) = y_1, \ w(t_2) = y_2, \ w^\Delta(t_2) = r_0,$$

for some fixed $\xi \in (-\infty, t_1)_{\mathbb{T}}$ such that $Card(\xi, t_1)_{\mathbb{T}} \geq 2$. Since $w^\Delta(t_2) = r_0 > r_k = x_k^\Delta(t_2)$ for all $k \in \mathbb{N}$, it follows that, for sufficiently large K, $x_K(t) - w(t)$ has GZ's at t_1 and t_2, and GZ's in $(t_1, \tau_1]_{\mathbb{T}}$ and $(t_2, \tau_2]_{\mathbb{T}}$. This is a contradiction to (TSC). Therefore, S is closed. Consequently, $S = \mathbb{R}$. Choosing $y_4 \in S$, we have a solution $y(t)$ of (5.1.8) satisfying

$$y(t_1) = y_1, \ y^\Delta(t_1) = y_2, \ y(t_2) = y_3, \ y^\Delta(t_2) = y_4.$$

This concludes Subcase (ii.2), which concludes Case (ii). The proof is complete. □

5.2 Right focal boundary value problems: uniqueness implies existence

In this section, we will concentrate on uniqueness of solutions implying their existence for right focal type boundary value problems for (5.0.3) of orders $n = 2, 3$. These results are due to Harris, Henderson, Lanz and Yin [39, 40].

Definition 5.2.1. *Given $t_1 \leq \cdots \leq t_n$ in $\mathbb{T}$, such that for at least one index $j \in \{1, \ldots, n-1\}$, $\sigma(t_j) < t_{j+1}$, boundary conditions of the form*

$$y^{\Delta^{i-1}}(t_i) = y_i, \quad 1 \leq i \leq n, \tag{5.2.1}$$

where $y_i \in \mathbb{R}$, $1 \leq i \leq n$, are called "right focal boundary conditions." The boundary value problem, (5.0.3)-(5.2.1) is called a "right focal boundary value problem on $\mathbb{T}$."

In particular, when $n = 2$, we are concerned with uniqueness of solutions implying their existence for the second order dynamic equation (5.1.2) (that is, $y^{\Delta\Delta}(t) = f(t, y(t), y^\Delta(t)))$, satisfying the right focal boundary conditions

$$y(t_1) = y_1, \quad y^\Delta(t_2) = y_2, \tag{5.2.2}$$

where $t_1 < t_2$ in $\mathbb{T}$, and $y_1, y_2 \in \mathbb{R}$.

And when $n = 3$, we are concerned with uniqueness of solutions implying their existence for the third order dynamic equation (5.1.4) (that is, $y^{\Delta^3}(t) = f(t, y(t), y^\Delta(t), y^{\Delta\Delta}(t)))$, satisfying the 2-point right focal boundary conditions, either given by

$$y(t_1) = y_1, \quad y^\Delta(t_2) = y_2, \quad y^{\Delta^2}(t_2) = y_3, \tag{5.2.3}$$

where $t_1 < t_2$ in $\mathbb{T}$, and $y_1, y_2, y_3 \in \mathbb{R}$, or given by

$$y(t_1) = y_1, \quad y^\Delta(t_1) = y_2, \quad y^{\Delta^2}(t_2) = y_3, \tag{5.2.4}$$

where $t_1 < t_2$ in $\mathbb{T}$, and $y_1, y_2, y_3 \in \mathbb{R}$, or satisfying the 3-point right focal boundary conditions given by

$$y(t_1) = y_1, \quad y^\Delta(t_2) = y_2, \quad y^{\Delta^2}(t_3) = y_3, \tag{5.2.5}$$

where $t_1 < t_2 < t_3$ in $\mathbb{T}$, and $y_1, y_2, y_3 \in \mathbb{R}$.

We will use extensively conditions (TSA), (TSB) and (TSCP) of Section 5.1, as well as a uniqueness condition on solutions of right focal boundary value problems, when such solutions exist.

(TSFC) For any $t_1 \leq \cdots \leq t_n$ in $\mathbb{T}$, such that for at least one index $j \in \{2, \ldots, n-1\}$, $\sigma(t_j) < t_{j+1}$, if $y(t)$ and $z(t)$ are solutions of (5.1.8) such that $(y(t) - z(t))^{\Delta^{i-1}}$ has a GZ at t_i, $i = 1, \ldots, n$, then $y(t) = z(t)$ on $\mathbb{T}$.

Because of uniqueness implies existence results for conjugate boundary value problems proven in the previous section, Section 5.1, we will also make application of those results for right focal problems primarily through a time scale Rolle's Theorem due to Bohner and Eloe [9].

Theorem 5.2.1. [[9], Bohner and Eloe] *If y has at least k GZ's on an interval $[a, b]_\mathbb{T}$, counting multiplicities, then y^Δ has at least $k - 1$ GZ's on $[a, b]_\mathbb{T}$, counting multiplicities.*

Remark 5.2.1. *It follows from Theorem 5.2.1 that condition (TSFC) implies condition (TSC) of Section 5.1.*

In addition, in the presence of uniqueness of solutions of the right focal boundary problems, we have the usual continuous dependence of solutions on right focal boundary conditions.

Theorem 5.2.2. *Assume that with respect to* (5.0.3), *conditions* (TSA), (TSB) *and* (TSFC) *are satisfied. Given a solution* $y(t)$ *of* (5.0.3) *on* $\mathbb{T}$, *an interval* $[a,b]_{\mathbb{T}}$, *points* $t_1 \leq \cdots \leq t_n$ *in* $[a,b]_{\mathbb{T}}$ *such that some* $\sigma(t_j) < t_{j+1}$, *and an* $\epsilon > 0$, *there exists a* $\delta(\epsilon, [a,b]_{\mathbb{T}}) > 0$ *such that, if* $|y^{\Delta^{i-1}}(t_i) - z_i| < \delta$, $i = 1, \ldots, n$, *then there exists a solution* $z(t)$ *of* (5.0.3) *satisfying* $z^{\Delta^{i-1}}(t_i) = z_i$, $i = 1, \ldots, n$, *and* $|y^{\Delta^j}(t) - z^{\Delta^j}(t)| < \epsilon$, *on* $[a,b]_{\mathbb{T}}$, $j = 0, \ldots, n-1$.

Our first uniqueness implies existence result for this section is for solutions of (5.1.2)-(5.2.2).

Theorem 5.2.3. [[39], Harris, Henderson, Lanz and Yin, Thm. 5] *Assume that with respect to* (5.1.2) *conditions* (TSA), (TSB), (TSFC) *and* (TSCP) *are satisfied. Then, given any* $t_1 < t_2$ *in* $\mathbb{T}$ *and* $y_1, y_2 \in \mathbb{R}$, *the boundary value problem* (5.1.2)-(5.2.2) *has a unique solution.*

Proof. We remark that the uniqueness of solutions is provided by condition (TSFC). So, let $t_1, t_2 \in \mathbb{T}$ and $y_1, y_2 \in \mathbb{R}$ be as stated.

We begin the proof by letting $z(\cdot\,; \xi, y_2)$ be the unique solution to the initial value problem (5.1.2) satisfying

$$z(t_2) = \xi, \qquad z^{\Delta}(t_2) = y_2. \qquad (5.2.6)$$

By condition (TSB), $z(\cdot\,; \xi, y_2)$ exists for all t in $\mathbb{T}$. The theorem will be proven if we can show that ϕ_{y_2} is surjective, i.e., Range $\phi_{y_2} = \mathbb{R}$.

First, it follows from the continuous dependence of solutions on initial conditions, as a consequence of (TSB), that ϕ_{y_2} is continuous.

Moreover, if $\phi_{y_2}(\xi) = \phi_{y_2}(\delta)$, then we have $z(t_1; y_2, \xi) = z(t_1; y_2, \delta)$ and $z^{\Delta}(t_2; y_2, \xi) - z^{\Delta}(t_2; y_2, \delta)$. By the uniqueness assumption (TSFC), $z(t; y_2, \xi) = z(t; y_2, \delta)$, for all $t \in \mathbb{T}$. As a consequences, $\xi = z(t_2; y_2, \xi) = z(t_2; y_2, \delta) = \delta$. Therefore, $\xi = \delta$ and ϕ_{y_2} is an injection.

Because ϕ_{y_2} is continuous and one-to-one, it follows that ϕ_{y_2} is a strictly monotone map with $\phi_{y_2}(\mathbb{R}) = (a,b)$ for some $a, b \in \mathbb{R} \cup \{\pm\infty\}$. The proof will be finished if we can argue that $a = -\infty$ and $b = +\infty$. We will show here that $b = +\infty$ as for the case that $a = -\infty$ is similar.

Suppose $b < \infty$. Then let $\{r_k\} \subset (a,b)$ be a strictly increasing sequence with $\lim r_k = b$, and let ξ_k be such that $\phi_{y_2}(\xi_k) = r_k$. Without loss of generality, we may assume that ϕ_{y_2} is increasing, as the argument for decreasing ϕ_{y_2} is similar. Then, $\xi_k < \xi_{k+1}$ for each $k \in \mathbb{N}$, and $\lim \xi_k = +\infty$.

Let z_k denote the solution of the initial value problem (5.1.2)-(5.2.6) with $\xi = \xi_k$. Condition (TSFC) implies that $z_k(t) < z_{k+1}(t)$ whenever $t \leq t_2$.

Now, let $t_0 \in \mathbb{T}$ be such that $t_0 < t_1$. If it is the case that the solutions $\{z_k\}$ are uniformly bounded on the time scale interval $[t_0, t_1]_{\mathbb{T}}$, then condition (TSCP) implies that the sequences $\{z_k\}$ and $\{z_k^\Delta\}$ converge uniformly on $[t_0, t_1]_{\mathbb{T}}$ to a solution $\tilde{z}$ of the dynamic equation (5.1.2). Evidently, $z_k(t_1)$ converges to $\tilde{z}(t_1) = b$ and $z_k^\Delta(t_1)$ converges to $\tilde{z}^\Delta(t_1)$ as $k \to \infty$. Since solutions of the dynamic equation (5.1.2) depend continuously on initial data, then it must follow that $z_k(t_2)$ converges to $\tilde{z}(t_2)$. However, this contradicts the observation that $z_k(t_2) = \xi_k \to \infty$ as $k \to \infty$. Hence, the increasing sequence $\{z_k\}$ must be unbounded on $[t_0, t_1)_{\mathbb{T}}$ and there is a point $\tau \in [t_0, t_1)_{\mathbb{T}}$ such that $z_k(\tau) \uparrow \infty$, as $k \to \infty$. And, by similar argument, we can also obtain that $\{z_k\}$ is unbounded on $(t_1, t_2]_T$.

Next, let w be the solution of the initial value problem for (5.1.2) satisfying

$$w(t_1) = b, \qquad w^\Delta(t_1) = 0.$$

Then, for K sufficiently large, $z_K - w$ must have GZ's in $(t_0, t_1]_{\mathbb{T}}$ and $(t_1, t_2]_{\mathbb{T}}$. If $z_K - w$ has generalized zeros at $\tau_0 \in (t_0, t_1]_{\mathbb{T}}$ and $\tau_1 \in (t_1, t_2]_{\mathbb{T}}$, then by Rolle's Theorem on time scales, $(z_K - w)^\Delta$ has a generalized zero in $(\tau_0, \tau_1)_{\mathbb{T}}$. In this case, condition (TSFC) implies that $z_K = w$ everywhere in $\mathbb{T}$, whenever K is sufficiently large, a contradiction. Hence, $b = +\infty$.

Thus, we showed that Range $\phi_{y_2} = \mathbb{R}$. And so, there exists some $\alpha \in \mathbb{R}$ such that $\phi_{y_2}(\alpha) = z(t_1; \alpha, y_2) = y_1$. Hence, we have a solution $z(t)$ of (5.1.2) satisfying the boundary conditions $z(t_1) = y_1$ and $z^\Delta(t_2) = y_2$. The proof is complete. $\qquad \square$

Our last three results of this section are uniqueness implies existence results for right focal boundary value problems for solutions of the third order dynamic equation (5.1.4). In particular, we establish such results, first for solutions of the 2-point problems (5.1.4)-(5.2.3) and (5.1.4)-(5.2.4), followed by such results for solutions of the 3-point problems (5.1.4)-(5.2.5).

Theorem 5.2.4. [[40], Harris, Henderson, Lanz and Yin, Thm. 8] *Assume that with respect to (5.1.4) conditions* (TSA), (TSB), (TSFC) *and* (TSCP) *are satisfied. Then, given any $t_1 < t_2$ in $\mathbb{T}$ and $y_1, y_2, y_3 \in \mathbb{R}$, the boundary value problem (5.1.4)-(5.2.3) has a unique solution.*

Proof. The uniqueness of solutions is a result of condition (TSFC). So, let $t_1, t_2 \in T$ and $y_1, y_2, y_3 \in \mathbb{R}$ be given. Let $z(t)$ be a solution of (5.1.4) satisfying the initial conditions

$$z(t_2) = 0, \qquad z^{\Delta}(t_2) = y_2, \qquad z^{\Delta\Delta}(t_2) = y_3.$$

Define

$$S = \{r \in \mathbb{R} \mid \text{there exists a solution } x(t) \text{ of (1) satisfying } x(t_1) = r,$$
$$x^{\Delta}(t_2) = z^{\Delta}(t_2), \ x^{\Delta\Delta}(t_2) = z^{\Delta\Delta}(t_2)\}.$$

We claim that $S = \mathbb{R}$. First, $S \neq \emptyset$ since $z(t_1) \in S$. Moreover, it follows from Theorem 5.2.2 that S is open.

Next, we show that S is also closed. Assume S is not closed. Then, there exist an $r_0 \in \overline{S} \setminus S$ and a sequence $\{r_k\} \subseteq S$ such that $\lim_{k \to \infty} r_k = r_0$. We may assume again that $r_k \uparrow r_0$. Then, for each $k \in \mathbb{N}$, let $z_k(t)$ be a solution of (5.1.4) satisfying $z_k(t_1) = r_k$, $z_k^{\Delta}(t_2) = z^{\Delta}(t_2)$, $z_k^{\Delta\Delta}(t_2) = z^{\Delta\Delta}(t_2)$.

Since $r_k < r_{k+1}$, from (TSB) and (TSC), we have $z_k(t) < z_{k+1}(t)$ on $(-\infty, t_2]_{\mathbb{T}}$. From the "compactness condition" (TSCP) and the fact that $r_0 \notin S$, there are points $\tau_1 < t_1 < \tau_2 < t_2$ in $\mathbb{T}$ such that $z_k(\tau_i) \uparrow \infty$, $i = 1, 2$.

Recall from Remark 5.2.1 that condition (TSC) also holds, and so from Theorem 5.1.5, solutions of 2-point conjugate boundary value problems for (5.1.4) exist and are unique. So, we let $w(t)$ be a solution to the 2-point conjugate boundary value problem of (5.1.4) satisfying

$$w(t_1) = r_0, \qquad w(t_2) = 0, \qquad w^{\Delta}(t_2) = x_k^{\Delta}(t_2) = z^{\Delta}(t_2).$$

Since $w(t_1) = r_0 > r_k = z_k(t_1)$, for all $k \in \mathbb{N}$, it follows that there exists a $k_0 \in \mathbb{N}$ and points $\gamma_1 < t_1 < \gamma_2 < t_2$ in $\mathbb{T}$ such that $(w - z_{k_0})$ has GZ's at γ_1 and γ_2. Thus, by the Rolle's Theorem on a time scale, there is an $\xi_1 \in \mathbb{T}$, $\gamma_1 \leq \xi_1 \leq \gamma_2$, at which $(w - z_{k_0})^{\Delta}$ has a GZ. We also have that $w^{\Delta}(t_2) = z_{k_0}^{\Delta}(t_2) = z^{\Delta}(t_2)$. Thus, by repeating the Rolle's Theorem, there is a point $\xi_2 \in \mathbb{T}$, $\xi_1 \leq \xi_2 \leq t_2$, such that $(w - z_{k_0})^{\Delta\Delta}$ has a GZ at ξ_2.

In conclusion, we have two solutions of (5.1.4), $w(t)$ and $z_{k_0}(t)$, that have the following GZ's: $w - z_{k_0}$ at γ_1, $(w - z_{k_0})^{\Delta}$ at ξ_1, and $(w - z_{k_0})^{\Delta\Delta}$ at ξ_2. Condition (TSFC) implies $w(t) = z_{k_0}(t)$ on $\mathbb{T}$. But this contradicts $r_0 \notin S$.

Therefore, S is closed and, consequently, $S = \mathbb{R}$. Then, by choosing $r = y_1 \in S$, we have a solution $y(t)$ of (5.1.4) satisfying

$$y(t_1) = y_1, \qquad y^\Delta(t_2) = y_2, \qquad y^{\Delta\Delta}(t_2) = y_3.$$

The proof is complete. □

Theorem 5.2.5. [[40], Harris, Henderson, Lanz and Yin, Thm. 9] *Assume that with respect to* (5.1.4) *conditions* (TSA), (TSB), (TSFC) *and* (TSCP) *are satisfied. Then, given any* $t_1 < t_2$ *in* $\mathbb{T}$ *and* $y_1, y_2, y_3 \in \mathbb{R}$, *the boundary value problem* (5.1.4)-(5.2.4) *has a unique solution.*

Proof. Again, the uniqueness of solutions is a result of condition (TSFC). From Theorem 5.2.4, there exists a unique solution $z(t)$ of (5.1.4), satisfying

$$z(t_1) = y_1, \qquad z^\Delta(t_2) = 0, \qquad z^{\Delta\Delta}(t_2) = y_3.$$

Define

$$S = \{r \in \mathbb{R} \mid \text{there exists a solution } x(t) \text{ of (5.1.4) with } x(t_1) = z(t_1),$$
$$x^\Delta(t_1) = r, \ x^{\Delta\Delta}(t_2) = z^{\Delta\Delta}(t_2)\}.$$

S is not empty because $z^\Delta(t_1) \in S$. Using a similar argument as in Theorem 5.2.4, but utilizing continuous dependence on boundary conditions instead of initial conditions, S is open.

We now claim that S is closed. Assuming the contrary, then, there exist an $r_0 \in \overline{S} \backslash S$ and a sequence $\{r_k\} \subseteq S$ such that $\lim_{k\to\infty} r_k = r_0$. Again, without loss of generality, let $r_k \uparrow r_0$. Then, for each $k \in \mathbb{N}$, there is a solution $z_k(t)$ of (5.1.4) that satisfies

$$z_k(t_1) = z(t_1), \qquad z_k^\Delta(t_1) = r_k, \qquad z_k^{\Delta\Delta}(t_2) = z^{\Delta\Delta}(t_2).$$

It follows from (TSFC) and the fact that $r_k < r_{k+1}$, we have

$$z_k(t) < z_{k+1}(t) \quad \text{on} \quad (t_1, t_2]_\mathbb{T}$$
$$z_k(t) > z_{k+1}(t) \quad \text{on} \quad (-\infty, t_1)_\mathbb{T}.$$

Also, from the condition (TSCP) and the fact that $r_0 \notin S$, $\{z_k(t)\}$ is unbounded on each compact interval in $\mathbb{T}$. Hence, there are points $\tau_1 < t_1 < \tau_2 < t_2$ in $\mathbb{T}$ such that $z_k(\tau_1) \downarrow -\infty$ and $z_k(\tau_2) \uparrow \infty$.

Next, let $w(t)$ solve the initial value problem for (5.1.4) satisfying,

$$w(t_1) = z_k(t_1) = z(t_1), \qquad w^\Delta(t_1) = r_0, \qquad w^{\Delta\Delta}(t_1) = 0.$$

Since $w^\Delta(t_1) = r_0 > z_k^\Delta(t_1)$, for all $k \in \mathbb{N}$, it follows that there exist points $\gamma_1 < t_1 < \gamma_2 < t_2$ in $\mathbb{T}$ and $k_0 \in \mathbb{N}$ such that $w(t_1) = z_{k_0}(t_1)$, and $(w - z_{k_0})$ has GZ's at γ_1 and γ_2. This means that $(w - z_{k_0})$ has three GZ's in $\mathbb{T}$. But applications of the time scale Rolle's Theorem and the uniqueness condition (TSFC) imply $w(t) = z_{k_0}(t)$ on $\mathbb{T}$. But this contradicts $r_0 \notin S$.

Therefore, S is closed. Consequently, $S = \mathbb{R}$, and then, by choosing $y_2 \in \mathbb{R}$, we have a solution $y(t)$ of (5.1.4) satisfying

$$y(t_1) = y_1, \qquad y^\Delta(t_1) = y_2, \qquad y^{\Delta\Delta}(t_2) = y_3. \qquad \square$$

Theorem 5.2.6. [[40], Harris, Henderson, Lanz and Yin, Thm. 10] *Assume that with respect to (5.1.4) conditions* (TSA), (TSB), (TSFC) *and* (TSCP) *are satisfied. Then, given any $t_1 < t_2 < t_3$ in $\mathbb{T}$ and $y_1, y_2, y_3 \in \mathbb{R}$, the boundary value problem (5.1.4)-(5.2.5) has a unique solution.*

Proof. From Theorem 5.2.5, there exists a solution $z(t)$ of (5.1.4) satisfying

$$z(t_2) = 0, \qquad z^\Delta(t_2) = y_2, \qquad z^{\Delta\Delta}(t_3) = y_3$$

Define

$$S = \{r \in \mathbb{R} \mid \text{there is a solution } x(t) \text{ of (5.1.4) satisfying } x(t_1) = r,$$
$$x^\Delta(t_2) = z^\Delta(t_2), \ x^{\Delta\Delta}(t_3) = z^{\Delta\Delta}(t_3)\}.$$

Then, $S \neq \emptyset$ since $z(t_1) \in S$. As in the previous theorems, S is an open subset of $\mathbb{R}$.

We next claim that S is closed. For contradiction purposes, assume S is not closed. Then, there exist an $r_0 \in \overline{S}\backslash S$ and a sequence $\{r_k\} \subseteq S$ such that $\lim_{k\to\infty} r_k = r_0$. Without loss of generality, let $r_k \uparrow r_0$. Then, for each $k \in \mathbb{N}$, there is a solution $z_k(t)$ of (5.1.4) with

$$z_k(t_1) = r_k, \qquad z_k^\Delta(t_2) = z^\Delta(t_2), \qquad z_k^{\Delta\Delta}(t_3) = z^{\Delta\Delta}(t_3).$$

From (TSFC) and the fact that $r_k < r_{k+1}$, we have

$$x_k(t) < x_{k+1}(t) \text{ on } (-\infty, t_2]_\mathbb{T}.$$

From (TSCP), there exist points $\tau_1 < t_1 < \tau_2 < t_2$ in $\mathbb{T}$ such that $x(\tau_i) \uparrow \infty$, $i = 1, 2$.

By Theorem 5.2.4, there exists a unique solution $w(t)$ of (5.1.4) satisfying

$$w(t_1) = r_0, \qquad w^\Delta(t_3) = 0, \qquad w^{\Delta\Delta}(t_3) = x_k^{\Delta\Delta}(t_3).$$

Since $w(t_1) = r_0 > z_k(t_1) = r_k$, for each $k \in \mathbb{N}$, it follows that there are points $\gamma_1 < t_1 < \gamma_2 < t_2$ in $\mathbb{T}$ and $k_0 \in \mathbb{N}$ such that $(w - x_{k_0})$ has GZ's at γ_1 and γ_2. By Rolle's Theorem on a time scale, there exists $\xi \in \mathbb{T}$, $\gamma_1 \leq \xi \leq \gamma_2$, where $(w - x_{k_0})^\Delta$ has a GZ. Hence, we have two solutions, $w(t)$ and $z_{k_0}(t)$, of (5.1.4) with the conditions where $(w - x_{k_0})$ has a GZ at γ_1, $(w - x_{k_0})^\Delta$ has a GZ at ξ, and $w^{\Delta\Delta}(t_3) = x_{k_0}^{\Delta\Delta}(t_3)$. If $\gamma_1 < \xi < \gamma_2$, uniqueness assumption (TSFC) implies that $w(t) = x_{k_0}(t)$. If $\gamma_1 = \xi$, uniqueness condition (TSFC) gives the same result. Either result yields a contradiction. Therefore, S is closed. Consequently, $S = \mathbb{R}$. Choosing $y_1 \in S = \mathbb{R}$, we have a solution $y(t)$ of (5.1.4) satisfying

$$y(t_1) = y_1, \qquad y^\Delta(t_2) = y_2, \qquad y^{\Delta\Delta}(t_3) = y_3.$$

The proof is complete.   □

5.3   Nonlocal boundary value problems: uniqueness implies existence

In this brief section, we will concentrate on uniqueness of solutions implying their existence for a nonlocal type boundary value problem for (5.0.3) of order $n = 2$. These results are due to Henderson, Tisdell and Yin [70].

In particular, in this chapter, we address the question of the uniqueness of solutions implying their existence for boundary problems for the second order dynamic equation (5.1.2), (that is, $y^{\Delta\Delta} = f(t, y, y^\Delta)$), satisfying the three-point boundary conditions,

$$y(t_1) = y_1, \ y(t_2) - y(t_3) = y_2, \tag{5.3.1}$$

where $t_1 \leq \sigma(t_1) < t_2 \leq \sigma(t_2) < t_3$ in $\mathbb{T}$, and $y_1, y_2 \in \mathbb{R}$.

Remark 5.3.1. *Sometimes the boundary value problem (5.1.2)-(5.3.1) is called a "nonlocal boundary value problem."*

We will use conditions (TSA), (TSB) and (TSCP) of Section 5.1, as well as a uniqueness condition on solutions of nonlocal boundary value problems, when such solutions exist.

(TSNC) Given points $r_1, r_2, r_3 \in (a, b)_\mathbb{T}$ with $r_1 \leq \sigma(r_1) < r_2 \leq \sigma(r_2) < r_3$, if $y(t)$ and $z(t)$ are solutions of (5.1.2) such that $y(t) - z(t)$ has a GZ at r_1, and $[y(r_3) - z(r_3)] - [y(t) - z(t)]$ has a GZ at r_2, then $y(t) = z(t)$, for all $t \in \mathbb{T}$.

Remark 5.3.2. *We notice, if (TSNC) is satisfied, then for any points $r_1, r_2, \in \mathbb{T}$ with $r_1 \leq \sigma(r_1) < r_2$, if $y(t)$ and $z(t)$ are solutions of (5.1.2)* such that $y(t) - z(t)$ has GZ's at r_1 and r_3, then $y(t) = z(t)$, for all $t \in \mathbb{T}$. That is, solutions of (5.1.2)-(5.1.3) are unique, hence exist, by Theorem 5.1.3.

And by the uniqueness of solutions as per (TSNC), we have the usual continuous dependence result, which we state without proof.

Theorem 5.3.1. *Assume that with respect to (5.1.2), conditions (TSA), (TSB) and (TSNC) are satisfied. Given a solution $y(t)$ of (5.1.2) on $\mathbb{T}$, an interval $[c,d]_{\mathbb{T}}$, points $t_1 \leq \sigma(t_1) < t_2 \leq \sigma(t_2) < t_3$ belonging to $(c,d)_{\mathbb{T}}$, and an $\epsilon > 0$, there exists a $\delta(\epsilon, [c,d]_{\mathbb{T}}) > 0$ such that, if $|x_i - t_i| < \delta$, $i = 1,2,3$, and $x_1 \leq \sigma(x_1) < x_2 \leq \sigma(x_2) < x_3$ belong to $(c,d)_{\mathbb{T}}$, and if $|y(t_1) - z_1| < \delta$ and $|y(t_2) - y(t_3) - z_2| < \delta$, then there exists a solution $z(t)$ of (5.1.2) satisfying $z(x_1) = z_1$, $z(x_2) - z(x_3) = z_2$, and $|y^{\Delta^i}(t) - z^{\Delta^i}(t)| < \epsilon$ on $[c,d]_{\mathbb{T}}$, $i = 0,1$.*

We invoke a shooting method to establish solutions of (5.1.2)-(5.3.1) exist under our uniqueness assumptions.

Theorem 5.3.2. *Assume that with respect to (5.1.2), conditions (TSA), (TSB), (TSNC) and (TSCP) are satisfied. Given points $t_1 \leq \sigma(t_1) < t_2 \leq \sigma(t_2) < t_3$ belonging to $\mathbb{T}$ and $y_1, y_2 \in \mathbb{R}$, there exists a unique solution of (5.1.2)-(5.3.1) on $\mathbb{T}$.*

Proof. Let $t_1 < t_2 < t_3$ in $\mathbb{T}$ and $y_1, y_2 \in \mathbb{R}$ be selected as in the statement of the theorem. Let $z(t)$ denote the solution of (5.1.2)-(5.1.3) satisfying

$$z(t_2) = y_2, \ z(t_3) = 0.$$

Next, define the set

$$S = \{y(t_1) \mid y(t) \text{ is a solution (5.1.2), and } y(t_2) - z(t_2) = y(t_3) - z(t_3)\}.$$

We observe that S is nonempty, since $z(t_1) \in S$. In addition, by applying Theorem 5.3.1, we conclude that S is an open subset of $\mathbb{R}$.

The remainder of the proof is devoted to showing that S is also a closed subset of $\mathbb{R}$. To that end, we assume for contradiction purposes that S is not closed. Then, there exist an $r_0 \in \overline{S} \backslash S$ and a strictly monotone sequence $\{r_k\} \subset S$ such that $\lim_{k \to \infty} r_k = r_0$.

We may assume without loss of generality that $r_k \uparrow r_0$. By the definition of S, we denote, for each $k \in \mathbb{N}$, by $u_k(t)$ the solution of (5.1.2) satisfying

$$u_k(t_1) = r_k, \text{ and } u_k(t_2) - u_k(t_3) = z(t_2) - z(t_3).$$

By (TSNC) and since $r_{k+1} > r_k$, we have

$$u_k(t) < u_{k+1}(t) \text{ on } (-\infty, x_2)_\mathbb{T}.$$

Consequently, from (TSCP) and the fact that $r_0 \notin S$, we may conclude that $\{u_k(t)\}$ is not bounded above at points of $(-\infty, t_1)_\mathbb{T}$ and at points of $(t_1, t_2)_\mathbb{T}$. That is, then there exist $\tau_1 < t_1 < \tau_2 < t_2$ such that

$$u_k(\tau_i) \uparrow +\infty, \quad i = 1, 2.$$

Now, let $w(t)$ be the solution of the initial value problem for (5.1.2) satisfying the initial conditions at t_1,

$$w(t_1) = r_0, \quad w^\Delta(t_1) = 0.$$

It follows that, for some K large, there exist points

$$\tau_1 < q_1 \leq \tau_2 < q_2 \leq \tau_3,$$

so that $u_K(t) - w(t)$ has GZ's at q_i, $i = 1, 2$, which contradicts Remark 5.3.2. Thus, S is also a closed subset of $\mathbb{R}$.

In summary, S is a nonempty subset of $\mathbb{R}$ that is both open and closed. We have $S = \mathbb{R}$. By choosing $r = y_1 \in S$, there is a corresponding solution $y(t)$ of (5.1.2) such that,

$$y(t_1) = y_1,$$
$$y(t_2) - y(t_3) = y_2.$$

The proof is complete.   □

5.4   Additional remarks

We remark that the results for this chapter have been restricted to dynamic equations of orders $n = 2, 3, 4$. It is not clear that the proofs for each of the main results reveal patterns to pursue for generalizations. Yet, the main results should motivate pursuit of generalizations.

Chapter 6

Postscript

Many of the methods employed in this book for uniqueness and existence of boundary value problems for ordinary differential equations have also been applied to boundary value problems that are described as being "between conjugate and right focal" boundary value problems for ordinary differential equations; namely, for an ordinary differential equation, if solutions of the right focal boundary value problems are unique, when solutions exist, then by applications of Rolle's Theorem, solutions of the "between" boundary value problems are unique, when solutions exist, whereas uniqueness of solutions of the "between" boundary value problems implies, by applications of Rolle's Theorem, that solutions of the conjugate boundary value problems are unique, when solutions exist.

For uniqueness and existence results for "between" boundary value problems for ordinary differential equations that are in the spirit of this book's results, we cite the works by Beverly [8], Chyan and Henderson [11], Ehme and Hankerson [17], Ehme and Lanz [19–21], Eloe and Henderson [28], Goecke and Henderson [31], Graef, Henderson, Luca and Tian [32], Henderson [47], Henderson and McGwier [68], and Peterson [99].

In addition, the methods of this book have been applied to establish uniqueness and existence results for Lidstone boundary value problems for ordinary differential equations; for a couple of those results, we cite the works by Davis and Henderson [16] and Ehme and Lanz [18].

Throughout this book, regardless of whether a result's conclusion is uniqueness or existence, the overriding hypothesis involves a uniqueness assumption on some solutions. So, natural questions to consider would include properties sufficient for the overriding uniqueness assumption to be satisfied.

Some of these properties can be stated in terms of monotonicity growth conditions on the nonlinearity in the differential equation; the classical papers by Lasota and Luczynski [86, 87] and the results in Kelley and Peterson [84, Theorems 7.23, 7.25, 7.26; pp. 329–331] provide typical such monotonicity conditions.

For some boundary value problems, the monotonicity conditions have been replaced by Lipschitz conditions on the nonlinearity in the differential equation, and then the Pontryagin Maximum Principle and optimal control theory have been applied to characterize maximal length intervals, in terms of the Lipschitz coefficients, on which solutions of the boundary value problems are unique, when solutions exist. Behind this latter approach on determining maximal intervals for uniqueness of solutions were the two papers on optimal intervals for nonoscillation of linear differential equations by Melentsova and Mil'shtein [91, 92]. In landmark papers, Jackson adapted some of the Melentsova and Mil'shtein techniques for nonlinear Lipschitz equations, first in determining optimal length intervals on which solutions of conjugate boundary value problems are unique [78], followed by another paper in which he adapted the techniques in determining optimal length intervals on which solutions of right focal boundary value problems are unique [79]. A number of papers exploiting Jackson's methods for certain boundary value problems for Lipschitz differential equations in determining maximal length intervals, in terms of the Lipschitz coefficients, on which solutions are unique, when solutions exist, include those by Clark and Henderson [14], Eloe and Henderson [24–27], Graef, Henderson, Luca and Tian [32], Hankerson and Henderson [38], Henderson [50–52, 61], and Henderson and McGwier [68].

Bibliography

[1] R. P. Agarwal, On multipoint boundary value problems for discrete equations, *J. Math. Anal. Appl.* **96**, No. 2 (1983), 520–534.

[2] R. P. Agarwal, Initial-value methods for discrete boundary value problems, *J. Math. Anal. Appl.* **100**, No. 2 (1984), 513–529.

[3] R. P. Agarwal, *Boundary Value Problems for Higher Order Differential Equations*, World Scientific Publishing Co., Inc., Teaneck, NJ, 1986.

[4] R. P. Agarwal, Compactness condition for boundary value problems, *Equadiff (Brno)* **9** (1997), 1–23.

[5] S. Agronsky, A. M. Bruckner, M. Laczkovich and D. Preiss, Convexity conditions and intersections with smooth functions, *Trans. Amer. Math. Soc.* **289**, No. 2 (1985), 659–677.

[6] B. Aulbach and S. Hilger, Linear dynamic processes with inhomogeneous time scale, *Nonlinear Dynamics and Quantum Dynamical Systems (Gaussig, 1990)*, Math. Res. **59**, Akademie-Verlag, Berlin, 1990, pp. 9–20.

[7] N. Azbelev and Z. Tsalyuk, On the question of the distribution of the zeros of solutions of a third order linear differential equation, *Mat. Sb. (N. S.)* **51** (1960), 475–486.

[8] R. Beverly, *Uniqueness Implies Existence for Fourth Order Boundary Value Problems*, M.S. Thesis, Auburn University, Auburn, June 1989.

[9] M. Bohner and P. W. Eloe, Higher order dynamic equations on measure chains: Wronskians, disconjugacy, and interpolating families of functions, *J. Math. Anal. Appl.* **246**, No. 2 (2000), 639–656.

[10] C. J. Chyan, Uniqueness implies existence on time scales, *J. Math. Anal. Appl.* **258**, No. 1 (2001), 359–365.

[11] C. J. Chyan and J. Henderson, Uniqueness implies existence for (n, p) boundary value problems, *Appl. Anal.* **73**, No. 3–4 (1999), 543–556.

[12] C. J. Chyan and W. K. C. Yin, A convergence theorem on a measure chain, *Math. Sci. Res. J.* **5**, No. 2 (2001), 15–22.

[13] S. Clark and J. Henderson, Uniqueness implies existence and uniqueness criterion for nonlocal boundary value problems for third order differential equations, *Proc. Amer. Math. Soc.* **134**, No. 11 (2006), 3363–3372.

[14] S. Clark and J. Henderson, Optimal interval lengths for nonlocal boundary

value problems associated with third order Lipschitz equations, *J. Math. Anal. Appl.* **322**, No. 1 (2006), 468–476.

[15] W. A. Coppel, *Disconjugacy*, Lecture Notes in Math., Vol. 220, Springer-Verlag, Berlin–New York, 1971.

[16] J. M. Davis and J. Henderson, Uniqueness implies existence for fourth-order Lidstone boundary value problems, *Panamer. Math. J.* **8**, No. 4 (1998), 23–35.

[17] J. Ehme and D. Hankerson, Existence of solutions for right focal boundary value problems, *Nonlinear Anal.* **18**, No. 2 (1992), 191–197.

[18] J. Ehme and A. Lanz, Uniqueness and existence for 2nth order Lidstone boundary value problems, *Panamer. Math. J.* **17**, No. 1 (2007), 75–81.

[19] J. Ehme and A. Lanz, Uniqueness implies existence for nonlinear conjugate-like boundary value problems, *Comm. Appl. Nonlinear Anal.* **14**, No. 3 (2007), 99–106.

[20] J. Ehme and A. Lanz, Uniqueness implies existence for various classes of nth order boundary value problems, *Int. J. Dyn. Syst. Differ. Equ.* **1**, No. 2 (2007), 98–101.

[21] J. Ehme and A. Lanz, Uniqueness implies existence of solutions for nonlinear focal-like boundary value problems, *Appl. Math. Lett.* **22**, No. 9 (2009), 1325–1329.

[22] P. W. Eloe, Difference equations and multipoint boundary value problems, *Proc. Amer. Math. Soc.* **86**, No. 2 (1982), 253–259.

[23] P. W. Eloe, A boundary value problem for a system of difference equations, *Nonlinear Anal.* **7**, No. 8 (1983), 813–820.

[24] P. W. Eloe and J. Henderson, Best interval lengths for some three point boundary value problems, *Differential Equations and Applications, Vol. I, II (Columbus, OH, 1988)*, Ohio University Press, Athens, OH, 1989, pp. 239–248.

[25] P. W. Eloe and J. Henderson, Optimal intervals for third order Lipschitz equations, *Differential Integral Equations* **2**, No. 4 (1989), 397–404.

[26] P. W. Eloe and J. Henderson, Uniqueness implies existence and uniqueness conditions for nonlocal boundary value problems for nth order differential equations, *J. Math. Anal. Appl.* **331**, No. 1 (2007), 240–247.

[27] P. W. Eloe and J. Henderson, Optimal intervals for uniqueness of solutions for nonlocal boundary value problems, *Comm. Appl. Nonlinear Anal.* **18**, No. 3 (2011), 89–97.

[28] P. W. Eloe and J. Henderson, Uniqueness implies existence and uniqueness conditions for a class of $(k+j)$-point boundary value problems for nth order differential equations, *Math. Nachr.* **284**, No. 2–3 (2011), 229–239.

[29] P. W. Eloe, J. Henderson and R. A. Khan, Uniqueness implies existence and uniqueness conditions for a class of $(k + j)$-point boundary value problems for n-th order differential equations, *Canad. Math. Bull.* **55**, No. 2 (2012), 285–296.

[30] P. W. Eloe, J. Henderson and R. A. Khan, Existence and uniqueness conditions for a class of $(k + 4j)$-point n-th order boundary value problems, *Nonlinear Dyn. Syst. Theory* **12**, No. 1 (2012), 49–62.

[31] G. M. Goecke and J. Henderson, Uniqueness of solutions of right focal problems for third order differential equations, *Nonlinear Anal.* **8**, No. 3 (1984), 253–259.

[32] J. R. Graef, J. Henderson, R. Luca and Y. Tian, Boundary-value problems for third-order Lipschitz ordinary differential equations, *Proc. Edinb. Math. Soc. (2)* **58**, No. 1 (2015), 183–197.

[33] M. Gray, *Uniqueness Implies Uniqueness and Existence for Nonlocal Boundary Value Problems for Third Order Ordinary Differential Equations*, Ph.D. Dissertation, Baylor University, Waco, May 2006.

[34] M. Gray, Uniqueness implies uniqueness and existence for nonlocal boundary value problems for third order ordinary differential equations, *Comm. Appl. Nonlinear Anal.* **13**, No. 4 (2006), 19–30.

[35] M. Gray, Uniqueness implies uniqueness for nonlocal boundary value problems for third order ordinary differential equations, *Dynam. Systems Appl.* **16**, No. 2 (2007), 277–284.

[36] D. Hankerson, *Boundary Value Problems for n-th Order Difference Equations*, Ph.D. Dissertation, The University of Nebraska-Lincoln, Lincoln, May 1986.

[37] D. Hankerson, An existence and uniqueness theorem for difference equations, *SIAM J. Math. Anal.* **20**, No. 5 (1989), 1208–1217.

[38] D. Hankerson and J. Henderson, Optimality for boundary value problems for Lipschitz equations, *J. Differential Equations* **77**, No. 2 (1989), 392–404.

[39] G. Harris, J. Henderson, A. Lanz and W. K. C. Yin, Second order right focal boundary value problems on a time scale, *Comm. Appl. Nonlinear Anal.* **11**, No. 4 (2004), 57–62.

[40] G. Harris, J. Henderson, A. Lanz and W. K. C. Yin, Third order right focal boundary value problems on a time scale, *J. Difference Equ. Appl.* **12**, No. 6 (2006), 525–533.

[41] P. Hartman, Unrestricted n-parameter families, *Rend. Circ. Mat. Palermo* **7**, No. 2 (1958), 123–142.

[42] P. Hartman, *Ordinary Differential Equations*, Wiley, New York, 1964.

[43] P. Hartman, On n-parameter families and interpolation problems for nonlinear ordinary differential equations, *Trans. Amer. Math. Soc.* **154** (1971), 201–226.

[44] P. Hartman, Difference equations: disconjugacy, principal solutions, Green's functions, complete monotonicity, *Trans. Amer. Math. Soc.* **246** (1978), 1–30.

[45] J. Henderson, Existence of solutions of right focal point boundary value problems for ordinary differential equations, *Nonlinear Anal.* **5**, No. 9 (1981), 989–1002.

[46] J. Henderson, Uniqueness of solutions of right focal point boundary value problems for ordinary differential equations, *J. Differential Equations* **41**, No. 2 (1981), 218–227.

[47] J. Henderson, Right $(m_1; \ldots; m_l)$ focal boundary value problems for third order differential equations, *J. Math. Phys. Sci.* **18**, No. 4 (1984), 405–413.

[48] J. Henderson, K-point disconjugacy and disconjugacy for linear differential

equations, *J. Differential Equations* **54**, No. 1 (1984), 87–96.

[49] J. Henderson, Existence and uniqueness of solutions of right focal point boundary value problems for third and fourth order equations, *Rocky Mountain J. Math.* **14**, No. 2 (1984), 487–497.

[50] J. Henderson, Best interval lengths for boundary value problems for third order Lipschitz equations, *SIAM J. Math. Anal.* **18**, No. 2 (1987), 293–305.

[51] J. Henderson, Boundary value problems for nth order Lipschitz equations, *J. Math. Anal. Appl.* **134**, No. 1 (1988), 196–210.

[52] J. Henderson, Optimality and existence for Lipschitz equations, *Internat. J. Math. Math. Sci.* **11**, No. 2 (1988), 267–274.

[53] J. Henderson, Existence theorems for boundary value problems for nth-order nonlinear difference equations, *SIAM J. Math. Anal.* **20**, No. 2 (1989), 468–478.

[54] J. Henderson, Focal boundary value problems for nonlinear difference equations. I, *J. Math. Anal. Appl.* **141**, No. 2 (1989), 559–567.

[55] J. Henderson, Focal boundary value problems for nonlinear difference equations. II, *J. Math. Anal. Appl.* **141**, No. 2 (1989), 568–579.

[56] J. Henderson, Boundary value problems for nonlinear difference equations, *Differential Equations (Colorado Springs, CO, 1989)*, Lecture Notes in Pure and Appl. Math. **127**, Dekker, New York, 1991, pp. 227–239.

[57] J. Henderson, Uniqueness implies existence for four-point nonlocal boundary value problems for second order equations, *Int. J. Appl. Math. Sci.* **1** (2004), 67–72.

[58] J. Henderson, Uniqueness implies existence for three-point boundary value problems for second order differential equations, *Appl. Math. Lett.* **18**, No. 8 (2005), 905–909.

[59] J. Henderson, Solutions of multipoint boundary value problems for second order equations, *Dynam. Systems Appl.* **15**, No. 1 (2006), 111–117.

[60] J. Henderson, Existence and uniqueness of solutions of $(k+2)$-point nonlocal boundary value problems for ordinary differential equations, *Nonlinear Anal.* **74**, No. 7 (2011), 2576–2584.

[61] J. Henderson, Optimal interval lengths for nonlocal boundary value problems for second order Lipschitz equations, *Commun. Appl. Anal.* **15**, No. 2-4 (2011), 475–482.

[62] J. Henderson and L. Jackson, Existence and uniqueness of solutions of k-point boundary value problems for ordinary differential equations, *J. Differential Equations* **48**, No. 3 (1983), 373–386.

[63] J. Henderson and A. M. Johnson, Uniqueness implies existence for discrete fourth order Lidstone boundary-value problems, *Proceedings of the Fourth Mississippi State Conference on Difference Equations and Computational Simulations (1999)*, Electron. J. Differ. Equ. Conf. **3**, Southwest Texas State University, San Marcos, TX, 2000, pp. 63–73 (electronic).

[64] J. Henderson, B. Karna and C. C. Tisdell, Existence of solutions for three-point boundary value problems for second order equations, *Proc. Amer. Math. Soc.* **133**, No. 5 (2005), 1365–1369.

[65] J. Henderson and C. J. Kunkel, Uniqueness of solutions of linear nonlocal

boundary value problems, *Appl. Math. Lett.* **21**, No. 10 (2008), 1053–1056.

[66] J. Henderson and D. Ma, Uniqueness of solutions for fourth-order nonlocal boundary value problems, *Bound. Value Probl.* **2006**, (2006) Art. ID 23875, 12 pp.

[67] J. Henderson and D. Ma, Existence of solutions for fourth order nonlocal boundary value problems, *Georgian Math. J.* **13**, No. 3 (2006), 473–484.

[68] J. Henderson and R. W. McGwier, Jr., Uniqueness, existence, and optimality for fourth-order Lipschitz equations, *J. Differential Equations* **67**, No. 3 (1987), 414–440.

[69] J. Henderson, A. C. Peterson and C. C. Tisdell, On the existence and uniqueness of solutions to boundary value problems on time scales, *Adv. Difference Equ.* **2004**, No. 2, 93–109.

[70] J. Henderson, C. C. Tisdell and W. K. C. Yin, Uniqueness implies existence for three-point boundary value problems for dynamic equations, *Appl. Math. Lett.* **17**, No. 12 (2004), 1391–1395.

[71] J. Henderson and W. K. C. Yin, Existence of solutions for third-order boundary value problems on a time scale, *Comput. Math. Appl.* **45**, No. 6-9 (2003), 1101–1111.

[72] J. Henderson and W. K. C. Yin, Existence of solutions for fourth order boundary value problems on a time scale, *J. Difference Equ. Appl.* **9**, No. 1 (2003), 15–28.

[73] J. Henderson and W. K. C. Yin, Two-point and three-point problems for fourth order dynamic equations, *Dynam. Systems Appl.* **12**, No. 1–2 (2003), 159–169.

[74] S. Hilger, Analysis on measure chains — a unified approach to continuous and discrete calculus, *Results Math.* **18**, No. 1–2 (1990), 18–56.

[75] L. Jackson, Existence and uniqueness of solutions of boundary value problems for third order differential equations, *J. Differential Equations* **13** (1973), 432–437.

[76] L. Jackson, Uniqueness of solutions of boundary value problems for ordinary differential equations, *SIAM J. Appl. Math.* **24**, No. 4 (1973), 535–538.

[77] L. Jackson, Boundary value problems for ordinary differential equations, *Studies in Ordinary Differential Equations*, Stud. in Math. **14**, Math. Association of America, Washington, D.C. (1977), pp. 93–127.

[78] L. Jackson, Existence and uniqueness of solutions of boundary value problems for Lipschitz equations, *J. Differential Equations* **32**, No. 1 (1979), 76–90.

[79] L. Jackson, Boundary value problems for Lipschitz equations, *Differential equations (Proc. Eighth Fall Conf., Oklahoma State Univ., Stillwater, Okla., 1979)*, Academic Press, New York, 1980, pp. 31–50.

[80] L. Jackson and G. Klaasen, Uniqueness of solutions of boundary value problems for ordinary differential equations, *SIAM J. Appl. Math.* **19**, No. 3 (1970), 542–546.

[81] L. Jackson and K. Schrader, Subfunctions and third order differential equations, *J. Differential Equations* **8** (1970), 180–194.

[82] L. Jackson and K. Schrader, Existence and uniqueness of solutions of

boundary value problems for third order differential equations, *J. Differential Equations* **9** (1970), 46–54.

[83] G. D. Jones, Existence of solutions of multipoint boundary value problems for a second order differential equation, *Dynam. Systems Appl.* **16**, No. 4 (2007), 709–711.

[84] W. G. Kelley and A. C. Peterson, *The Theory of Differential Equations. Classical and Qualitative*, Second edition, Universitext, Springer, New York, 2010.

[85] G. Klaasen, Existence theorem for boundary value problems for nth order ordinary differential equations, *Rocky Mountain J. Math.* **3** (1973), 457–472.

[86] A. Lasota and M. Luczynski, A note on the uniqueness of two point boundary value problems, I, *Zeszyty Nauk. Uniw. Jagiello. Prace Mat.* **12** (1968), 27–29.

[87] A. Lasota and M. Luczynski, A note on the uniqueness of two point boundary value problems, II, *Zeszyty Nauk. Uniw. Jagiello. Prace Mat.* **13** (1969), 45–48.

[88] A. Lasota and Z. Opial, On the existence and uniqueness of solutions of a boundary value problem for an ordinary second order differential equation, *Colloq. Math.* **18** (1967), 1–5.

[89] D. Ma, *Uniqueness Implies Uniqueness and Existence for Nonlocal Boundary Value Problems for Fourth Order Differential Equations*, Ph.D. Dissertation, Baylor University, Waco, December 2005.

[90] J. Mawhin, Functional analysis and nonlinear boundary value problems: the legacy of Andrzej Lasota, *Ann. Math. Sil.* **27** (2013), 7–38.

[91] Yu. A. Melentsova and G. N. Mil'shtein, An optimal estimate of the interval on which a multipoint boundary value problem possesses a solution, *Differentsial'nye Uravneniya* **10** (1974), 1257–1265 (in Russian).

[92] Yu. A. Melentsova and G. N. Mil'shtein, On the optimal estimate of the interval of nonoscillation for linear differential equations with bounded coefficients, *Differentsial'nye Uravneniya* **17**, No. 12 (1981), 2160–2175 (in Russian).

[93] J. S. Muldowney, A necessary and sufficient condition for disfocality, *Proc. Amer. Math. Soc.* **71**, No. 1 (1979), 49–55.

[94] J. S. Muldowney, On disconjugacy and k-point disconjugacy for linear ordinary differential operators, *Proc. Amer. Math. Soc.* **92**, No. 1 (1984), 27–30.

[95] I. P. Natanson, *Theory of Functions of a Real Variable*, Frederick Ungar Publishing, New York, 1955, (translated from the Russian by L. F. Boron).

[96] Z. Opial, On a theorem of O. Aramă, *J. Differential Equations* **3** (1967), 88–91.

[97] A. C. Peterson, A theorem of Aliev, *Proc. Amer. Math. Soc.* **23** (1969), 364–366.

[98] A. C. Peterson, Green's functions for focal type boundary value problems, *Rocky Mountain J. Math.* **9**, No. 4 (1979), 721–732.

[99] A. C. Peterson, Existence–uniqueness for focal-point boundary value prob-

lems, *SIAM J. Math. Anal.* **12**, No. 2 (1981), 173–185.

[100] A. C. Peterson, Existence and uniqueness theorems for nonlinear difference equations, *J. Math. Anal. Appl.* **125**, No. 1 (1987), 185–191.

[101] D. Peterson, *Uniqueness, Existence and Comparison Theorems for Ordinary Differential Equations*, Ph.D. Dissertation, University of Nebraska, Lincoln, May 1977.

[102] K. Schrader, Uniqueness implies existence for solutions of nonlinear boundary value problems, *Abstracts Amer. Math. Soc.* **6** (1985), 235.

[103] T. Sherman, Properties of solutions of nth order linear differential equations, *Pacific J. Math.* **15** (1965), 1045–1060.

[104] T. Sherman, Conjugate points and simple zeros for ordinary linear differential equations, *Trans. Amer. Math. Soc.* **146** (1969), 397–411.

[105] G. Vidossich, On the continuous dependence of solutions of boundary value problems for ordinary differential equations, *J. Differential Equations* **82**, No. 1 (1989), 1–14.

Index

Printed in the United States
By Bookmasters